DISCARDED

AMERICAN ECONOMIC GROWTH:
The Historic Challenge

Edited by
William F. Donnelly, S.J.
University of Santa Clara

LAMAR UNIVERSITY LIBRARY

MSS Information Corporation
655 Madison Avenue, New York, N.Y. 10021

This is a custom-made book of readings prepared for the courses taught by the editor, as well as for related courses and for college and university libraries. For information about our program, please write to:

MSS INFORMATION CORPORATION
655 Madison Avenue
New York, New York 10021

MSS wishes to express its appreciation to the authors of the articles in this collection for their cooperation in making their work available in this format.

Library of Congress Cataloging in Publication Data

Donnelly, William F   comp.
  American economic growth.

  CONTENTS: Murray, G. S. Forty years ago: The great depression comes to Arkansas.--Schatz, A. W. The Anglo-American trade agreement and Cordell Hull's search for peace, 1936-1938.--Bennett, J. D. Roosevelt, Wilkie, and the TVA. [etc.]
  1. United States--Economic conditions--Addresses, essays, lectures. I. Title.
HC103.D7   330.9'73'008   73-10339
ISBN 0-8422-5110-3
ISBN 0-8422-0309-5 (pbk.)

Copyright © 1973
by
MSS INFORMATION CORPORATION
All Rights Reserved.

CONTENTS

Preface 5

PART ONE: THE TREE IS PLANTED — From the Colonies to the Civil War 6

Fish and Flour for Gold: Southern Europe and the Colonial American Balance of Payments
  JAMES G. LYDON 7

Goals and Enforcement of British Colonial Policy, 1763-1775
  NEIL R. STOUT 20

Speculations on the Significance of Debt: *Virginia*, 1781-1789
  MYRA L. RICH 30

The Great Migration to the Mississippi Territory, 1798-1819
  CHARLES D. LOWERY 47

Secretary Taney and the Baltimore Pets: A Study in Banking and Politics
  FRANK OTTO GATELL 67

Coal-burning Locomotives: A Technological Development of the 1850's
  EDWARD F. KEUCHEL 90

The Economics of Industrial Slavery in the Old South
  ROBERT S. STAROBIN 102

Facilities for the Construction of War Vessels in the Confederacy
  WILLIAM N. STILL, JR. 137

PART TWO: GROWING STRONG — From the Civil War through World War I 157

Water, Land, and People in the Great Valley
  PAUL S. TAYLOR 158

The Troy Case: A Fight Against Discriminatory Freight Rates
  KENNETH R. JOHNSON 178

Origins of the American Tobacco Company
  PATRICK G. PORTER 190

Poverty in the Urban Ghetto
  JOHN F. BAUMAN 208

Who Killed the Aldrich Plan?
 ANDREW GRAY 216

PART THREE: HARVEST TIME — From the 1920's to the Post-War Age ... 238

The National Association of Manufacturers and Labor Relations in the 1920s
 ALLEN M. WAKSTEIN 239

The Harlan County Coal Strike of 1931
 TONY BUBKA 253

Forty Years Ago: The Great Depression Comes to Arkansas
 GAIL S. MURRAY 270

The Anglo-American Trade Agreement and Cordell Hull's Search for Peace 1936-1938
 ARTHUR W. SCHATZ 292

Roosevelt, Wilkie, and the TVA
 JAMES D. BENNETT 311

Some Observations on Rationing
 CHARLES F. PHILLIPS 320

Charting a Course between Inflation and Depression
 BARTON J. BERNSTEIN 332

Eisenhower and Ezra Taft Benson: Farm Policy in the 1950s
 EDWARD L. SCHAPSMEIER and FREDERICK H. SCHAPSMEIER 344

PREFACE

When a student wants to find out the why and wherefore of the American scene, he must eventually take a look at the economic side of the nation's history. Sometimes he'll find that a particular individual made the difference. On other occasions the push and pull arises from foreign or outside factors. In most cases, however, a bit of educated reflection will discover some common denominator in those economic dimensions that attend the major turns and events in the American Republic. What this means is that other factors may make the world go round, but economic considerations keep it spinning.

American Economic History is a fascinating discipline, rich in empirical evidence and the expertise to promote a nation's growth. While still something of an infant in comparison to more senior colleagues in Europe or the East, the United States manages to attain a sufficient measure of political and material success to keep her envied throughout the world. Add to this the advantages of a free and democratic society, and the resultant product is eminently worthy of a serious and detailed inquiry.

What is truly unfortunate, however, is to miss the genuine interest and dynamism that should accompany economic investigations. The articles included in this collection have been chosen on that basis, for their inherent interest and their facile capacity to bring the economic past to the concerned attention of the contemporary student. It is to be hoped too that the future will see that attention intensified with a will to bring to completion those concepts that must underlie the nation's progress as she moves along the highways and freeways yet to come.

Any collection, of course, must remain incomplete. Numerous articles have necessarily been excluded and spatial considerations limit the subjects treated. Those presented, however, are superior in quality and were chosen to exemplify one or another essential facet that accompanies national growth. Taken together they account for the American present, for political and industrial prominence and the high standard of living most of us enjoy. Hopefully students will find from these efforts another chance to come alive intellectually and cull from this limited offering those principles of economics which will profitably serve their own lives and the future of the Republic as well.

Special thanks are to be given to the publishers and authors of the articles presented, and for the kind permission to include their research here. The experience over several years of challenging student inquiry also demands an expression of gratitude for the stimulation thus created and the friendships that ensued; and my assistant, Hubert W. Jansen, is to be especially commended.

PART ONE: THE TREE IS PLANTED

From the Colonies to the Civil War

John Smith and his Jamestown companions were not the first Europeans to take up residence in America, but they were the earliest Europeans to make a success of it. Horrendous problems and discouragement were commonplace, for them as well as their countrymen to the north; but before long they had a surplus and a thriving foreign trade brought new wealth and a variety of commercial opportunities abroad. At home, the greatest advantage was the land, the resources contained within it, a favorable climate, good navigable rivers, and a beneficent ocean rich in cod and haddock and the profitable returns of an extensive whaling industry. Understandably, things went well for almost one hundred years until conflicting interests clashed head on, with a subsequent tightening of British policy which led to the clash of arms and a definitive break in the Mercantilist relations of colonies and mother country across the sea.

Out of that conflict came a new nation inspired with a democratic spirit, endowed with vast new tracts of land, and even more heavily weighed down with financial problems and a huge amount of debt. At the same time the underlying strength was there and new lands and new people combined to build foundations for the striking growth to come. Concerns of finance and banking were contained, though not really solved, while invention and technology sparked the efforts to build manufacturing and a strong industrial base. An adequate supply of labor, in the South especially, was never really attained, so the institution of slavery was to be found in industry and agriculture alike. With emotions and economic interests so thoroughly entwined, this and other problems led irrevocably to the tragedy of Civil War. Even then some commerce still endured, and both belligerent sections survived it all. What is even more striking is the fact that this relatively short span of one hundred and fifty years saw a wilderness come to life and deep roots planted. Some day they would bear abundant fruit, and John Smith and the original colonists would find that remarkable in the extreme.

By James G. Lydon

Fish and Flour for Gold: Southern Europe and the Colonial American Balance of Payments*

In August, 1754 the schooner *Jolly Robin* sailed out of the Piscataqua River with a mixed cargo of fish and lumber, bound for Bilbao in northern Spain. Her cargo was disposed of there through the house of Gardoqui for £255, sterling. Of this sum, £34 were returned to Portsmouth in cash and £101 were transferred to London in bills of exchange. The remainder of the proceeds purchased a cargo of salt which sold at Piscataqua for £154, more than the cost of the original cargo plus all expenses. A second voyage was made in 1755, repeating this pattern. This time 43 per cent of the returns came in cash (£59) and bills of exchange (£79) and again the remaining funds bought a salt cargo which repaid the first costs plus expenses. From these voyages, the owners of *Jolly Robin* realized £273, sterling, in cash and bills.[1]

These two transactions, involving relatively small sums, are not especially significant. Yet, if they represent a pattern of trade repeated almost endlessly by a large number of vessels for more than a hundred years, their relationship to the over-all problem of the colonial specie shortage is evident. The question to be answered is: Did this market consistently provide returns to American merchants of 40 per cent or more in bills and cash?

Southern Europe offered a very attractive market for the goods of North America. At Bilbao, Lisbon, Cadiz, Barcelona, Leghorn

* Part of the research for this article was supported by a summer grant from the Baker Library and Business History groups of the Harvard University Graduate School of Business Administration.
[1] "Schooner Jolly Robin's book of Acco'ts. for Voyages as Settled per Dan'l Staniford," Heard Collection, AQ 1 (Baker Library, Harvard University Graduate School of Business Administration).

and other ports, American fish, wheat, flour, lumber, corn, rice, and other products found outlet. In return, trading vessels brought back aromatic wines from Oporto, Sherez, Madeira, and the Canaries or solar salt from Setubal and the Isle of May. Chests of oranges, limes, currants, and raisins came from Cadiz and Alicant and sundry other goods — olives, anchovies, and capers or perhaps some Leghorn hats or Barcelona handkerchiefs. Luxury goods were occasionally imported — Bilbao mirrors, "moracker" leather chairs, Turkey carpets, Italian marble for a mantlepiece, and even a box of vermicelli. Wine and salt, however, were the "blue chips" of the trade. These and the two great colonial staples, fish and grain, must attract our attention.

Colonial trade from North America to Southern Europe has been relatively neglected by historians. Third in tonnage, behind the trades to the West Indies and the British Isles, its volume greatly exceeded that employed in slaving, yet historians have been more attracted by man's inhumanity to man. Naturally, trade with Great Britain has demanded attention and sheer volume has emphasized West Indian commerce. Though neglected, the trade to Southern Europe has not been ignored. Almost all historians who have examined colonial trading patterns mention — in some cases stress — that trade to Southern Europe was quite significant as a source of returns.

Seventeenth- and eighteenth-century English and American writers were well aware of its importance. Jeremiah Dummer, when agent for Massachusetts Bay, noted: "Salt is used partly for saving provisions, but principally for curing fish, which is — the principal branch of returns made from the Continent to Great Britain by way of Spain, Portugal, and the Straits." [2] In his *The Importance of the British Plantations in America to this Kingdom*, Fayrer Hall, referring to Pennsylvania, stated: "They send great quantities of corn [wheat] to Portugal and Spain, frequently selling the ship as well as the cargo; and the produce of both is sent thence to England." [3] Somewhat later, John Huske, in *The Present State of North America*, wrote: "It is from our American Colonies — we have our Silver and Gold either by their trade with foreigners in America, or by way of Spain, Portugal, and Italy, in payment for their immense quantities of Fish, Rice, &c." [4]

---

[2] Curtis P. Nettels, *The Money Supply of the American Colonies before 1720* (Madison, 1934), pp. 96–97.
[3] Fayrer Hall, *The Importance of the British Plantations in America to this Kingdom* (London, 1731), p. 98.
[4] John Huske, *The Present State of North America*, quoted in Raymond McFarland, *The History of the New England Fisheries* (New York, 1911), p. 98.

Unfortunately, most of these writers offer few specifics to support their commentaries and, in essence, it is the volume of the traffic that demonstrates its importance. If fish and flour exports were exchanged for salt and wine, costing only 60 per cent of the value of the outward cargoes, the larger the volume the greater the returns to England. Perhaps the best approach is to examine the two colonial staples mainly involved, as to their entrance into this trade, the volume exported to Iberia and the Straits, and their significance in the balance of payments.

Dried codfish, heavily salted, kept well in warm climates and was in great demand in Catholic Europe. Lenten and Friday fasts and abstinences created a need for this product. In 1755 Bilbao, the entrepot for Madrid and the central plain required 70,000–80,000 quintals of cod a year, while the whole Iberian market consumed 300,000 quintals yearly.[5] Even then, there were Englishmen who complained that Spanish Catholics did not follow their religious requirements as they should. Since only merchantable fish of the first and second grades were salable in these markets, we have some clue to the volume of this trade. When figures for total fish caught are available, a rule of thumb that 50 per cent were of merchantable grades can be employed to estimate shipments to Southern Europe. Fortunately, scattered statistics are to be found among Port Office Records and tonnage volume is also significant. New England's exports of fish can be approximated for several periods.

English fishermen quite early exploited the waters off New England, well aware of the gold to be gained by carrying the catch to Spain and Portugal. Massachusetts, plagued by a sickly economy in the late 1630's, made efforts to encourage the fishery as a source of returns to alleviate the specie shortage. Several government measures stimulated fishing. As early as 1641, some 300,000 fish were sent to market,[6] and fishing had soon become a major factor in the Massachusetts economy. By 1661–1662, Boston sent more shipping to the area south of Cape Finisterre than to the West Indies.[7] In four months in 1686, Boston shipped more than 11,000 quintals of fish to Iberia.[8] By the turn of the century, the city's

---

[5] William Douglas, *Summary, Historical and Political . . . of the British Settlements in North America* (2 vols., London, 1755), vol. I, p. 295.

[6] James K. Hosmer (ed.), *John Winthrop's Journal: A History of New England, 1630–1649* (2 vols., New York, 1908), vol. II, p. 42.

[7] The Names of such Ships and Masters that have come in and gone out of our Harbors & Given bond for His Majesty's Customs, August 16, 1661 to February 25, 1662" (Photostat at the Mass. Historical Society).

[8] Abstracts of English Shipping Records Relating to Massachusetts Ports — from Original Records in the Public Record Office, London (Mass. Historical Society, cited hence as Photostats, MHS). Part I, p. 28 indicates that 500 tons of fish were shipped to Bilbao. Two

exports there had risen to 25,000 and after 1713, 60,000 quintals or more of merchantable cod went to these markets from Salem and Boston each year.[9] Fayrer Hall, who may have exaggerated his figures, states that by 1731 New England was exporting 230,000 quintals to Southern Europe. A wartime decline in the fisheries ended in 1747 and seven years later Boston and Salem dispatched almost 6,000 tons of shipping to the south of Europe.[10] In that year, 1754, Boston alone exported 26,000 quintals and the following year more than 48,000.[11] Salem tonnage to Southern Europe was twice that of Boston, and New Hampshire, Maine, and Nova Scotia also swelled fish shipments. The volume of the trade, though it naturally fluctuated, increased fairly steadily through the period.

To a very considerable extent, the value of the New England fishery depended upon these exports to Southern Europe. Though varying in price, the fish sold there generally were worth 50 per cent more than the refuse fish sent to the West Indies. From the statistics available and by use of mercantile correspondence, waste books, etc., the value of these fish exports can be estimated. The merchantable cod taken in 1641 were worth about £1,500, sterling. Boston's known exports in 1686 can be valued at £5,500; by 1700 these fish shipments had reached £15,000 and after 1713 their value rose steadily. Hall's figures for 1731 make New England's exports south of Finisterre £172,500.[12] In the later period fish prices rose so the returns were proportionately higher. The larger the volume, the greater were the returns to England. Though the percentage of the catch cleared to the West Indies rose somewhat in these years, the portion shipped to Southern Europe was always in total more valuable.

Both Philadelphia and New York, jealous of New England's fishery returns and her dominance of their trade, sought to enter this lucrative traffic. Governor Thomas Dongan of New York noted in 1685 the expenditure of £2,500 to develop "fishing" and the Duke of York was "especially desirous" of advancing this business in his

---

other cargoes cleared to Portugal and the Canary Isles. The estimate is 550 tons of fish or 11,000 quintals.
[9] Nettels, *Money Supply*, p. 79; Photostats, MHS, Part I (Entrances and Clearances, 1686–1717) and Part II (Entrances and Clearances, 1717–1755).
[10] Hall, *Importance of British Plantations*, p. 102; Photostats, MHS, Part II.
[11] Photostats, MHS, Part II.
[12] Hosmer (ed.), *John Winthrop's Journal*, vol. II, p. 42. Ralph G. Lounsbury, *The British Fishery at Newfoundland, 1634–1763* (New Haven, 1934), p. 58 indicates that 100 dried cod equalled a quintal. Several sources price New England fish at 32 rials or 16 shillings at this time. Half the fish were merchantable. See note 8 above for volume of fish shipments; John Hull's Journal, 1685–1689 (Baker Library) prices merchantable fish at ten shillings/qtl., so 11,000 qtls. equalled £5,500. For 1700 shipments see Nettels, *Money Supply*, p. 79. The 1731 figure is from Hall's *Importance of British Plantations*, p. 102.

colony.[13] But hopes for a fishery proved fleeting and New York and Philadelphia later exchanged West Indian produce for fish in New England and Newfoundland in order to "pursue a valuable trade with Spain, Portugal, and Italy, where they chiefly obtained money or bills of exchange in return."[14]

Through the indulgence of the English government, return cargoes of salt could be carried directly to New England. This privilege was most important in the development of Southern Europe as a source of colonial returns. The bulk of the west-bound cargoes was in salt. Cheaper than the fish sent to purchase it, salt was a basic necessity for the fisheries. For example, in 1686, Isle of May salt sold at Boston for 10 to 12 shillings per hogshead, while a "kintal" or hundredweight of merchantable fish ranged at about the same price.[15] Thus a hundredweight of fish purchased more than five times its weight in salt, *at Boston*. Fish was dearer in Europe and the salt even cheaper. The returns remaining after the purchase of a salt cargo were what made this trade so attractive.

In the case of the other major import — wine — the situation differed only slightly. Harold Innis, in his excellent work on the cod-fisheries, states that between 1720 and 1750, "Ships to Lisbon and Alicante, carrying fish — could stow only a quantity of wine equal to one half the sale price of the fish and were compelled to take the remainder in money."[16] This type of compulsion certainly did not discourage shipments to that area. Cargo space was the basic limitation on both salt and wine shipments. When cheaper Spanish wines were laded, there was a balance in favor of the colonials but more expensive vintages, such as Madeira, were apparently exchanged evenly against the eastbound cargoes. In any case, salt ladings were much more common.[17]

Thus, the fish trade to Southern Europe offered an excellent source of returns to cover accounts payable in Britain. Notarial records for Boston, as early as the 1640's, show innumerable debts paid English merchants through delivery of cod for those markets.[18]

---

[13] Great Britain Public Record Office, *Calendar of State Papers, Colonial Office Series*, vol. VIII (1675-1676), p. 409. Henceforth cited *CSPC*.

[14] Thomas Pownal, *The Administration of the British Colonies* (5th ed., 2 vols., London 1774) vol. I, pp. 256-57.

[15] Photostats, MHS, Part I, p. 28. Prices for these goods may be found in John Hull's Journal, 1685-1689 (Baker Library).

[16] Harold A. Innis, *The Cod Fisheries: The History of an International Economy* (New Haven, 1940), p. 159.

[17] As an example, of 90 vessels entering Boston from Southern Europe between November 27, 1752 and December 3, 1755, 71 carried only salt, 3 salt and wine, and 16 wine only.

[18] *Boston Records*, XXXII (A Volume Relating to the Early History of Boston containing the "Aspinwall Ntarial Records" from 1644 to 1651) (Boston, 1903).

In the late seventeenth century, it was noted as "the only great stapple which the Country produceth for forraine parts and is so benefitiall for making returns for what we need." [19] John Higginson of Salem, writing his brother in England at the turn of the century, said: "Places proper in Europe to make returns to England from, and are much improved for that end, from hence, are Bilboa, Cadiz, Oporto, and the Streights." [20] Merchant's directions for the disposal of their fish in Europe repeated a monotonous refrain, "and remit the produce." The produce commonly ranged from 30 to 50 per cent of the sale price. Often the total proceeds of a cargo were remitted, the vessel returning in ballast or plying European trade routes to earn a return lading.

As for the second great food staple — grain, New Englanders very early supplied part of Southern Europe's demand for wheat and corn. By the spring of 1648 wheat had become scarce at Massachusetts Bay and John Winthrop noted: "Our scarcity came by occasion of our transporting much to the West Indies and the Portugal and Spanish Islands." [21] Two years later, Edward Johnson wrote in his *Wonder Working Providence*: "Portugal hath had many a mouthful of bread and fish from us in exchange of their Madeara liquor, and also Spain." [22] Black rust destroyed New England wheat and soon the middle colonies were shipping to the Iberian market. Jacob Leisler, who married into a prosperous wine business, was captured by the Barbary pirates on a voyage there in 1678.[23] In the same year Sir Edmund Andros estimated that New York exported 60,000 bushels of grain yearly.[24] These shipments were largely indirect. Yankee merchants held New York in economic thralldom down to Queen Anne's War, trading European goods for grain and draining money from that colony. This slavery was not quiescently accepted. As early as the 1680's a 10 per cent duty was levied by New York on European goods entering the colony indirectly. Coercive tariffs were employed under Governors Edward Hyde Cornbury and Robert Hunter and, after 1715, wine imported indirectly paid double

---

[19] Samuel Eliot Morison, *The Maritime History of Massachusetts, 1783–1860* (Cambridge, Mass., 1921), p. 13.
[20] James Duncan Phillips, *Salem in the Eighteenth Century* (Boston, 1937), pp. 77–78, citing a letter from John Higginson to Nathaniel Higginson in London, Salem, 1700.
[21] Hosmer (ed.), *John Winthrop's Journal*, vol. II, p. 341.
[22] J. Franklin Jameson (ed.), *Johnson's Wonder Working Providence, 1628–1651* (New York, 1910), p. 247.
[23] Lawrence A. Leder, *Robert Livingston and the Politics of Colonial New York* (Chapel Hill, 1961), p. 17n. Robert N. Toppan and A. T. S. Goodrick, *Edward Randolph, Including His Letters and Official Papers* (7 vols., Boston, 1898–1909), vol. IV, pp. 254, 258.
[24] George Louis Beer, *The Old Colonial System* (2 vols., New York, 1912), vol. II, p. 345n.

duties.²⁵ Pennsylvania too encouraged direct trade to Southern Europe by economic pressures.

Well before 1700 New York had managed to establish contacts with the Wine Islands. The letters of William Bolton, an English factor at Madeira, mentioned 9 vessels from the colony trading there between 1695 and 1700, in comparison to 4 from Pennsylvania and 10 from New England. Early in 1700 he wrote: "This place hath been abundantly supply'd with all sorts of Provisions, soe that since Wheate cannot be shipped from England — because of the coming of soe many vessels from the Westerne Colonies." ²⁶

Pennsylvania, as noted, was also involved. Bolton, in 1697, referred to a Pennsylvania vessel in from Cadiz, carrying 1,800 pieces-of-eight, as well as a ketch which arrived in 1700 with 150 moyos of wheat "which causeth that noe Corne sels." ²⁷ Other grain producing colonies sought outlets in Southern Europe before the turn of the century. In 1671 Governor William Berkeley of Virginia requested liberty for the colony to ship pipestaves, timber, and corn to European markets, obviously south of Cape Finisterre.²⁸ A petition, submitted to the Commissioners sent to Virginia as a result of Bacon's Rebellion, asked permission to send wheat and other products to the Wine Islands and return directly with salt.²⁹ Permission to import salt directly was never extended to the southern colonies.

Thus, grain exportation began quite early but apparently down to 1700 the major advantages of the trade accrued to New England. By that date, however, both Philadelphia and New York were extending their trade in grain, though much of their produce was carried by Yankee shippers throughout the colonial period.

Benzanson, Gray, and Hussey note in their study of prices in colonial Pennsylvania that by 1700 the colony's bread and flour had flooded the West Indian market.³⁰ When crops were good, this was evidently a continuing phenomenon. Fortunately, the Treaty of Methuen, negotiated by Britain in 1703, allowed an outlet in Portugal for the surplus. Soon Philadelphia merchants were talking of shipping into the Straits.³¹ The rapid growth of trade to Lisbon caused Isaac Norris of Philadelphia to write in 1711: "I am apt to

---

[25] Nettels, *Money Supply*, p. 107; A. A. Giesecke, *American Commercial Legislation before 1789* (New York, 1910), pp. 22–24.
[26] Andre L. Simon (ed.), *The Bolton Letters, 1695–1714* (London, 1928), p. 156.
[27] Ibid., p. 180. A moyo equals 23.0 bushels of grain, total then was 3,454 bushels.
[28] Beer, *Old Colonial System*, vol. II, p. 114.
[29] Ibid., p. 143.
[30] Anne Bezanson, Robert D. Gray, and Miriam Hussey, *Prices in Colonial Pennsylvania* (Philadelphia, 1935), p. 9.
[31] Arthur L. Jensen, *The Maritime Commerce of Colonial Philadelphia* (Madison, 1963), p. 57.

think the country has, within ten or twelve years, increased to near ten times its then produce of corn, wheat especially. The market of Lisbon has been of great advantage to us." [32] In these years New York shipping to Europe rose sharply and when the war ended, it increased again.[33]

In 1714 Governor Hunter complained that the New York wine market had been glutted for a year or more.[34] By 1712 Maryland too had opened a small trade to Lisbon in grain, returning wine, cash, and bills. Joshua Gee reported to the Board of Trade in 1716 that Pennsylvania had lately "shipt large quantities of corn to Portugal and other parts of Europe, to put themselves in a capacity of purchasing in England cloathing and other necessaries which they want." [35] By the second decade of the century Pennsylvania imported more than 250 pipes of wine yearly.[36] Quite significantly, the Crown granted both Pennsylvania and New York the right to import salt directly in 1726 and 1730 respectively.

The decade of the 1730's saw this traffic assume greater proportions. Fayrer Hall noted of Pennsylvania in 1731: "Whatever they sell their Wheat, Flour, and Bread for in *Portugal* or *Spain*, is generally sent to *London,* which cannot amount to less than 25,000 l per *Annum."* [37] He estimated that indirect trade to Southern Europe via Newfoundland and the Canaries remitted an additional £8,000. His figures may have been exaggerated, though in the average year between 1729 and 1736 some twenty-one vessels cleared for Southern Europe from Philadelphia. It is also important to note that the total English balance against Pennsylvania for 1731 was but £31,474.[38]

Philadelphia mercantile correspondence indicates the growing importance of this traffic in the 1730's with such comments as: "Wheat, flour and bread are like to be very dear and scarce — most of our wheat being already sent up the Straits." and "the Lisbon trade and Cadiz likewise, has answered better of late than the West Indies." and again, "Phil'a is as much or more oblidged to Lisbon than any port whatsoever, London excepted." [39] In 1731 it was

[32] Bezanson, *et als, Prices,* p. 9, citing Isaac Norris to Joseph Pike, June, 1711.
[33] *CSPC,* vol. XXIX (1716–1717), p. 256.
[34] *Ibid.,* vol. XXVIII (8/1714–12/1715), p. 15.
[35] *Ibid.,* vol. XXIX (1/1716–7/1717), p. 271.
[36] Mary A. Hanna, *The Trade of the Delaware District before the Revolution* (Northampton, Mass., 1917), p. 265n.
[37] Hall, *Importance of British Plantations,* p. 98.
[38] *Ibid.;* Charles Whitworth, *The State of the Trade of Great Britain in its Imports and Exports Progressively from the Year 1697 to 1773* (London, 1776), p. 67.
[39] Bezanson, *et als., Prices,* p. 23, citing John Reynell to Richard Deeble in Plymouth, England, Philadelphia, February 25, 1735; Jensen, *Maritime Commerce,* p. 59, citing John Reynell to Michael Dicker, Philadelphia, November 19, 1736, and William Till to Lawrence Williams, Philadelphia, September 11, 1740.

claimed that Pennsylvania sent more than 55,000 bushels of grain to Ireland and the south of Europe.[40] By the middle of the 1730's, more than one-fifth of the tonnage cleared from Philadelphia went to the Iberian area. Between August, 1735 and August, 1736 New York and Philadelphia cleared 69 vessels for Southern Europe. (New York, 31; Philadelphia, 38.) [41]

War soon depressed the trade but in the late forties it revived again. In March, 1747 Samuel Powell, Jr. of Philadelphia wrote: "Flour is now upon the start again. — we have now a call for Lisbon. Most of our two-decked vessels are going thither, which will raise the price." [42] In the first post-war years the trade flourished and the Pennsylvanians cleared more than thirty vessels a year to Iberia, the Wine Islands and up the Straits.[43] A temporary decline occurred in 1754 and then war intervened again.

Bezanson, Gray, and Hussey, in commenting on Philadelphia's traffic with Southern Europe, state: "Approximately one third of the trade might be said to have facilitated the settlement for British imports." [44] Purchases of island wine apparently balanced with exports thence, thus, the returns came from grain sold at Lisbon and other mainland ports. More than half of the vessels commonly returned directly to Philadelphia. Salt filled the holds of most of these ships and again the price relationship of wheat to salt favored the colonials. In 1751–1752, for example, a bushel of wheat bought four or five bushels of salt.[45] Once more, this ratio is based upon American prices where wheat was cheaper and salt dearer than in Europe.

There are indications that grain exports to Southern Europe — and fish also — were, at times a losing venture. It is possible that, when Polish, English, or Sicilian competition for the grain market kept prices low, shipments were still made in order to secure cash and bills of exchange to pay English creditors. When debts were pressing, a cargo sent off to Spain or Portugal could answer such demands. On the other hand, merchants who engaged in several trading patterns could employ the returns from Southern Europe to cover expansion in other directions.

Two other colonial exports were also important in the trade with Southern Europe, lumber and rice. Wood products were constantly

---

[40] Hanna, *Trade of the Delaware District*, p. 264–65.
[41] Statistics taken from the *Pennsylvania Gazette* (Philadelphia), August 7, 1735 to August 2, 1736.
[42] Bezanson, *et als.*, *Prices*, p. 35, citing Samuel Powell, Jr. to Gabriel Manigault, Philadelphia, March 16, 1747.
[43] Jensen, *Maritime Commerce*, p. 57. Virginia D. Harrington, *New York Merchant on the Eve of the Revolution* (New York, 1935), p. 290.
[44] Bezanson, *et als.*, *Prices*, p. 29.
[45] *Ibid.*, pp. 237–44.

in demand in Southern Europe. Pipestaves were traded there even before 1640 and remained a salable commodity through the whole period. In 1671, John Mason noted that New Hampshire shipped there 20,000 tons of deal and pipestaves and ten shiploads of masts.[46] Lumber and ship timber often went as deck cargo and sold well. At the turn of the century it was reported that a lumber cargo from Portsmouth to Cadiz produced more than three times its original cost in New England.[47] Between 1752 and 1756 Boston shipped, in the average year, 76,000 pipe and barrel staves and 36,000 board feet of lumber south of Cape Finisterre. In addition there were large amounts of ship timber, planks, and finished wood products such as oars, desks, bookcases, and early in 1756, five prefabricated house frames.[48]

Rice had evidently found some outlet in Portugal before its enumeration in 1705. Though some was smuggled after that date, it did not go to Southern Europe in large amounts until its removal from the enumerated list in 1730. Almost at once these shipments became a significant factor in South Carolina's trade. Between 1730 and 1739 almost 88,000 barrels of rice was sent to ports south of Cape Finisterre, with a value exceeding £150,000.[49] Southern Europe evidently continued to take about one-fifth of the crop. Malaachi Postlethwayt in his *Universal Dictionary of Commerce* (1751) informs us that in a good year shipments to Lisbon and other ports reached 10,000 barrels.[50] These exports much more than paid for the tawny red Madeira so popular in the homes of the South Carolina planters. As with the other products sent to Iberian ports, the additional returns from the sale of the rice were remitted to England and spent for English manufactures.

Shipments to Southern Europe were not confined to fish, grain, lumber, and rice. A wide variety of other goods was sent there, including tobacco, dyewoods, hats, candles, beeswax, and feathers. In the 1720's Thomas Amory even shipped a billiard table to one of his correspondents.

It is very difficult to estimate the amount of the remittances from Southern Europe sent to England or of the cash brought to America. Yet, the value of the goods exported and those imported can be

---

[46] *CSPC.*, vol. VII (1669–1674), p. 294.
[47] Eleanor L. Lord, *Industrial Experiments in Colonial America* (Baltimore, 1898), p. 106.
[48] Photostats, MHS, Part II (1717–1755).
[49] Malachi Postlethwayt, *Universal Dictionary of Trade and Commerce* (2 vols., London, 1751–1755), vol. I, p. 362. See also *South Carolina Gazette*, 1731–1738. Average price of rice in this era is considered as 38 shillings/bbl.
[50] Postlethwayt, *Universal Dictionary*, vol. I, p. 362.

compared in the periods for which Port Office Records are available. Between May 1, 1686 and September 29, 1686, approximately 550 tons of fish, plus an undetermined amount of lumber and other goods, left Boston for Southern European ports, while 400 tons of salt and 201 pipes of wine entered.[51] Utilizing these figures and employing Boston prices for all three products, the value of the fish exceeds that of the salt and wine by £918.[52] Using the same approach with Salem's statistics for July, 1716 through June, 1717, when over 67,000 quintals of fish and sundry other goods cleared and 1,660 hogsheads of salt and 108 pipes of wine entered, the balance on the fish side of the ledger amounts to £41,784 sterling.[53] The significance of these returns from Southern Europe is properly highlighted when this figure is compared to the trade imbalance for *all* New England, in 1717, of only £73,103.[54]

Another area remains for brief consideration, the transferral of these credits to England. Colonial shipments were commonly consigned to English or local merchants in Spanish, Portuguese, and Mediterranean ports. Through innumerable merchant houses, such as Hill & Lamar or Patrick Joyce & Co. at Madeira; Parminter & Barrow, Lory & Michell, or the Gardoquis at Bilbao; Anderson & Co. and Macky & Smith at Gibraltar; William Gibbs, and Parr & Bulkeley at Lisbon; Michael and Richard Harris at Barcelona; cargoes were sold and ladings purchased for the colonial merchants. These firms, often branches of English houses, also handled the transfer of the credits. Traffic in colonial staples made up a very important part of their total business. In 1717 there were ninety such English firms located in Lisbon alone.[55] Bills of exchange were sent by them to London and then were followed by shipments of bullion to cover these demands. Spain and Portugal had made the exportation of specie a capital offense but it had become so routine that on one occasion English merchants charged with this crime presented common usage as their defense.[56] Generally the law was loosely enforced but when it was tightly applied the bullion was smuggled out to England. A British newspaper, commenting on Spanish bullion exports early in the eighteenth century, stated that it was "always strictly enjoyn'd but the execution Universally

---

[51] Photostats, MHS, Part I (1686–1717).
[52] Prices are taken from John Hull's Journal, 1685–1689 (Baker Library).
[53] Salem statistics are to be found in Photostats, MHS, Part II (1717–1756). Prices were found in mercantile correspondence.
[54] Whitworth, *State of the Trade of Great Britain*, p. 63.
[55] V. M. Shillington and A. B. Chapman, *The Commercial Relations of England and Portugal, 1487–1807* (London, 1907), p. 217.
[56] *Ibid.*, pp. 239–41, 249.

omitted." [57] Thus the colonial credits amassed in Southern Europe reached the hands of the English creditors.

From one-third to one-half of the value of the goods shipped to Southern Europe was converted into cash and bills of exchange. An educated guess indicates that these returns covered perhaps a third of the imbalance with England when nothing interfered with the trade. That being the case, the availability of the Iberian markets and those in the Mediterranean assumes much more significance. If Southern European trade helped so materially with the burden of the balance of payments, any dislocation of this trade should have been reflected by fluctuations in the colonial economy and by depressed commodity prices.

Variations in crop production in America and Europe and the coincident competition or lack of it was naturally important. Colonial newspapers and mercantile correspondence demonstrate a continued awareness of the factors effecting the markets in Southern Europe. Wars and rumors of war were almost as important. In 1717 Timothy Fitch wrote a London correspondent: "I'm Solicitous how to make you some remittances, Wee Heurd Rumours of some jealousies conceived of Spain, which makes us desirous of a further Aco't before wee proceed farr in loading fish." [58] Fitch was also concerned by the threat of pirates from Sallee. More than twenty years later, in 1741, Reverend John White of Gloucester complained of the failure of the fishery, laying it to the war with Spain and fear of war with France, which had lowered the price of fish and increased that of salt and other necessities.[59]

Periodic depressions between 1713 and 1740 have an uncanny relationship with the various Anglo-Spanish crises of this era. Major fluctuations occurred in 1719, with the War of the Quadruple Alliance, in 1727, when minor hostilities broke out between the two powers, and again at the end of this period, with the advent of the new war. Seizures of English vessels in the West Indies between these dates are reflected by less violent fluctuations. The rise and fall of wholesale commodity prices show a neat correlation with these diplomatic crises. There appears to be a definite relationship between the colonial depressions of this era and this trading pattern.[60] As another example, it is interesting that when the

[57] *The Mercator* (London), Number 74 (1713), p. 2.
[58] Letter Book of Timothy Fitch, 1714–1717 (New England Historic and Genealogical Society), letter dated May 3, 1717.
[59] John J. Babson, *Notes and Additions to the History of Gloucester* (Salem, 1891), second series, p. 123.
[60] Fluctuations in the economy are noted by Richard Pares, *War and Trade in the West Indies* (Oxford, 1936), pp. 14–16; George F. Warren, Frank A. Pearson, and Herman M. Stoker, *Wholesale Prices for 213 Years, 1720–1932* (Ithaca, 1932), pt. II, p. 201; Earl J.

various movements to organize banks developed in Massachusetts they coincided with low points in the trade to Southern Europe. Periodic demands for paper money may also have had some relationship to the availability of these markets. A detailed examination of the factors effecting this trade may well expose the viability of these associations.

Other tangential questions also require examination and explanation. For example: Did English mercantilists consciously encourage the development of Colonial trade with Southern Europe? How significant was protection against the Barbary pirates to its growth? What was its relationship to other trading patterns? In what way did strictures placed on it after 1763 contribute to imperial tensions? Is its collapse in the 1780's related to the depression of that era? And these are but a few.

Colonial trade to Southern Europe, from the 1640's to the American Revolution and after, was then a very important factor in the growth of the colonial economy. Perhaps a third or more of the adverse balance of payments was covered by Spanish, Portuguese, and Mediterranean returns when those markets were available. In this light the burden placed on the colonial economy by the mercantilist system appears to have been much less onerous than some historians have considered it.

To return to the schooner *Jolly Robin* and her voyages to Bilbao, we can suggest that they represent a trading pattern which had been established for over a hundred years; which was not confined to fish cargoes alone; which commonly saw a 40 per cent return in cash and bills; and which was extremely significant in the development of the American colonial economy.

Hamilton, *War and Prices in Spain, 1651–1800* (Cambridge, Mass., 1947), p. 176; Bezanson, et al., *Prices*, pp. 9–13, 315–16; Arthur H. Cole, *Wholesale Commodity Prices in the United States, 1700–1861* (Cambridge, Mass., 1938).

Goals and Enforcement of British Colonial Policy, 1763-1775

BY NEIL R. STOUT

GREAT Britain came out of the Seven Years War in 1763 with a huge national debt, a vastly enlarged empire, and a load of new responsibilities. These were fundamental considerations in shaping Britain's policies toward her North American colonies between 1763 and 1775. The 'salutary neglect' of the past half century, in the British government's view, could have no place in the new imperial system. During the Seven Years War British officials began to rethink their country's colonial policies and by 1763 they had projected certain goals and enforcement procedures. Three goals are apparent: (1) the empire must be defended; (2) the colonies must be placed under stricter central control; and (3) the colonies must be made to yield increased revenues.

The first goal required maintenance of sufficient military forces to discourage French incursions on the extended empire. France was not decisively beaten in the Seven Years War, a war which weaknesses in imperial defenses between 1750 and 1755 had helped to start. The military estimates for the British government's 1763 budget were, therefore, the largest in the nation's peacetime history. Over 8,000 soldiers were to be stationed in North America, more than twice as many as in 1754.[1] The largest single element of the Royal Navy outside of Britain's home waters was to be the twenty-one ship North American squadron covering the American coast from Florida to Nova Scotia, with smaller naval squadrons assigned to Newfoundland, the Caribbean, and the Gulf of Mexico.[2] Although their primary mission was defense, these military units were quickly made instruments of colonial policy as well. The other two goals,

---

[1] Bernhard Knollenberg, *Origin of the American Revolution: 1759-1766* (New York, 1960), p. 88. Knollenberg, however, does not believe the troops' purpose was defensive.

[2] Admiralty to Secretary of State, 5 January 1763. British Public Record Office [PRO] S.P. 42/43.

AMERICAN NEPTUNE, 1967, vol. 27, pp. 211-220.

more control over the colonies and more revenues from them, are inextricably combined; for increased revenue collections depended upon better political control, and, given Britain's financial problems, closer control had to be paid for through colonial revenues. The first goal was also tied to the latter two, for the colonies were expected to help pay for their defense, and the military was ordered to assist British political and revenue officials in the colonies.

This neat package did not come as a sudden revelation to George Grenville when he became Prime Minister in April 1763.[3] Early in 1759 the Board of Trade had collected from its files of the past twenty years all letters relating to illicit trade in the colonies and sent them to the Board of Customs Commissioners for comment. The Customs Commissioners replied that their own files also revealed 'great Difficulties and Doubts... in the execution of the Acts of Trade... and of his Majesty's Revenue of Customs.' The Customs Commissioners grouped the problems under three headings. First, much foreign molasses was illicitly imported from the French Caribbean islands because, as the Commissioners saw it, the six-penny duty on foreign molasses was uncollectible and simply made smuggling profitable. Second, evasions of the Navigation Acts, while not as flagrant as evasions of the molasses tax, could be halted completely only by strengthening the customs service and taking prosecutions out of colonial courts, which were 'not so impartial as in England.' Third, the Commissioners admitted that wartime trade between the British North American colonies and French Caribbean colonies was carried on with impunity, helping to keep the French in the war.[4]

Early in 1760 a committee of the Privy Council reported widespread colonial trade with the enemy French and alleged that the charter colonies of Rhode Island and Connecticut were virtually independent of British rule.[5] Meanwhile, complaints flooded in from admirals and generals that the French Caribbean islands were being sustained by British-American provisions carried in British-American ships. From 1759 on, the Royal Navy begged for orders to seize ships engaged in trade with the French under bogus flags of truce or through the neutral Spanish port of Monte Christi. Finally the navy began making seizures without orders from London. Customs officers in such ports as Boston and New York tried to make trade with the enemy unprofitable by collecting, for the first

[3] For an excellent discussion of this point, see Thomas C. Barrow, 'Background to the Grenville Program, 1757-1763,' *William and Mary Quarterly*, 3d Ser., XXII (1965), 93-104

[4] Customs Commissioners to Board of Trade, 10 May 1759. PRO T. 1/392.

[5] W. L. Grant and James Munro, eds., *Acts of the Privy Council of England, Colonial Series* (6 vols., London, 1908-1912), IV, 443-447. Dated 17 February 1760.

time, the six-penny duty on foreign molasses, the chief commodity traded for American provisions. (Out of their collection efforts arose the famous writs of assistance case in Boston.) The Royal Navy assisted the customs collectors by intercepting merchant ships and escorting them to the customs houses before they could unload their cargoes clandestinely.[6] In August 1760 Secretary of State William Pitt at last took official notice of the colonies' trade with the enemy in a sharply worded circular letter to the colonial governors ordering them to report the extent of illicit trade in their provinces and their measures to stop it. Many governors, no doubt motivated mainly by a desire to placate Pitt, replied that the problem had already been solved by the Royal Navy.[7] Whether the navy had, in fact, halted the trade is unimportant; the significance lies in the fact that the Royal Navy claimed success and was backed in its claims by governors and customs men. Furthermore, collections of the molasses duty rose markedly during 1760 and 1761, indicating that the navy and the customs service had at last put some teeth into the Molasses Act of 1733.[8] Here, then, was a possible solution for Britain's colonial revenue problems.

It should not be surprising that George Grenville's first step in building his new imperial system was to continue, formalize, and strengthen naval enforcement of the trade and revenue laws. The Act of 3 George III, Chapter 22, introduced while Grenville was still First Lord of the Admiralty in the Bute administration, became law in April 1763, shortly after he formed his own ministry. Among its provisions, the act stated that 'for the more effectual prevention of the infamous practice of smuggling, it may be necessary to employ several of the ships and vessels of war belonging to his Majesty.'[9] This innocuous-sounding clause was the key to Grenville's enforcement program. By the first of May 1763, less than a month after Grenville became Prime Minister, officers of warships being sent out to America were receiving Treasury warrants making them

[6] The problem of trade with the enemy is discussed at length by many authors. These are two of the best. Richard Pares, *War and Trade in the West Indies 1739-1763* (Oxford, 1936), pp. 356-487. George Louis Beer, *British Colonial Policy 1754-1765* (New York, 1907), chs. vi, vii.

[7] Gertrude S. Kimball, ed., *Correspondence of William Pitt* . . . (2 vols., New York, 1906), II, 320-321, 344, 348-363, 373-382, 388, 401-403.

[8] Beer, *British Colonial Policy*, pp. 115-116.

[9] 3 Geo. III, c. 22, did not actually authorize the Treasury to issue customs deputations to navy officers. It did not need to, since the Treasury had had this power for a century under 14 Car. II, c. 11. The real importance of the Act of 3 Geo. III, so far as the Royal Navy was concerned, was its provision that seizures made by naval vessels should be divided according to wartime prize law, half to the seizing ship and half to the king, rather than in the usual one-third each to the seizer, the king, and the local royal governor. This provision caused much strife between the Royal Navy and the governors.

deputy customs officers.[10] The Admiralty and Treasury hurriedly drew up instructions for the navy officers in their new role as customs men.[11] On 9 July 1763 Secretary of State Lord Egremont informed the colonial governors of the navy's enforcement mission and bluntly ordered them to co-operate.[12]

Meanwhile, the Treasury took steps to reform the shore-based customs service. In May 1763 it ordered the Board of Customs Commissioners to find ways to improve collections in America and the West Indies, which 'amounts in no degree to the sum which might be expected of them.'[13] In July the Treasury ordered all American customs officers to their posts, a measure most disconcerting to many of the absentees, who had thought their offices were sinecures. Most important, the Treasury ordered strict enforcement of the Molasses Act of 1733 and a halt to the old and perfectly logical practice, worked out between merchants and customs men, of collecting only a penny or two of the six-penny-per-gallon duty on foreign molasses.[14] Customs officers like Surveyor General John Temple opposed enforcement of what had always been a bad law, but the Grenville government knew what it was doing. Its first aim was to impress both the colonists and its own appointees that it would tolerate no laxness in collection of revenue. The duties themselves could be, and soon were, adjusted to make them revenue-producing rather than protective.

In October 1763, just as Admiral Lord Colvill's North American squadron and the absentee customs officials were reaching their posts in America, a Treasury memorandum to the Privy Council outlined a comprehensive enforcement program. First noting that colonial revenues had not increased, despite the colonies' commercial growth, and returned less than a quarter of the cost of collecting them, the memorandum outlined the steps already taken to strengthen the revenue service and halt illicit trade. The Treasury said other measures would be needed, not only because 'the military establishment necessary for maintaining these colonies requires a large revenue to support it,' but also because the colonies were growing so rapidly in territory and population that Britain might soon become unable to impose her will upon them. Governors and customs men were to be held strictly to their duty, and the army and navy were to assist them and protect them 'from the violence of any desperate and

---

[10] List of deputations in Customs to Admiralty, 10 September 1765. PRO Adm 1/3866.
[11] Admiralty to Secretary of State, with enclosures, 27 May 1763. PRO S.P. 42/64. See also Customs instructions to navy officers, Customs to Admiralty, N.D. PRO Adm 1/3866.
[12] Printed in New-York Historical Society *Collections*, LV (1922), 222-225.
[13] Treasury to Customs, 21 May 1763. PRO T. 11/27.
[14] Treasury to Customs, 25 July, 3 September, 10 November 1763. PRO T. 11/27.

lawless persons who shall attempt to resist the due execution of the laws.' The Treasury memorandum added: 'The advantages of a sea guard, more especially in those parts, are sufficiently obvious. We depend upon it as the likeliest means for accomplishing these great purposes, and ... earnestly wish that the same may not only be continued but even extended and strengthened as far as the naval establishment will allow.' The Royal Navy, the nation's bulwark in war, was now to help solve Britain's colonial problem.[15]

News of closer control of their trade staggered the American merchants, already in the grip of a post-war recession. Massachusetts' Governor Francis Bernard said it caused more alarm than did the French capture of Fort William Henry in 1757, and other commentators bore him out.[16] The *Providence Gazette* announced that Rhode Islanders could look after their own affairs, 'without the concurrent Assistance of *Swaggering Soldiers* or insulting *Captains Bashaws*, I mean Captains of War Ships.'[17] For the first time in years, colonial legislatures and merchants' associations energetically protested to Parliament against the scheduled renewal of the Molasses Act. Then the shippers turned to the old tactics that had worked so well in the past to make customs officers interpret the trade and revenue laws reasonably. First they tested the new customs officers' vigilance by attempting to continue their old practices; then, if caught, harassed them with legal obstacles, arrest on trumped-up charges, and mob violence. Concerted opposition came whenever a port first felt the pinch of the new enforcement procedures—in Boston and New York it came in late 1763; in Charleston, where the molasses trade was unimportant, it did not come until 1767.[18]

The Royal Navy proved equal to the challenge, although it received, at first, very little aid from the Royal Governors, with whom it had a dispute over distribution of shares from captured contraband;[19] nor from vice-admiralty court judges like Richard Morris of New York and John Andrews of Rhode Island, who served the shippers first and the king sec-

[15] *Acts of the Privy Council, Colonial Series*, IV, 217-220. Dated 5 October 1763. Note that the memorandum was written before any reports on the navy's success could have come from North America, proving the government's faith in the Royal Navy.

[16] Quoted in Frederick B. Wiener, 'The Rhode Island Merchants and the Sugar Act,' *New England Quarterly*, III (1930), 468.

[17] 3 December 1763.

[18] For 1763, see letters to Admiralty of Admiral Lord Colvill (PRO Adm 1/482), and Captains Thomas Bishop (Adm 1/1424), John Brown (Adm 1/1494), James Hawker (Adm 1/1898), and Archibald Kennedy (Adm 1/2012). For Charleston in 1767, see Captain James Hawker to Admiralty, 2 June 1767. PRO Adm 1/1899.

[19] See, e.g., Admiral Lord Colvill to Admiralty, 21 December 1763, 22 January 1764. PRO Adm 1/482.

ond;[20] nor from the regular customs officers, who accused the navy men of poaching on their domain.[21] But the British government consistently backed the navy against both merchants and royal officials. The government confirmed the navy's share of seizures and told the governors to quit carping.[22] A new vice-admiralty court with jurisdiction over all America gave navy officers the option of prosecuting their seizures in a court not automatically prejudiced against them, and the old courts had their procedures and fee lists carefully examined.[23] The shore-based customs officers were ordered to assist the navy in prosecutions and to provide bail and bond money for navy officers who, like Captain John Brown, had been sued or thrown in jail by local merchants.[24]

Government directives accomplished some of these changes; many more came with the passage of the Revenue Act of 1764, better known as the Sugar Act.[25] It was well designed to make smuggling (and honest trade) more difficult and less profitable, and to remove hostile (and often salutary) impediments from the enforcement process. All its many controversial provisions were based on reports from the field, many of them directly traceable to Royal Navy officers' complaints made during their first four months on the American station. Admiral Lord Colvill, for example, asked that the navy's share of seizures be confirmed against the royal governors' demands, that all trade with the tiny French islands of St. Pierre and Miquelon be outlawed, and that a new vice-admiralty court for all America be located at Halifax.[26] The troublesome loading and unloading provisions and requirement that a bond be posted for non-enumerated cargoes seem almost like answers to letters written by Captains Archibald Kennedy and James Hawker in December 1763.[27] Customs officers had long wanted more protection from damage suits; the

---

[20] Carl Ubbelohde, *The Vice-Admiralty Courts and the American Revolution* (Chapel Hill, 1960), pp. 94-97. Charles M. Hough, ed., *Reports of Cases in the Vice-Admiralty Court of the Province of New York* (New Haven, 1925), pp. 215-218. Admiral Lord Colvill to Admiralty, 26 December 1763. PRO Adm 1/482. Captain John Brown to Admiralty, 16 May 1764. PRO Adm 1/1494.

[21] Admiral Lord Colvill to Admiralty, 12 January 1765. PRO Adm 1/482. John Temple to Treasury, 10 September 1764. PRO T. 1/429. Ubbelohde, *Vice-Admiralty Courts*, p. 78.

[22] 4 Geo. III, c. 15, section XLII, further clarified by 5 Geo. III, c. 45, section XXXI. Customs to Treasury, 22 November 1766. PRO T. 1/453.

[23] Ubbelohde, *Vice-Admiralty Courts*, esp. ch. 2. Order in Council of 6 June 1764 printed in *Acts of the Privy Council, Colonial Series*, IV, 670.

[24] Captain John Brown to Admiralty, 16 May 1764. PRO Adm 1/1494. Captain Archibald Kennedy to Admiralty, 20 November 1764. PRO Adm 1/2012. Treasury to Customs, 28 January 1765. PRO T. 11/27. Admiralty to Lord Colvill, 7 March 1765. PRO Adm 2/538.

[25] 4 Geo. III, c. 15.

[26] Sections XXXV, XLI, XLII. Admiral Lord Colvill to Admiralty, 25 October, 21 December 1763. PRO Adm 1/482.

[27] Sections XXI, XXIII, XXIX, XXX. Kennedy to Colvill, 26 December 1763 and Hawker to Colvill, 12 December 1763, enclosed in Admiral Lord Colvill to Admiralty, 22 January 1764. PRO Adm 1/482.

navy reinforced their plea when Captains Hawker and Bishop were threatened with damage judgments and Captain Brown was actually thrown in jail by owners of a New York ship he had seized. One result was the much-maligned 'probable cause' provision of the Sugar Act.[28] Finally, numerous sections of the act allowed examination of papers and cargoes 'within two leagues of shore,' pointing directly to naval enforcement.[29] The Royal Navy must share the credit—or blame—for the Sugar Act.

Purely as a revenue measure, the Sugar Act was a success, the only successful tax Parliament ever imposed on the American colonies. The molasses tax, especially after it was lowered to a penny per gallon in 1766 and imposed on both British and foreign molasses, yielded enough revenue to allow the customs service to pay its expenses for the first time, with enough left over to help support other parts of British colonial officialdom.[30] The Royal Navy made the Sugar Act a success. Molasses, and to a lesser extent wine, is too cheap, bulky, and hard to conceal to lend itself to smuggling past even minimally alert customs. But until the sea guard was established to supplement the regular customs service, there was little risk in ignoring the customs houses. The colonial merchant considered it a sinful waste to pay a tax he could avoid, but when the risk went up he found it paid to obey the law. The Royal Navy added to the risks. The scanty records that survive show that the navy made slightly over half the seizures of the whole American customs establishment, but seizures are not nearly as important as the navy's assistance in the collection of duties.[31] For example, the log of H.M.S. *Sardoine* for 28 June to 18 September 1764 shows that she examined at least two hundred vessels in the Delaware River below Philadelphia. Of these only five were seized, and one was later released; but every vessel carrying a cargo that lent itself to evasion of duties, such as wine from Madeira or molasses from St. Eustatius, had a navy man or two put on board her to make sure she proceeded directly to the customshouse, where the regular customs officers collected the lawful duties.[32] Where there was a constant naval patrol, as in the Boston and New York areas, collections under the Sugar Act were high. Where it was intermittent, as in Rhode Island and Maine, collections rose

[28] Sections XLV-XLVII.
[29] Sections XXV, XXIX, XXX, XXXIII.
[30] Oliver M. Dickerson, *The Navigation Acts and the American Revolution* (Philadelphia, 1951), 86, 201. Oliver M. Dickerson, 'The Use Made of the Tax on Tea,' *New England Quarterly*, XXXI (1958), 232-243.
[31] Dickerson, *Navigation Acts*, p. 202. Edward Channing, *A History of the United States*, III (New York, 1912), 89-90. Dora Mae Clark, 'The American Board of Customs Commissioners, 1767-1783,' *American Historical Review*, LV (1939-1940), 800, 805.
[32] Log of H.M.S. *Sardoine*, 28 June-18 September 1764. PRO Adm 51/859.

and fell with the arrival and departure of the warships.³³ Equally important was the Royal Navy's protection of shore-based officials; mobs were not nearly as effective when the customs men had an armed, floating refuge.³⁴

Had the British government been content with the molasses tax and the other duties imposed by the Sugar Act, the history of British North America might well have been very different. Merchants grumbled, but paid, and the economic catastrophe they predicted did not materialize. But Parliament passed the Stamp Act, which united Americans more than anything else during the colonial period. A major reason for the Stamp Act's failure was that the British government, fearing no trouble, made no adequate provisions to enforce it. The Royal Navy, had it received proper instructions, might have made the Stamp Act work. No stamps under navy protection were ever destroyed. In Rhode Island, Customs Collector John Robinson took refuge on board H.M.S. *Cygnet* during the Stamp Act crisis and refused to issue ship clearances until the Rhode Islanders guaranteed him safe conduct. Had he been so instructed by the Treasury, Robinson could have held out until the Rhode Island shippers agreed to purchase stamped clearances.³⁵ In New York, Captain Archibald Kennedy refused to honor unstamped clearances and kept the port bottled up for a month, while local shippers first threatened his property and at last contritely begged him to lift his blockade. Kennedy acted on his own, he received no backing from his government, and no other navy captain went so far as he; but he showed that Britain, with a little attention to naval enforcement, could have forced compliance with the Stamp Act (or perhaps have brought on the American Revolution ten years early).³⁶

The British government paid much greater attention to enforcement in its next attempt to tax the colonies, the Townshend Acts of 1767. A final reform of the American customs establishment saw the formation of an American Board of Customs Commissioners, located in Boston, which could give closer attention to colonial revenue collection. The customs

---

³³ Dickerson, *Navigation Acts*, p. 186. Captain Philip Durell to Customs, 14 August 1766. PRO T. 1/453. Admiral Lord Colvill to Admiralty, 12 January 1765. PRO Adm 1/482. See also John F. Walzer, Colonial Coastwise Trade, 1763-1775 (unpublished M.A. thesis, U. of Wisconsin, 1960), Appendix E, taken from PRO Customs 16/1.

³⁴ Commodore James Gambier to Admiralty, 6 November 1770. PRO Adm 1/483. William Parker, et al., eds., *Archives of the State of New Jersey, First Series* (33 vols., Newark, Trenton, Paterson, and Somerville, 1880-1928), X, 209-216, 290. Clark, 'American Board of Customs,' *American Historical Review*, LV (1939-1940), 801.

³⁵ Edmund S. and Helen M. Morgan, *The Stamp Act Crisis* (Chapel Hill, 1953), ch. 9, esp. pp. 149-153.

³⁶ Ibid., pp. 163, 166-167. Neil R. Stout, 'Captain Kennedy and the Stamp Act,' *New York History*, XLV (1964), 44-58.

commissioners gave early notice of their intention to enforce the laws strictly when they seized John Hancock's sloop *Liberty* for the technical violation of loading without a permit, but only after the battleship *Romney* was sent to assist them. The *Liberty* seizure touched off a riot that forced the customs commissioners to flee to *Romney*, where they stayed until British troops occupied Boston in October 1768. The *Liberty* riot is often given as the reason for the dispatch of the troops, but the British government had already decided upon military occuptation of the town before the riot occurred.[37] During the winter of 1768-1769 Boston was graced by the presence of four regiments of redcoats and nearly half the North American naval squadron. Here, indeed, was military enforcement of Britain's colonial policy; yet revenue collections were no better than before the American Board of Customs Commissioners took over.[38]

There are two reasons for revenues not increasing under the Townshend program: first, the Sugar Act was already about as well enforced as it needed to be by the Royal Navy and the old customs establishment, and the American Board of Customs Commissioners did not enhance their efficiency; second, the Townshend taxes tremendously increased the problems of enforcement. Tea, the only important item taxed under the Townshend Acts, lent itself well to smuggling unless extremely close vigilance was exercised. The British government refused to spend enough money to provide this sort of vigilance. Commanders of the North American squadron repeatedly asked for more ships, particularly small, fore-and-aft-rigged coast guard cutters, and for more sailors to man the ships they had. The government always ignored these requests.[39] British consuls frequently reported American ships leaving Holland or Germany with large cargoes of tea, but these ships always seemed to arrive in America in ballast.[40] Neither the navy nor the shore-based customs service made any significant seizures of contraband tea, while collections of the tea tax declined from £9,723 in 1768 to £1,677 in 1772.[41]

Parliament had made the molasses tax enforceable by cutting it, and

[37] Secretary of State Lord Hillsborough to General Thomas Gage, 8 June 1768. PRO C.O. 5/86.

[38] In 1767, the first full year of the one-penny duty on molasses, collections under the Sugar Act came to £33,844. From 1768 to 1774, when the Customs Commissioners were active, collections under the Sugar Act averaged £33,104 per year. Dickerson, *Navigation Acts*, p. 201. See also Clark, 'American Board of Customs,' *American Historical Review*, LV (1939-1940), 804-805.

[39] Commodore Samuel Hood to Admiralty, 8 May 1769. PRO Adm 1/483. Commodore James Gambier to Admiralty, 6 November, 4 December 1770. PRO Adm 1/483. Admiral John Montagu to Admiralty, 22 March 1773. PRO Adm 1/484. Admiralty to Montagu, 15 June 1773. PRO Adm 2/548. Neil R. Stout, 'Manning the Royal Navy in North America, 1763-1775,' AMERICAN NEPTUNE, XXIII (1963), 174-185.

[40] Extract of letters from consuls enclosed in Commodore James Gambier to Admiralty, 18 November 1770. See also Gambier to Admiralty, 5 December 1770, 27 February 1771. PRO Adm 1/483. Treasury to Admiralty, 8 November, 4 December 1770. PRO Adm 1/4286.

[41] Channing, *History of U. S.*, III, 128.

thus cutting the profits from smuggling. It tried to do the same thing for tea with the infamous Tea Act of 1773. But Americans had also learned something from the Sugar Act. The Sons of Liberty in every port from Boston to Charleston swore that they would not again compromise principles for low import duties. Boston dumped the low-duty tea overboard, and her punishment led to the Revolution.

The blockade of Boston under the provisions of the Boston Port Act was Britain's last attempt to enforce her colonial policy before war broke out, but the blockade also caused the final breakdown of her enforcement procedures. With the army and navy occupied in Boston, not only tea but arms and ammunition poured into other parts of America. The scanty British forces could not even protect the king's property—the patriots captured at least seventy-two cannon from forts in Rhode Island, Maine, and even Charlestown, Massachusetts, during the last four months of 1774.[42] When hostilities broke out a few months later, the patriots were about as well prepared with materials of war as the British troops were, all because Britain refused to give up either her goals or her economy drive when they were mutually contradictory.

In the last analysis, the new enforcement procedures laid down in 1763 failed, for the goals of British policy were not realized. Their ultimate failure, however, resulted from their initial success. The strengthened customs service and especially the sea guard established by the Royal Navy did, in fact, successfully impose unwanted taxes on American imports. Faced for the first time in decades with the reality of British power, Americans reacted with economic sanctions, force, and constitutional arguments to thwart an extension of that power. They defeated the Stamp Act because Britain had failed to learn the lesson of enforcement from the Sugar Act. The Townshend taxes failed because their enforcement was too expensive. The Tea Act failed because American radicals feared it could be enforced, and they precipitated a crisis to head it off. The final attempt to enforce Britain's colonial policy, the Intolerable Acts, failed because enforcement was too little and too late. The goals of British policy in 1763 did not represent a departure from Britain's previous theory of empire. The real departure was the attempt to enforce those goals. This is what caused Americans to reassess their place in the British empire.

[42] Admiral Samuel Graves to Admiralty, 8 August, 31 August, 3 September, 23 September, 15 December 1774, 8 January, 19 March, 23 September 1775. PRO Adm 1/485.

SPECULATIONS ON THE SIGNIFICANCE OF DEBT
Virginia, 1781-1789

by MYRA L. RICH

CHRONIC debt was a by-product of colonial Virginia tobacco argriculture and of the economic relationship between the Chesapeake colonies and Great Britain. Recent studies have indicated the necessity of credit in an economy based on commercial agriculture,[1] and have pointed out that debt in Virginia was not a class phenomenon. More significant is the growing recognition that debt was not limited to the dealings with Britain's merchants, but that most of Virginia's debts were contracted and held among the colonists.[2] Virginians themselves knew they were indebted, to Britain and to each other, and variously attributed their indebtedness, "the greatest of imaginable hardships," to the "amazing scarcity of specie,"[3] to "dissipation and extravagance," and to the want of "industry and frugality."[4]

Virginians' attitudes were based on accurate observation. England intended to drain the colonies of specie, both by promoting colonial consumption of British goods and by forbidding coinage. The ensuing problems of colonial trade and the need for efficiency in marketing colonial tobacco had favored the development of the consignment system based on credit between planter and British merchant. The resulting availability of credit and of British manufactured goods, and Virginians' natural desire to live like gentlemen, had induced many of them to purchase more than they could afford.

But even without these consumption patterns, the state's rudimentary financial system would have forced them to seek credit and to accumulate debt. Virginians had no medium of exchange. Their shortage of specie was aggravated by seasonal fluctuations in trade, since the little specie they possessed entered the state in payment for their exports. Forced to turn to currency substitutes, Virginians during the colonial period paid their debts with tobacco notes, a form of commodity money in which each note

---

[1] Samuel M. Rosenblatt, "The Significance of Credit in the Tobacco Consignment Trade: A Study of John Norton & Sons, 1768-1775," *William and Mary Quarterly*, 3rd ser., XIX (1962), 383-399; Emory G. Evans, "Planter Indebtedness and the Coming of the Revolution in Virginia," *William and Mary Quarterly*, 3rd ser., XIX (1962), 511-533.

[2] Robert E. and B. Katherine Brown, *Virginia, 1705-1786: Democracy or Aristocracy?* (East Lansing, 1964), especially chapter V, "Debtors, Creditors and Paper Money."

[3] Rockingham County, Petition to the House of Delegates, June 5, 1784, Virginia State Library.

[4] *Virginia Journal and Alexandria Advertiser*, April 5, 1787.

represented a certain number of hogsheads of tobacco of a stipulated grade. During the Revolutionary War the state itself issued notes to buy supplies and to pay the military; and such of the state certificates as remained in circulation after the war were used as a medium of exchange.[5] Further, an act of Assembly made debts assignable, so that one man's note could be passed by its holder to another man in payment of an obligation.[6] None of these substitutes was entirely adequate, and the state would not resort to emissions of paper money. No banking system existed to make capital available, nor had the Assembly produced an adequate code of commercial laws to give form to business dealings.[7]

In the absence of these possibilities citizens relied on a network of private credit, which represented loans among friends and the remnants of other private business dealings. This use of credit as a medium of exchange was common during the colonial period,[8] but in Virginia, during the Confederation, it was less successful owing to the dislocation of the tobacco trade with England during and after the Revolution. The network of credit was further complicated by the presence of unpaid prewar debts to the mother country, one of the subjects mentioned in the peace treaty, and by the debts resulting from continued postwar trade with England. To make matters worse, the county courts, which should have enforced the few existing laws and aided in the settlement of debts and other commercial disagreements, were inadequate and overburdened, and frequently reluctant.

Virginians' observations of their economic affairs produced a body of public and private writings expressing the difficulties of payment and collection of debts, the scarcity of money, the importance of debtor legislation, and sanctity of contract. The complex situation which they described had formidable consequences for Virginia. An investigation of the patterns of holding debts and the methods used to settle them indicates that indebtedness severely affected commercial relationships of merchants and planters and posed serious problems for Virginia's General Assembly. More impor-

---

[5] E. James Ferguson, "Currency Finance: An Interpretation of Colonial Monetary Practices," *William and Mary Quarterly*, 3rd ser., X (1953),153-180. See also William Z. Ripley, *The Financial History of Virginia, 1609-1776*, in *Columbia College Studies in History, Economics and Public Law*, IV (1893-1894), 145-162.

[6] William W. Hening, *The Statutes at Large: Being a Collection of all the Laws of Virginia* (Richmond, Philadelphia, and New York, 1809-1823), XII, 358-359. The fact that this statute was passed in the midst of Virginia's paper-money fever indicates that the Assembly might have regarded it as a means of evading a paper emission while still providing a medium of exchange.

[7] J. S. Davis, *Essays in the Earlier History of American Corporations* (Cambridge, 1917), II, 37.

[8] Charles Grant, *Democracy in the Connecticut Frontier Town of Kent* (New York, 1961), especially chapter 5, "Threats to Economic Opportunity: Debt at Kent and Her Neighboring Towns."

tant, it had repercussions for Virginia's long-term relationship to Great Britain and to the new union.

I

While the absence of money complicated the process of transacting business in Virginia and led to business dealings on credit, the most important impact of the credit system for both merchants and planters was on the settlement of their accounts. Thomas Rutherfoord, an eighteen-year-old Irishman who had come to Virginia to make his fortune, discovered that he could not dispose of his goods at all unless he extended credit. Then, to his chagrin, he found that it was "impossible to force a sale here of anything without a risque of being never paid for it."[9] Dealing on credit made the realization of profits nearly impossible: "To do much business here one year requires an age to wind up."[10]

The collection of delinquent accounts was one of the worse problems that merchants faced in Virginia. The moment Thomas Rutherfoord opened his store he became a creditor, and his own statements indicate that most of his profits were on paper. His letters and those of other merchants, as well as Virginia's county court records, make it clear that being a creditor in a society of reluctant debtors was not at all conducive to making money.[11]

Rutherfoord's one escape was that employed by the other retail houses which sold almost exclusively on credit. The merchants regarded the extension of credit as bad business practice, but it was the only answer to their problems. If merchants like Rutherfoord did not extend credit they could not sell at all, for few customers could offer cash payments. As a result, both retail and wholesale storekeepers, though offering to receive payment in "cash, good bills of exchange on Europe, or country produce," usually accepted personal notes also in payment for their goods.[12]

Though the merchants complained that they were forced to extend credit and vowed that sales for cash were the only proper way to do business, their protestations were exaggerated. They were willing to gamble that they would ultimately collect what was owed them, and in the meantime to mark

---

[9] Thomas Rutherfoord to Hawksley & Rutherfoord, April 20, 1786, Thomas Rutherfoord Letterbook, Virginia Historical Society.

[10] Thomas Rutherfoord to Henry Westray, September 26, 1786. To speed things along Rutherfoord opened a second store and divided his remaining goods in half, hoping that they would sell more quickly and enable him to return home (Thomas Rutherfoord to Hawksley & Rutherfoord, December 30, 1786).

[11] None of Rutherfoord's account books exist, but his many letters attest to the likelihood of paper profits.

[12] It is fruitless to attempt any kind of generalization on the basis of merchants' account books, for there are too few of them. But in the Maryland and Virginia Accounts at the Library of Congress the books preserved record mainly IOU's and payments in kind.

their profits on their books. Because they preferred to sell rather than to let their goods sit on their shelves, they gave credit freely.[13]

The problems of Virginia's financial system weighed as heavily on the planters as the merchants. If the merchants were distressed over nonpayment of accounts, the planters were equally troubled about the difficulties of purchasing the goods they desired and making good their obligations. The problem of planter debt had plagued Virginia from its days as a colony and continued well into the 1790's. Many of the planters had been debtors before the war; but if a planter had not been in debt in 1776, he was sure to have achieved that status by 1786 or 1787.

The volume of planters' debts and merchants' complaints after peace was established reflected the fact that Virginians had resumed trade with Great Britain. After the war, as a matter of convenience, the planters continued to send their crops to England, despite attempts to initiate new commercial contacts. Virginia's marketing connections had been firmly established with English merchants. The planters needed to trade with a country that would not demand specie payment, and British houses could supply them with the credit they required. Because Virginians could not supply their own credit and marketing facilities overnight, and because it would require time to establish trade with other countries, most planters found it easier to return to the British market.

Virginians also returned to England for trade because of their predilection for British goods. They had manufactured some of their own clothing during the war, but Jefferson accurately predicted that "such [is] our preference for foreign manufactures, that be it wise or unwise, our people will certainly return as soon as they can, to the raising raw materials, and exchanging them for finer manufactures than they are able to execute themselves."[14] Virginia's preferences for British goods and credit were well known to the merchants of England, who were more than willing to encourage the former colonists to return to old connections.[15]

---

[13] For example, William Allason, a Scottish merchant living at Falmouth, who did business in Fauquier County, made routine and frequent appearances in the county court to settle his accounts receivable. His name appears in the court records 33 times in 8 years, and his own correspondence indicates in one instance an effort of 25 years to settle one debt (William Allason Letterbooks, Virginia State Library, and Fauquier County Court Records, Virginia State Library).

[14] Thomas Jefferson, *Notes on Virginia*, ed. William Peden (Chapel Hill, 1955). p. 164.

[15] One feature of the reopening of British connections after the war was the willingness of British merchants to continue to serve as agents for the planters. Another was a group of new merchants who appeared in Virginia, seeking to fulfill the increased postwar demands of Virginia planters and to receive their tobacco. Some were native Americans, but most were British. Their presence complicated the financial picture in the state by adding another load of debts to those incurred before the war. More important, the payment of prewar debts could be legally avoided; the payment of new ones could not.

The linking of British merchants and planters was not the only creditor-debtor relationship in Virginia society. As Richard Pares has pointed out with respect to the West Indies, there existed an internal network of debt which transcended occupational or professional lines.[16] As often as they were in debt to the British, Virginians were in debt to one another.

It was a rare person who was ever exclusively a debtor or creditor. This dual role was characteristic of a large segment of Virginia society, though obviously some bore more of the burden of debt than others. Very often the man who made every effort to collect the debts due him did so because he owed money to other people who were in turn pressing him for payment. The fact that a man was often both a debtor and a creditor complicates the problem of assigning people to economic classes or groups, and means that within both the mercantile and planter groups there existed more complex arrangements than would at first appear.

Even the greatest of Virginians were involved in the network of debt. George Washington wrote of his precarious circumstances, which had been consistently poor since the war. In 1785 he assured John Mercer "that the disclosure I made you of my circumstances was candid; and that it cannot be more disagreeable to you to hear, than it is to me to repeat that my wants are pressing, some debts which I am really ashamed to owe, are unpaid."[17] Still later, in a letter to George William Fairfax, Washington revealed his dual status as both creditor and debtor when he related that the "ungenerous not to say dishonest practices of most of my debtors who pay me with a shilling or six pence in the pound" had rendered him unable to pay his own debts.[18]

The interlocking connections between creditors and debtors, and the fact that assigned debts circulated as currency severely complicated the settlement of debts, for a suit always involved many beside the original parties. In 1786 Yates & Lovell, a Fredericksburg firm, were being sued for debt; but they were also the creditors of Barbour & Johnston, who in turn were creditors in their own right. Yates & Lovell wrote Barbour & Johnston that "the great scarcity of money with us, and several demands on us for large sums induce us to write you & we hope you will assist us immediately with

---

[16] Richard Pares, "Merchants and Planters," *Economic History Review*, Supplements, 4 (1960). See also Pares, *A West India Fortune* (London, 1950), especially Chapter XI and Appendix.

[17] George Washington to John Mercer, December 20, 1785, *The Writings of George Washington*, edited by J. C. Fitzpatrick (Washington, 1931-1944), XXIX, 363.

[18] George Washington to George William Fairfax, June 30, 1786, *ibid.*, XXVIII, 471-478. In another instance Washington wrote that his "numerous . . . demands for cash, when I find it impossible to obtain what is due to me, by any means, have caused me more perplexity, and gives me more uneasiness than I have ever experienced before from the want of money" (George Washington to Charles Lee, April 4, 1788, *ibid.*, XXIX, 459-460).

what money you can raise, for we assure you we are in immediate want & your compliance will oblige."[19] Unfortunately Barbour & Johnston were unable to pay immediately, for they themselves possessed no money. Thus on May 10, 1787, Yates & Lovell were "sorry to hear we are to wait for a payment from you till you obtain a judgment against Mr. Sandford.... We must assure you it is by no means our wish to bring suit for what you owe us, on the contrary we solemnly declare nothing could be more disagreeable to us." Yet Yates & Lovell were in a difficult position. Owing "large sums," they were in the unpleasant predicament of "relying on the promises of our debtors to assist us."[20]

The complexities of debt and the difficulties of settlement were further exemplified by the activities of James Hunter who had been an active merchant in Virginia at the time of the Revolution, but had overextended and fallen upon hard times. Hounded by his creditors, he turned to those who owed him money, only to find that they were unable to pay. Sometime in 1787 he wrote his wife from Portsmouth saying, "I have been very unhappy in not being able to collect a little money here, I shall strain every Nerve to do so, but really the Distress is amazing—what to do I know not—Adam [a relative who was aiding him in his collections] is returned from Richmond and no money—good god what a world."[21] Like so many other Virginians, Hunter depended on his position as creditor to obtain relief from his debts. To his dismay he received half a dozen letters from people indebted to him substantiating their inability to make payment.

Without money, and without any particular inclination or ability to pay, planters allowed their debts to drag on until finally they were brought to court and forced to pay in produce or property. But even resorting to the courts did not always succeed. In King William County by a strange chance the courthouse burned to the ground the night before the opening session of the county court. Of course all the records burned with it, thereby absolving many King William residents of their obligations. Some thought the deed had been "designedly done."[22]

---

[19] Yates & Lovell to Barbour & Johnston, January 11, 1786, Barbour Family Papers, Alderman Library, University of Virginia.

[20] Yates & Lovell to Barbour & Johnston, October 25, 1786. In this case Yates & Lovell, under pressure from their creditors, dissolved their partnership in August 1787, and sued Barbour & Johnston for the money owed them. Because of crowded calendars in Virginia's courts they did not obtain a judgment until April 1791 at which time they were awarded £740 specie. There is no indication that the judgment was ever executed (Records of the Fredericksburg District Court, April 30, 1791, in the Barbour Papers).

[21] James Hunter to Marianna Hunter, ——— 1787, James Hunter Papers, Alderman Library.

[22] John Dawson to James Madison, June 12, 1787, Madison Papers, Library of Congress. Francis Corbin wrote to Arthur Lee that the King William incident had been followed by a similar burning in New Kent County (Francis Corbin to Arthur Lee, August 8, 1787, Lee Papers, VII, 143, Houghton Library, Harvard University). John Price Posey, a former member of the House of Delegates, was hanged for setting fire to the New Kent clerk's office (*Virginia Gazette and Weekly Advertiser* [Richmond], August 23, 1787; January 31, 1788).

II

These internal complications were an integral part of Virginia's economic life, and had been so since colonial days. But the added problems of the early years of independence threatened the already precarious network of obligations within the state. One of these problems was the postwar depression. Causes and conditions varied from state to state, but in Virginia one reason for the crisis—evidenced by an extreme shortage of currency and a drop in tobacco prices—was Virginians' overindulgence in British goods after the war.

During the years of fighting Virginia's foreign trade had been severely curtailed, both by the presence of enemy ships and by the state's demands for supplies for the Continental army. The needs of the Revolution dominated all others, and Virginians were expected to sacrifice to gain their freedom from Britain. In keeping with this spirit a man was not to purchase anything that he could possibly produce for himself. "A Landholder" said that there was "no country upon earth which produces so many of those necessaries, or in which they can be procured so easily, as in Virginia. An industrious Virginian can raise not only beef, bacon, mutton, poultry, wheat, barley, oats, hay, rye, corn, hemp, cotton, flax, wool and horses for his own use and consumption, but also for sale." He did not have to use foreign liquors, for his own "apple and peach trees afford him liquors in abundance; or should these fail, his barley might supply the loss. His dairy supplies him with milk and butter for breakfast instead of the luxuries of tea and coffee."[23]

But Virginians longed for the textiles, dishes, books, liquor, and other manufactured goods of Great Britain. They had been accustomed to using them since their earliest days as colonists, and during the eighteenth century their importations from the mother country had frequently reached the level of extravagance. When the war ended, Virginians reasserted these desires. Once again they resumed trade with English merchants, and took advantage of liberal British credit and temporarily high prices for their tobacco to purchase more than they could afford. A correspondent of the *Virginia Independent Chronicle* suggested that Virginians "have religiously preserved the fancies, prejudices, and customs which they [the British] had stamped on our minds for the produce of their manufactures."[24]

---

[23] *Virginia Gazette or American Advertiser* (Richmond), May 11, 1782.
[24] *Virginia Independent Chronicle* (Richmond), November 14, 1787. This opinion was widely held in Virginia and appears repeatedly not only in letters to the newspapers but also in private correspondence.

Virginians' extravagances contributed to the postwar depression first by increasing the burden of debt borne by the citizens. Not only did they have to face the repayment of debts incurred before the war, but they piled new debts on top of old, decreasing their ability to pay and decreasing the faith of their creditors in them. Moreover the increased importations created an unfavorable balance of trade in Virginia, causing specie to flow out of the country and leaving unpaid balances behind. Continued commerce with Britain would "inevitably either drain our country of specie, or to the amount of this balance bring us yearly in debt. . . . The terms upon which a country in its infancy, must carry on a commercial intercourse with one long established, are not equal; since produce and manufactures, bear so small a proportion to each other."[25] As Virginians' purchases increased, the British merchants took control of the market. They took specie when they could get it, or commandeered tobacco in payment, and put themselves in a position to drop tobacco prices since the balance of trade flowed in their favor and Virginians had to sell in order to pay their debts.[26]

Virginia's depression resulted also from the machinations of Robert Morris, a Philadelphia financier, who became Secretary of Finance to the Continental Congress in 1776 and returned to his private business after the cessation of hostilities. At the behest of the French Farmers General, the government-sponsored monopoly of tobacco purchases, he entered into a contract to supply France with Chesapeake tobacco. Armed with a substantial advance of money from the Farmers General, he cornered the market for Southern tobacco.

As Jefferson explained it, Morris's profits arose because as "sole purchaser of so great a proportion of the tobacco made, [he] had the price in his own power. A great reduction in it took place; and that not only on the quantity he bought, but on the whole quantity made. The loss to the states producing the article did not go to cheapen it for their friends here [France]. Their price was fixed. What was gained on the consumption was to enrich the person purchasing it."[27] In other words, Morris gained on the gap between the price at which he purchased the tobacco and the price at which France had contracted to buy it. The lower he could push the American price, the greater would be his profit.

Morris's efforts were successful, not only because the Farmers General

---

[25] Letter signed "Amicus," *Virginia Independent Chronicle*, June 13, 1787.

[26] A letter from Thomas Jefferson to Lucy Ludwell Paradise, dated August 27, 1786, sets out the gradual progress of Britain's stranglehold on Virginia (*The Papers of Thomas Jefferson*, edited by Julian P. Boyd [Princeton, 1950———], X, 304-305, hereafter cited as *Jefferson Papers*).

[27] Thomas Jefferson to Mortmorin, July 23, 1787, *Jefferson Papers*, XI, 615.

gave him the money for his initial purchases, but also because his considerable commercial experience enabled him to organize a dozen or more Virginia and Maryland mercantile firms as agents to support and carry out his monopoly.[28] By the first months of 1786 the double monopoly of the French tobacco trade had produced chaos in the Virginia market. Prices of Virginia tobacco dropped from a postwar high of 40 shillings per hundred-weight to 30 in March 1785 and to 20 by March 1786. They remained around the low of 20 shillings at least until the end of 1787.[29]

The reduction in price and, more important, the glut of the French market by Morris' shipments which curtailed demand, seriously affected Virginia planters and merchants. The Morris contract had thrown the tobacco trade into "agonies," reducing the value of the staple as a medium of exchange in Virginia and abroad; and "the commerce between America and this country [France], so far as it depended on that article [tobacco], which was very capitally too, was absolutely ceasing."[30]

Most Virginians agreed with George Washington, who wrote to incoming governor Edmund Randolph that "Our affairs seem to be drawing to an awful crisis: it is necessary therefore that the abilities of every man should be drawn into action . . . to rescue them if possible from impending ruin."[31] Not only did planters suffer from the drop in prices, but the state also felt the effects. Citizens who had no specie and who faced low prices for their staple contributed sparingly and reluctantly to the state at tax time. Madison told Jefferson in December 1786 that "Our Treasury is empty, no supplies have gone to the federal treasury, and our internal embarrassments torment us exceedingly."[32]

A by-product of economic instability and specie shortage was fear—fear of continued or more severe depression, of a paper emission and devaluation,

---

[28] For details of the Morris monopoly see Frederick Nussbaum, "American Tobacco and French Politics," *Political Science Quarterly*, XL (December 1925) 497-516; Alan Schaffer, "Virginia's 'Critical Period'," in *The Old Dominion*, edited by Darrett B. Rutman (Charlottesville, 1964), 152-178. The story is best told by Morris himself and his agents in their letters to him and to each other. The most revealing letters are those Morris wrote to Tench Tilghman of Baltimore, in the Ford Collection, New York Public Library.

[29] Prices are taken from the *Virginia Journal and Alexandria Advertiser* where they were reported quite regularly between March 1785 and December 1787. The newspapers provide the best available consistent report of prices. For corroboration one is compelled to rely on quotations given by Virginians in their private correspondence. For example, James Madison to Thomas Jefferson, March 18, 1786, in which Madison reports that the price of tobacco was 22 shillings at Richmond. *Jefferson Papers*, IX, 334. The quotation of 40 shillings was expressed by Jefferson in a letter to John Adams, July 9, 1786, *Jefferson Papers*, X, 105-107.

[30] Thomas Jefferson to John Adams, July 9, 1786, *Jefferson Papers*, X, 105-107.

[31] George Washington to Edmund Randolph, November 19, 1786, *Writings of Washington*, XXIX, 77.

[32] James Madison to Thomas Jefferson, December 4, 1786, *Jefferson Papers*, X, 576.

and of the impossibility of collecting debts. Alexander Hamilton, an agent for one of the largest Scottish houses operating in Virginia and Maryland, told his Glasgow office in 1784 that "collecting debts in this world was at all times a very fatiguing as well as disagreeable business, it is now greatly more so."[33] He was amazed that "the enormous debts owed by the United States to Britain, did not make every body in trade cautious of trusting such large sums to people many of whom had no other way of paying for them . . . without selling their little pieces of land and a few Negroes . . . for which little or no ready money can be got."[34] Years later Hamilton had nothing better to report: "The debts come in very slow and grow dayly worse. . . . The Sums thus lost are very considerable and unless something is done . . . there is no knowing to what extent [more debts] will arise."[35]

If prices continued depressed, tobacco would be unacceptable in payment of debts. A paper emission and subsequent devaluation would increase the difficulties of creditors who would in effect be paid at a discount. Moreover the prevailing sense of uncertainty, the instability of the state and confederation governments, added to the pressure. Small wonder that creditors rushed to collect what they could in order to protect and fortify themselves against the possibility of future hardships.

Another complication was Virginia's need for revenue which necessitated a system of state taxes. Under the best of circumstances these would have been paid entirely in specie, but during the money shortage of the 1780's the state had to accept commodities instead. The revenue acts passed between 1780 and 1786 specified that land taxes might be paid partly in state bills of credit, partly in specie. All other taxes—on cattle, horses, carriages—might be paid in specie, tobacco, hemp, or flour.[36] The Revenue Act of 1787 limited the means of payment: citizens were entitled to use either government warrants or specie; but an additional act of the same legislative session provided that tobacco would be an acceptable substitute.[37]

The provisions for the acceptance of commutables were, of course, less effective after the fall in the market price of tobacco. Madison admitted as much in 1787 when he told Washington that "in admitting Tobacco for a

---

[33] Alexander Hamilton to James Brown & Co., March 10, 1784, in Glassford & Co. Account Books, Letters (1784-1790), vol. 34, Library of Congress.

[34] Alexander Hamilton to James Brown & Co., November 24, 1784.

[35] Alexander Hamilton to James Brown & Co., April 5, 1788. See also William Allason to John Lilly, May 24, 1788, in which Allason speaks of being retired from trade except with his "Brother in Falmouth who lives there as much to be in the way of looking after old balances as to any gain that might be made from keeping store."

[36] For example, Hening, *Statutes*, X, 202-203.

[37] Hening, *Statutes*, XII, 412-432, 455-457.

commutable we [the Assembly] perhaps swerved a little from the line in which we set out. I acquiesced in the measure myself as a prudential compliance with the clamours within doors & without, and as a probable means of obviating more hurtful experiments."[38] The price problem led to the Revenue Act of 1788, which required payment in specie or equivalent warrants, but reduced all taxes by one-third "in proportion to the losses sustained by the public, by receiving tobacco at prices exceeding the real value."[39] All revenue acts in Virginia which allowed payment in tobacco had stipulated a valuation for the crop. But as the price of tobacco declined, the Assembly probably could no longer accept the staple as a commutable at the prices that had been set in previous revenue acts because they were higher than the market price. Rather than allowing future revenue collection to depend on the fluctuation of the market, delegates apparently preferred to reduce the stated amount of the tax and have it paid in specie and warrants which had a more stable value than tobacco.

Madison had correctly estimated the importance of the "clamours" from the people of Virginia. Petitions poured into the Assembly and the governor's office, some from the inhabitants of a county protesting jointly, others from sheriffs begging not to be punished for their failure to collect the taxes in their districts. During and after the war Virginians complained that "the mode of Taxing us all in Cash, when there is so little Cash in the Country will certainly lay us under the Greatest Distresses."[40] In December 1786 the inhabitants of Pittsylvania County suggested that their "present distresses" might be "relieved, and the credit of the State restored" if the Assembly would issue paper money, "and if tobacco should be made receivable, in payment of a certain part of the public taxes."[41] Far more numerous were the petitions of county sheriffs, who were held responsible by the state for uncollected taxes. In December 1787 the Augusta County sheriff reported "not one-third of the taxes for 1783-84 collected, and no purchasers will attend sales of property determined upon." From Berkeley came a similar report: "Taxes cannot be collected for scarcity of money and the impossibility to sell the property of debtors."[42]

---

[38] Such as paper money. James Madison to George Washington, December 7, 1786, Madison Papers, Library of Congress.
[39] Hening, *Statutes*, XII, 707-708.
[40] Culpeper County petition, November 15, 1783, Virginia State Library.
[41] Pittsylvania petition in *Journals of the House of Delegates of the Commonwealth of Virginia, 1776-1790* (Richmond, 1828), December 7, 1786. See also petitions for Westmoreland, Richmond, and Northumberland counties, November 1787, in Virginia State Library.
[42] *Calendar of Virginia State Papers and Other Manuscripts*, edited by William P. Palmer (Richmond, 1883-1884), IV, 337. Ten such petitions appeared on the same date. See also petitions of the sheriffs of Powhatan, Chesterfield and Lunenberg, March 6, 1787; from the sheriffs of

Because of the scarcity of money, the requirement of state taxes had a destabilizing effect on the network of internal debts. Virginians might not have pressed one another for collection if no "outside" influence had demanded specie. The imposition of taxes forced men to press their private claims, causing weaker men to go bankrupt; and ended with general evasion of taxes or payment in kind.

III

Depression, specie shortage, and state taxes all contributed to the pressure on Virginians to exercise their claims as creditors before they were beset by others to pay their debts. But the most discussed and feared pressure on Virginia's internal network of debt was, like state taxes, an "external" factor: the peace treaty with Great Britain which required that all debts remaining unpaid from before the war be settled. The prewar debts owed in Britain added to Virginia's economic instability during the Confederation had made the planters reluctant to pay the debts they accrued after the war. The treaty posed a deadline for Virginians since the British (and later the federal government) would insist on enforcement. More important, the payment of the British debts would affect the network of debt held within the state, for when a man owed money to a British merchant he would spare no pains to collect from his own debtors in Virginia in order to make payment.

When the colonies declared their independence Virginians already owed some 20 million pounds sterling in Britain.[43] After the war they had little desire to repay it, because of their antipathy to England. Furthermore they were in no position to do so, for the payment of British debts imposed a strain on their financial system which threatened to topple it from its already precarious position. The impending possibility of payment produced two parallel movements in Virginia. One provided a temporary means of evading payment to Britain; the other sought to find a permanent solution to the problems of internal debt that plagued the state.

From the outset of treaty negotiations Virginians had regarded the payment of British debts with suspicion, realizing that such a provision would subject Virginia to extra hardship. During the negotiations Madison wrote Edmund Randolph that his "apprehension from the article in favor of

---

Isle of Wight, April 7, 1786, Loudoun and Princess Anne, April 18, 1786, Mecklenburg, August 2, 1786, and Nansemond, September 10, 1786 in the Executive Papers, Virginia State Library; and *The Letters and Papers of Edmund Pendleton*, edited by David John Mays (Charlottesville, 1967), II, 483-485, 491-492.

[43] This is Jefferson's estimate (*Jefferson Papers*, X, 27).

British creditors correspond with those entertained by all whose remarks I have heard upon it. My hope is that in the definitive treaty the danger may be removed by a suspension of their demands for a reasonable term after peace."[44] Madison hoped in vain, for the treaty in final form provided for full repayment in sterling of all prewar debts. As attempts were made to comply with the treaty, pressure on Virginia increased. In September 1785 Edmund Randolph voiced the sense of impending peril when he wrote to Madison that "the scarcity of money in Virginia can hardly be conceived ... and the approach of british debts thickens the horror of the prospect."[45]

Spurred by reluctance to fulfill the provisions of the treaty, Virginia's General Assembly passed legislation to avoid payment. During the war the Assembly had provided for payment of the debts out of confiscated British property and occasional payments by the planters into the state loan office.[46] But in 1780 the Assembly repealed the part of this act which "enabled persons owing money to a subject of Great Britain to pay the same or any part thereof into the publick loan office, and obtain certificates of such payment in the name of the creditor."[47] In substitution the Assembly provided for payment according to a more equitable scale of depreciation, adjusting for the wartime inflation in the value of the state's currency. Wherever judgments had been obtained, executions were suspended until December 1783.[48] Before the suspension period had expired the Assembly acted again, this time to repeal the suspension and affirm that "no debt or demand whatsoever, originally due to a subject of Great Britain, shall be recoverable in any court in this commonwealth."[49] This act was reinforced at the next session of Assembly but a new and crucial distinction between pre- and postwar debts was added: all postwar debts incurred after January 1782 had to be paid, while debts incurred before that date remained exempt. These acts further aided Virginia citizens by providing that debts might be paid in property at stated rates rather than in cash, even at the depreciated rate.[50]

Finally in October 1787, under pressure from England, Congress, and

[44] James Madison to Edmund Randolph, April 8, 1783, in Edmund C. Burnett, ed., *Letters of the Members of the Continental Congress* (Washington, 1921-1938), VII, 126-127. See also Edmund Randolph to James Madison, March 29, 1783, in the Madison Papers, Library of Congress.

[45] Edmund Randolph to James Madison, September 12, 1785, Madison Papers, Library of Congress.

[46] Hening, *Statutes*, IX, 377-380.
[47] Hening, *Statutes*, X, 227.
[48] Hening, *Statutes*, X, 471-478.
[49] Hening, *Statutes*, XI, 75-76.
[50] Hening, *Statutes*, XI, 176-180.

Virginia's statesmen, the General Assembly repealed "so much of all and every act or acts of assembly as prohibits the recovery of British debts," in compliance with the terms of the British treaty. This was clear enough, but the Assembly, still reluctant to subject Virginians to this burden, threw an impediment in the way of collection. The terms of the act were to be suspended until the governor of the state had been officially notified of Britain's fulfillment of the terms of the treaty—that is, until Britain had given up the Northwest posts and either returned or made compensation for the slaves taken from Virginians during the war.[51]

But though the Assembly successfully passed legislation to avoid repayment of debts (and was safe in so doing until the new government after 1789 could enforce the treaty provisions), this was obviously only a temporary evasion of the issue. Debts held within the state continued to crowd court calendars. The second and more direct solution to the general problem—an alteration of Virginia's commercial and judicial system—seemed the only hope of improvement. Two approaches were available to the Assembly: to facilitate repayment of debts within the state through an emission of paper money, and to reorganize the state's system of courts so as to insure prompt repayment and equal justice for creditor and debtor. These were by no means interchangeable solutions. An emission of paper money would have reached the heart of Virginia's problem while a reformation of the judicial system merely smoothed the process of adjustment to the scarcity of money without alleviating it. Nevertheless, Virginians regarded both as preferable to payment to Britain and as providing more long-lasting benefits to the state.

Of the two solutions the paper money alternative appealed to some (Patrick Henry was one of its most vigorous advocates), but despite the agitation for such a solution that swept the country in the late 1780's Virginia's Assembly rejected it. They felt that paper money, because it did not represent "real" wealth, would "destroy all public and private confidence," by devaluating the currency "whilst it weakens the industry and virtue of individuals" by making it easier for debtors to repay their debts. By increas-

---

[51] Hening, *Statutes*, October 1787, XII, 528. A complicating factor was the attitude that people were wrong to incur debt in the first place and, therefore, should not complain when they had to face the day of reckoning. Richard Henry Lee told Richard Lee that "Those evils which happen independently of us, we must bear with manly firmness; and those which flow from our own misconduct we have no right to complain of. Among the last is the pressure of private debt, which almost universally arises from idleness and extravagence; one or both—This will be corrected and remedied by industry and economy" (Richard Henry Lee to Richard Lee, September 13, 1787, *The Letters of Richard Henry Lee*, edited by James Curtis Ballagh [New York, 1911-1914], II, 436-437). Jefferson himself wanted the debts repaid, though he favored a moderate plan of installments (Thomas Jefferson to James Madison, April 25, 1786, *Jefferson Papers*, IX, 433-435).

ing the quantity of money in circulation, a paper emission would benefit the debtor and hurt the creditor, perhaps risking inflation and leaving the "state exposed to all the evils of a general rapacious speculation."[52] James Currie wrote Jefferson that a paper emission "will certainly continue the delusion we are under in regard to our own Finances, and procrastinate the period when we ought and from dire necessity must live in every respect more conformable to our Situation as an infant Republick."[53]

Instead the Assembly preferred the slower and more cumbersome method of working for a "uniform, equal, and speedy administration of justice." The courts were so far behind in their business that "it frequently happens, that a man is reduced to poverty and distress by his creditors, when he has much more due to him than he owes."[54] One of many public calls for judicial reform reminded Virginians that "when justice is speedily and equally administered, the same law that obliges me to pay what I owe, obliges every man to pay what is owing to me. . . . And do you propose to support a law for establishing the Circuit Courts, and for lengthening the terms of all Courts, so as they may regularly go through their dockets at least twice a year?"[55]

In response to the problems caused by "delay of justice," the Assembly undertook to reform the judicial system. At their October 1785 session the delegates tried to reorganize the county courts, acknowledging that "the methods hitherto established for the administration of justice within this commonwealth have proved ineffectual."[56] The act set aside certain special sessions of the county courts for the trial of suits, hoping that this would ensure prompt action by the courts.

In October 1786 two more acts attempted to facilitate recovery of debts. One was the act "to empower securities to recover damages in a summary way." It gave some protection to the "many persons" who had "been reduced from affluence to poverty, by securityships," that is, by their agreement to guarantee payment of someone else's debt. Having made the payment and seeking to recover from the person they had helped, these "securities" were very often unable to get their money due to the "insolvency of the principal

---

[52] Prince George County, Instructions to Delegates, *Virginia Gazette and Weekly Advertiser* (Richmond), June 28, 1789.

[53] James Currie to Thomas Jefferson, October 17, 1786, *Jefferson Papers*, VIII, 642. See also Madison's notes for a speech against paper money before the House of Delegates, November 1786, in *The Writings of James Madison*, edited by Gaillard Hunt (New York, 1900-1910), II, 279-281.

[54] Prince George County, Instructions, *Virginia Gazette and Weekly Advertiser*, June 28, 1789.

[55] Letter signed "Aristides," *Virginia Independent Chronicle* (Richmond), March 21, 1787.

[56] Hening, *Statutes*, XII, 32-36.

... or a tardy administration of justice ... whereby the said securities have been involved in great inconveniences, and often times in manifest ruin."[57] The second act passed at this session provided for summary judgment in petitions for the recovery of small amounts, between 25 and 100 shillings, or between 200 and 800 pounds of tobacco.[58]

Yet in the minds of many citizens these procedural reforms constituted an evasion of duty by the members of the Assembly whose task should have been to effect a complete revision of the judicial system. Instead the members had passed a few half-hearted measures and voted down a more thoroughgoing circuit court bill which would have taken considerable pressure off the county courts and accelerated the judicial process. The obvious solution was to elect a new set of delegates who would pass the judicial reforms that many felt were a necessity.[59]

The Assembly did act in the fall of 1787, but apparently nothing they did was sufficient. Even the plan to provide district courts, which became law, was inadequate, for outlying areas still did not have conveniently located courts. Moreover, in the areas which did have courts, a debt of less than thirty pounds still was forced into the county courts where collection proceeded at the same slow pace.[60] Not until October 1792 was any comprehensive legislation passed—at this session Virginia got its first bankruptcy law and thorough provision for judgments and executions.

Until 1792 things went on as usual. William Allason, writing to Thomas McCulloch, expressed concern at the "reprehensible" conduct he saw around him. "Times are greatly altered here from what you knew them, in order to come at a little money we are oblig'd to wade through a Law Suit that is generally tedious, and consequently very expensive by the Lawyers unseasonable requisitions, altho' we have four times as many of that profession as is realy necessary."[61]

By the time the ratification convention gathered in the summer of 1788, Virginia had not effected any permanent remedies. The Assembly had rejected paper money, and had failed to accomplish the necessary reform

---

[57] Hening, *Statutes*, October 1786, XII, 268-270. This act was first suggested in 1779 under Jefferson's administration and was incorporated into the revised code of laws that was a pet project of Jefferson.

[58] Hening, *Statutes*, XII, 353-354.

[59] See the Madison correspondence for discussion of the progress of judicial reform in the state. Resistance to its adoption was a source of great discouragement to Madison and his friends. See, for example, Joseph Jones to James Madison, May 30, 1786; James Madison to Edmund Pendleton, January 9, 1787, in the Madison Papers, Library of Congress.

[60] Hening, *Statutes*, XII, 467-474.

[61] William Allason to Thomas McCulloch, March 4, 1791, Allason Letterbooks, Virginia State Library.

of the courts. Moreover, their successful but temporary evasion of the payments to British creditors would end with ratification of the Constitution because it would empower the federal government to enforce the treaty provision requiring payment of prewar debts. For these reasons one consequence of the presence of debt in Virginia was the reluctance of many citizens to endorse the constitution for fear of further disrupting the debt-ridden economy.[62]

More important, the chronic debt and the underlying conditions that created it imprisoned Virginia in a colonial relationship to Britain until the 1790's (when the French Revolution and European demand for grain temporarily liberated them) and to the Northern states thereafter. Because of the scarcity of specie and their commitment to land, to agriculture, and hence to regular borrowing, Virginia never accumulated capital or developed a class of native merchants who might in time have supplied it. Even the brief period of self-sufficiency during and immediately after the Revolution failed to provide the necessary impetus for Virginians to create their own capital and commercial services. Rather, they continued to rely on those sources of capital that were readily available: the merchants of Britain and the North. To this extent they perpetuated their colonial dependency even though nominally and politically they had become a free people.

---

[62] See, for example, Patrick Henry, speech to the Virginia Ratifying Convention, June 23, 1787, in *Debates and Other Proceedings of the Convention of Virginia* . . . , David Robertson, comp. (Richmond, 1805), p 414. See also Edmund Randolph to James Madison, October 23, 1787, Madison Papers, Library of Congress.

THE GREAT MIGRATION TO THE MISSISSIPPI TERRITORY, 1798-1819

By CHARLES D. LOWERY

Historically, Americans have been an inveterately restless people. From the earliest times restlessness and mobility have been major characteristics of American society and character. The tenuous footholds at Jamestown and Plymouth had hardly been secured when restive settlers began pushing into a hostile interior. And while it took almost two centuries for the population of the new country to push westward to the Alleghenies, the march to the Pacific was accomplished in a little over fifty years. The lure of a vacant continent, with its promise of new life and unlimited opportunities, was too strong for countless Americans to resist.

The Frenchman Alexis de Tocqueville, who published a remarkably penetrating study of American society after a tour in the early 1830's, commented on the "strange unrest" of the American:

> In the United States a man builds a house in which to spend his old age, and he sells it before the roof is on; he plants a garden and lets it just as the trees are coming into bearing; he brings a field into tillage and leaves other men to gather the crops... ; he settles in a place, which he soon afterwards leaves to carry his changeable longings elsewhere.... Death at last overtakes him, but it is before he is weary of his bootless chase of that complete felicity which forever escapes him.[1]

In few periods of American history has the population been as mobile as the period of which Tocqueville

---

[1] Alexis de Tocqueville, *Democracy in America* (2 vols.; New York, 1945), II, 136-137.

wrote. The first half of the nineteenth century was a time of feverish movement and unprecedented expansion in the United States. During these decades immigrants surged through the passes of the Alleghenies in a seemingly endless flow to inundate the choice lands of a waiting continent. As one contemporary observed, it seemed as if old America was folding up and moving west.[2]

The Great Migration, the appellation given to this mass westward movement, began after the Revolution and continued until mid-nineteenth century, by which time the peopling of the eastern half of the continent had been completed. While it may be said that the Great Migration began when settlers started pushing into Kentucky and Tennessee in the years immediately after the Revolution, it was not until the late 1790's that migration commenced on a general scale. In 1800 Kentucky and Tennessee were the only two states west of the mountains. In 1820, when the general depression following the Panic of 1819 ended the first phase of the migration, there were eight western states and the Missouri Territory was ready for statehood. The total population of the West had grown from 386,000 in 1800 to 2,216,000 in 1820.[3] This surging tide of westward moving pioneers constituted the greatest population movement the nation had ever known.

During the two decades 1800-1820 immigrants moved largely two into general areas—the Old Northwest and the Old Southwest. It is the purpose of this paper to examine some broad aspects of the migration to and settlement of Alabama and Mississippi, a major portion of the Old Southwest.

[2] William H. Sparks, *The Memories of Fifty Years* (Macon, Georgia, 1870), 27.
[3] *Fourth Census of the United States*, 1.

The Alabama-Mississippi area was opened to settlement in 1798 with the organization of the Mississippi Territory. Within a very short time settlers were pouring into the Territory in large numbers. During the peak years of movement immediately after the War of 1812, the migration assumed such momentous and unprecedented proportions that some observers began to predict the depopulation of the older states. By 1820 some 220,000 persons were living in the new states of Alabama and Mississippi. Seldom in history has an area been populated so rapidly.

The forces behind the migration to the Old Southwest were many. The economic force was of unquestionable importance. Economic opportunities for the average man had diminished in the Upper South as the available supply of good land dwindled. As a result of successive planting of tobacco for more than a century, primitive agricultural methods, and the temptation characteristic of staple producing areas to "butcher" the soil, land that had not been exhausted was difficult to find anywhere in the Upper South after the Revolution.[4] In the 1790's and early 1800's traveler after traveler wrote of the worn fields and abandoned farms of this section. The Duc de la Rochefoucauld-Laincourt, who traveled through Virginia and Maryland in 1795-1796, was appalled at the condition of agriculture. Thousands of once fertile acres, he observed, had been completely exhausted by continuous growth of tobacco.[5] A settler in the Virginia Piedmont was greeted in 1800 by a "scene of desolation that baffles all description—farm after farm . . . worn out, washed and

---

[4] Avery O. Craven, *Soil Exhaustion as a Factor in the Agricultural History of Virginia and Maryland* (Urbana, Illinois, 1926), 72 ff; Lewis C. Gray and Esther K. Thompson, *History of Agriculture in the Southern United States to 1860* (2 vols.; Washington, 1933), II, 606-609.

[5] Craven, *Soil Exhaustion*, 82-83.

gullied, so that scarcely an acre [can] be found in a place fit for cultivation."⁶ Other areas of the Upper South were described as "dreary and uncultivated wastes" of "barren and exhausted soil, half clothed negroes, lean and hungry stock," dilapidated fences and decaying houses.⁷ The extent of soil exhaustion in Virginia, the Carolinas and Georgia undoubtedly prompted hundreds of families to abandon their farms and move to the rich, virgin lands of the Southwest.⁸

Another factor contributing to the economic distress of the Upper South and giving impetus to westward expansion was the precipitate decline after the Revolution of the markets for Southern staples, especially tobacco and rice. During the colonial period these staples had enjoyed a guaranteed and relatively stable market in England. This was no longer the case after the Revolution. England imposed stiff duties on these commodities, and increased competition from other tobacco and rice producing countries, along with the general market displacement and price decline occasioned by the Napoleonic wars, caused widespread distress among Southern planters.⁹

An even more important force behind the migration to the Southwest was the rapid expansion of cotton culture. Before Eli Whitney invented a workable cotton gin in 1793 the only kind of cotton that could be grown profitably in this country was the long staple variety, the growth of which was limited largely to the sea-islands and swampy lowlands of South Carolina and Georgia. The cotton gin made possible the development

---

⁶ *Farmers' Register*, I, 150. Quoted in Craven, *Soil Exhaustion*, 83.
⁷ *American Farmer*, I, 99. Quoted in Craven, *Soil Exhaustion*, 84.
⁸ J. G. de Roulhac Hamilton (ed.), *The Papers of Thomas Ruffin* (4 vols.; Publications of the North Carolina Historical Commission; Raleigh, 1918-1920), I, 226-227.
⁹ Gray and Thompson, *History of Agriculture*, II, 595 ff.

of the Cotton Kingdom. Short staple or upland cotton, which heretofore could not be grown profitably because of the difficulty of separating the seed from the fiber, could be grown throughout most of the Lower South. The high price commanded by the commodity in an English textile industry that had recently adopted the machinery needed to produce cotton cloth mechanically caused a flurry of excitement in the Southern states. Plantation owners, hard pressed by the uncertain tobacco market, were quick to see the fortunes awaiting those who switched to the promising new crop. By 1812, when war temporarily halted westward expansion, the cotton frontier had progressed across the Piedmont and was beginning to advance into the fertile Southwest.[10]

But economic factors alone do not explain why the Great Migration occurred. It was something more positive than sheer economic necessity that sent settlers pouring into the Mississippi Territory. Immigrants were not driven into this region; they were lured. They might have remained in the older settled areas and improved economic conditions by adopting scientific farming techniques, as many of those who remained ultimately did. But the prospects of a new life in the Southwest held greater attraction. They were drawn to the new region by the promise of greater opportunities, by the promise of something, often intangible, better than what they left.

The surprising dimensions of the migration to the Southwest can be explained only in light of the fact that this region, like the West in general, had assumed a mythical and symbolic character. It had become in the popular mind the new Canaan—an infinite expanse of fertile country "like the land of promise, flowing with

[10] *Ibid.*, II, 673 ff.

milk and honey," where "soft zephyrs gently breathe on sweets, and the inhaled air gives a voluptuous glow of health and vigor, that seems to ravish the intoxicated senses."[11]

Witness, for example, the glowing terms used to describe the region. The Tennessee Valley was advertised as "Happy Valley," and Huntsville was described as a town possessing every advantage which a superb location, rich soil and superior climate could offer. The Alabama River basin was called the "Acadia of Southern America." Mississippi was known as the "Garden of America."[12] Numerous immigrants set out for the region believing that it was a new Garden of Eden. One immigrant en route to the Southwest, who spoke of his destination with glowing anticipation and enthusiasm, was certainly convinced of this: "It is a wide empty country," he said, "with a soil that yields such noble crops that any man is sure to succeed, go where he will."[13] The immigrant from Maryland who wrote in extravagant praise of his new home was no less convinced: "The crops [here] are certain and want of the necessaries of life never for a moment causes the heart to ache—abundance spreads the table of the poor man, and contentment smiles on every countenance."[14] If the region was often viewed through the golden mist of Utopian fantasy, its attraction was in no wise diminished.

It was widely believed that in this undeveloped

[11] Gilbert Imlay, *A Topographical Description of the Western Territory of North America* (London, 1792), 39.
[12] Albert B. Moore, "Rummaging in Alabama's Background," *Alabama Review*, IV (July, 1951), 181; Samuel R. Brown, *The Western Gazeteer: Or Emigrant's Directory* (Auburn, New York, 1817), 15; Huntsville *Republican*, November 17, 1819; Albert B. Moore, *History of Alabama* (Tuscaloosa, 1934), 76.
[13] Captain Basil Hall, *Travels in North America in the Years 1827-28* (3 vols.; Edinburgh, 1829), III, 133. See aso Adam Hodgson, *Letters from North America* ... (2 vols.; London, 1824), I, 113-114, 141.
[14] *Niles' Register*, XIII, 38.

and almost empty land unsurpassed opportunities for self-advancement awaited the individual. He must display industry, courage, and resourcefulness, but if he possessed these qualities, then security, independence, and prosperity were within his reach. As the idyllic land where opportunities for wealth and happiness were unbounded, the Southwest symbolized the promise of American life.[15] Thousands of immigrants who pushed into the Mississippi Territory thought that they moved forward to a rendezvous with some untransacted destiny. And in their westward trek they not only affirmed their belief in the American dream, but also revealed that the supreme confidence, the inveterate optimism, and the innate restlessness that first brought many colonists to America were becoming stronger and more widespread national characteristics.

During the first phase of the Great Migration, which began about 1798 and continued through 1819, two fairly distinct waves of immigrants flowed into the Alabama-Mississippi area. The first wave appeared shortly after the Territory opened and subsided when the War of 1812 broke out. The second wave developed after the war, reached its peak in the years 1818-1819, and receded after the Panic of 1819. In the period 1798-1812 the flow of immigrants was steady but, by comparison with the period 1815-1819, unspectacular. In the earlier period settlers moved into three general areas—the Natchez District, the lower Tombigbee River basin, and the Tennessee Valley.

Of these three regions, the Natchez District received the heaviest influx of settlers during the early period. In 1798 this region had a combined free and

---

[15] See, for example, the letter of William Roane to Thomas Ruffin, August 13, 1819, in Hamilton (ed.), *The Papers of Thomas Ruffin*, I, 194.

slave population of some 4,500 persons.[16] Two years later the counties of Adams and Pickering, into which the Natchez District had been divided in 1799, contained a total population of 4,446 white persons and 2,995 slaves.[17] During the next decade these counties increased by approximately 3,500 whites and 6,000 Negroes. By 1811 a tier of five new counties lying north and south of Adams County and extending eastward to the present Alabama line had been created,[18] and the total population of the counties embraced within the present state of Mississippi was 31,306 persons, 14,706 of whom were Negroes.[19]

The settlements along the lower Tombigbee grew rather slowly during this same period. In 1800 the total white and Negro population of Washington County, which included the Tombigbee settlements and until 1809 extended westward to the Pearl River, numbered some 1,250 persons.[20] Ten years later the county had grown to only 2,010 whites and 910 Negroes.[21] To the south of Washington County lay the newly created county of Baldwin, which had a population of 667 whites and 760 Negroes.[22] Thus, while the total population of the Mississippi portion of the Territory increased by almost 27,000 persons during the period 1798-1810, the settlement in south Alabama increased by less than 3,000.

The population growth of the Tennessee Valley was far more impressive. In 1804 the northern boundary of the Mississippi Territory was extended from the

---

[16] Arthur P. Whitaker, *The Mississippi Question, 1795-1803, A Study in Trade, Politics and Diplomacy* (Gloucester, Massachusetts, 1962), 276, note 24.
[17] *Second Census of the United States*, 83.
[18] The counties of Jefferson and Claiborne, created in 1802 out of Pickering county are not considered here as new counties.
[19] *Third Census of the United States*, 83.
[20] *Second Census of the United States*, 85-86.
[21] *Third Census of the United States*, 83.
[22] *Ibid.*

line of 32° 28′ to the Tennessee border. Settlers soon began to push into the Tennessee Valley, especially the Huntsville area. In 1808 Madison County was created, and within two years it had a population of 4,699 persons, one-fifth of whom were slaves.[23]

Until the outbreak of war in 1812 migration to the Mississippi Territory continued on a modest level. During the war Indian hostility and general economic displacement acted as an effective dike to hold back the flow of immigrants. Migration, except to the Tennessee Valley, where the effects of the Creek Indian War were not pronounced, slowed to a trickle. With the restoration of peace, however, the flood-gates were opened, and a veritable immigrant torrent inundated the Territory.

A number of developments explain the remarkable increase in immigration after the Treaty of Ghent. During the war that portion of Spanish West Florida situated between the Perdido and Pearl rivers fell to American forces. Interior planters were thereby assured of unmolested river shipments directly to the sea. Andrew Jackson's victory over the Creek Indians in the Battle of Horseshoe Bend not only restored peace to the Territory, but also opened for settlement an extensive tract of land in the Alabama interior. Moreover, his victory revealed that the Southern Indians were too divided and weak to resist for long the powerful forces driving them from their ancestral lands. Also, travel into the Territory had been facilitated by improvements in existing roads and the opening of new ones. These factors, along with an excellent cotton market and a general period of optimism and speculation, gave tremendous impetus to the migration.

[23] *Ibid.*; Edward Chamber Betts, *Early History of Huntsville, Alabama, 1804-1870* (Montgomery, 1909), 6-14.

By horse, by wagon, by boat, and by foot the flood of humanity poured in. Roads leading into the Territory were filled with a seemingly endless succession of settlers. Travelers moving northward against the current might travel for days without losing sight of immigrant trains. One traveler passing through Alabama in late 1816 counted almost 4,000 immigrants in nine days of travel.[24] More than one traveler observed that the movement was more like the march of an army of occupation than the migration of settlers. So great was the influx that corn, which seldom sold for more than fifty cents per bushel in the Territory, brought as much as four dollars, and bacon and other foodstuff were practically unobtainable.[25]

If many visionary Americans were enthralled as they contemplated this pioneer army moving westward at what seemed to them the command of destiny, others saw reason for considerable alarm. For a time it seemed as if the older states might be depopulated.[26]

A lawyer from the Piedmont section of North Carolina wrote in 1817, as the migration was nearing its peak:

> You, Sir, can't conceive of the anxiety and confusion that pervades all ranks of people in this section of country to remove to the Alabama. ... The *Alabama Feaver* [sic] rages here with great violence and has *carried off* vast numbers of our citizens. I am apprehensive, if it continues to spread as it has done, it will almost depopulate the country. There is no question that this feaver is contagious. . . , for as soon as one neighbor visits another who has just returned from the Alabama he immediately discovers the same symptoms which are exhibited by the person who has seen the allureing [sic] Alabama.[27]

[24] *Niles' Register*, XI, 336.
[25] *Ibid.*, XII, 341.
[26] Sparks, *The Memories of Fifty Years*, 27.
[27] James Graham to Thomas Ruffin, August 10, November 9, 1817, in Hamilton (ed.) *Papers of Thomas Ruffin*, I, 193-194, 197-199.

John Randolph of Roanoke, the brilliant but erratic leader of the Old School Republicans, viewed with alarm the exodus of people from the Virginia Tidewater. Touring his home region after the war with Great Britain, he wrote:

> What a spectacle does our lower country present! Deserted and dismantled country-houses, once the seats of cheerfulness and plenty, and the temples of the Most High ruinous and desolate, 'frowning in portentous silence upon the land. . . .' I do not wonder at the rage for emigration. What do the bulk of the people get here that they cannot have for one fifth the labor in the western country. . . ? In a few years more, those of us who are alive will move off to Kaintuck or the Massissippi [sic], where corn can be had for six-pence a bushel and pork for a penny a pound.[28]

The Richmond *Enquirer* in 1816 lamented the spectacular increase in migration: "At no period since the settlement of the Western Country has the tide of population set stronger . . . [that] way than at present. But a few years more and the pivot on which the union will balance will be the Allegheny Mountains, or west of them."[29] In August, 1817, the Charleston *Reporter* declared that the flow of population to the Southwest had reached unprecedented proportions.[30] Several months later a traveler reported that between Fredericksburg and Richmond he met large bodies of settlers moving to Alabama.[31] In November, 1815, a committee of the North Carolina legislature reported that during the preceding twenty-five years more than 200,000 people had left the state to settle in the Mississippi Valley, and

---

[28] Quoted in Ulrich B. Phillips, *American Negro Slavery* (New York, 1918), 183.

[29] Richmond *Enquirer*, November 21, 1816.

[30] Charleston *Reporter*, August 6, 1817.

[31] William T. Harris, *Remarks Made During a Tour Through the United States of America in the Years 1817, 1818, 1819* (Liverpool, 1819), 21.

a committee of the Virginia House of Delegates made a similar complaint the following year.[32]

By the close of 1819, when a serious panic and the beginning of a depression slowed westward expansion, both Alabama and Mississippi had come into the Union. In the decade 1810-1820 the population of Mississippi grew from 16,600 whites and 14,706 Negroes to 42,176 whites and 33,272 Negroes, a total increase of more than 44,000 persons.[33] Thus, within the decade Mississippi's population more than doubled, and the increase over 1800 was more than tenfold.

The growth of Alabama was far more remarkable. In the decade 1810-1820 Alabama's population grew from 6,422 whites and 2,624 Negroes to 99,198 whites and 47,665 Negroes, a total increase of 137,817 persons.[34] Within a decade her population increased more than sixteenfold and more than one hundredfold since 1800.[35] Seldom in history has the growth of an area been more phenomenal.

While the entire Territory was populated with surprising rapidity, the growth of one area, the Tennessee Valley, was without parallel. In 1810 there was but one county in this region, and its population, free and slave, was less than 5,000 persons. A decade later six

---

[32] Niles' Register, IX, 165; Thomas Perkins Abernethy, *The South in the New Nation, 1789-1819* (Baton Rouge, 1961), 471.

[33] *Fourth Census of the United States*, 30.

[34] *Ibid.*, 28-29. It is necessary to supplement the incomplete Federal census of 1820 with "Alabama Census Returns, 1820" in *Alabama Historical Quarterly*, XI, 337-338.

[35] In the decade 1810-1820 the flow of population into Alabama was greater than into Mississippi because more land had been cleared of Indian title in the former. In the years 1814-1816 the Creek, Choctaw, Cherokee and Chickasaw Indians relinquished title to vast tracts of choice land in Alabama. No comparable cessions were made in Mississippi until after 1819. An immigrant coming into the Southwest in the period 1815-1819 had a much wider choice of land in Alabama than in Mississippi, where he was restricted to a relatively narrow band of territory in the southern portion of the state.

additional counties had been created. The combined population in 1820 of the seven counties bordering the Tennessee River was 58,540 persons, 18,618 of whom were slaves.[36]

No other general region within the Territory grew with such phenomenal rapidity. The tier of five counties bordering the Mississippi River had in 1820 a combined free and slave population of 33,148 persons, but the growth of these counties had been gradual and much less spectacular than the growth of the Tennessee Valley.[37] Some individual counties did show remarkable increases, however. Take, for example, the counties of Tuscaloosa, Montgomery, and Monroe, all in Alabama. In 1818 there were less than 300 persons in Tuscaloosa County. Two years later there were 5,069 whites and 2,253 slaves living there. Montgomery grew almost as rapidly. Immigrants did not begin to push into that county until about 1817, but within three years time 3,827 whites and 2,602 slaves had settled there. And Monroe County, organized in 1815 and located in the southern part of the state, had a population of 4,511 whites and 3,695 slaves by 1820.[38]

By way of concluding, some general observations concerning the settlement of the Territory should be made. The heaviest flow of migration occurred during boom years. There was a notable diminution during hard times. In periods of depression few immigrants could raise the money to finance their move. Cash was required to purchase lands, acquire supplies and tide the families over until crops could be raised and sold.

The flow of population, which followed generally the lines of least resistance and greatest opportunity,

[36] *Fourth Census of the United States*, 28-29.
[37] *Ibid.*, 30.
[38] *Ibid.*, 28-29.

tended also to follow isothermal lines. Hence the states of Georgia and Tennessee contributed to the Territory the overwhelming majority of settlers. The Tennessee Valley was peopled principally by families from Tennessee who had easy access to the area by the Tennessee River. The southern counties of the Territory were settled largely by Georgians, who could travel along the Federal Road all the way from Georgia to Natchez.[39]

The Carolinas contributed the next greatest number of settlers. Coming primarily from the Piedmont area, these immigrants settled generally in the region drained by the Tombigbee and Black rivers, though they did not settle here exclusively.[40] Virginia, Kentucky and Maryland contributed relatively few settlers. However, many of the Virginia immigrants were men of some wealth and ability who were able to exert considerable influence on the life and institutions of the Southwest.[41] Like the Carolinians, immigrants from Kentucky, Maryland, and Virginia settled throughout the Territory, but there were discernible concentrations of Virginians in central Alabama and along the Mississippi River.[42]

It should be noted also that the advance of population into the Mississippi Territory possessed many of the characteristics generally ascribed to the westward movement. There were distinct stages of population advance and, indeed, institutional development which, in varying degrees, conform to the accepted patterns of the frontier in general. In the vanguard of the move-

[39] Thomas Perkins Abernethy, *The Formative Period in Alabama, 1815-1828* (Montgomery, 1922), 32; Abernethy, *The South in the New Nation*, 468-469; William O. Lynch, "The Westward Flow of Southern Colonists before 1861," *The Journal of Southern History*, IX (August, 1943), 315-317.
[40] Abernethy, *The South in the New Nation*, 469; Lynch, "The Westward Flow of Southern Colonists," 316.
[41] Raleigh *Register*, January 15, 1819; Lynch, "The Westward Flow of Southern Colonists," 316; Savannah *Republican*, March 5, 1819.
[42] Abernethy, *The South in the New Nation*, 469.

ment into the Southwestern frontier were the traders and trappers from Georgia and the Carolinas who entered the Indian country well before the Revolution and prepared the way for those who followed.[43]

In the wake of the Indian traders, the herdsmen, who subsisted primarily upon a grazing and hunting economy, followed. The importance of this group in the opening and development of the Southwest has often been overlooked. When the Mississippi Territory was organized there were many herdsmen already living in the area. In the ensuing years their numbers increased. Many persons well acquainted with the life and economy of the Territory commented upon the extensive herds of cattle and droves of swine that ranged the swamps, canebreaks and prairies of the region.[44] In time, as agricultural settlers pushed in and the open range dwindled, the herdsmen moved further west or relocated in the barren pine belts.

Contrary to general opinion, the inhabitants of the pine barrens were not all poor whites. The great error that contemporary travelers and a later generation of writers have made concerning the piney woods folk has been to consider them agriculturalists. This they were not. It is understandable that travelers, accustomed to measuring affluence in terms of cotton and slaves, were notably unimpressed by the small corn, pumpkin and pea patches maintained by these people. What they in

---

[43] Avery Craven, "The 'Turner Theories' and the South," *The Journal of Southern History*, V (August, 1939), 301-302.

[44] *Ibid.*, 302; Frank L. Owsley, "The Pattern of Migration and Settlement on the Southern Frontier," *The Journal of Southern History*, XI (May, 1945), 149-153. William H. Sparks, a jurist who dwelt in the Natchez District, frequently traveled through the piney woods section of Mississippi during the territorial period. He attested to the fact that the counties lying east of the Pearl River and west of the Tombigbee settlements sustained immense herds of cattle and that the people of this region derived their living almost exclusively from stock-tending and hunting. The piney woods sections of Alabama were populated by a similar group. Sparks, *The Memories of Fifty Years*, 331.

their haste often failed to see were the extensive herds of cattle and droves of hogs which provided, in a great many cases, a comfortable living. And a later generation of writers, judging these people by the standards of a booming cotton economy, has failed to recognize their true frontier character. They were the inhabitants of a sort of inner frontier which, with the exception of the presence of Indians, was as much a frontier as the western edge of settlement. That the great pine belts were frontier regions is evidenced by the fact that they were remote and sparsely populated, well stocked with game and remained, for the most part, a portion of the public domain until after the Civil War.[45]

Finally came the agricultural settlers. These may be grouped into two broad categories—the nonpermanent and the permanent. The great majority of those immigrants falling within the nonpermanent category were poor farmers. While this type of immigrant might be found moving into the Territory throughout the period under consideration, he was more apt to come during the early territorial period, when there was still plenty of good quality public land on which to squat, than during the later period when most of the good land had been closed to public entry.

A great many of the nonpermanent settlers were squatters who settled on the public domain and moved on when the land was sold. However, it would appear that a very substantial number did acquire title to the land. But they did not retain the title for very long. These were men with "the West in their eyes." They could not resist the lure of the greener forests and richer valleys of the ever advancing frontier.[46] Drawn west-

[45] Owsley, "Settlement on the Southern Frontier," 155.
[46] Timothy Flint, a New England schoolteacher who lived in the Southwest during the early 1820's, observed that innumerable immigrants

ward by some El Dorado of the imagination, they sold their land and moved with the frontier. They became habitually rootless and nomadic as they followed the frontier into Alabama and Mississippi, and thence across the Mississippi River to Texas and beyond. The wife of one immigrant moving into the Southwest, provoked by her husband's ceaseless wandering, explained this propensity. "It is all for the mere love of moving," she said. "We have been doing so all our lives—just moving from place to place—never resting—as soon as ever we get comfortably settled, then it is time to be off to something new."[47]

Not all the nonpermanent settlers were subsistence farmers. Some were men of fairly substantial means. One such individual, who exemplifies very well the restless character of the nonpermanent settler, was Gideon Lincecum who, among other things, was a planter, physician and teacher. After perhaps a half-dozen moves up and down the Carolina-Georgia frontier, Lincecum was struck with the "Alabama fever." Crossing the Georgia frontier, he spent a season in the Creek Indian territory and then pushed through several hundred miles of wilderness to the little settlement of Tuscaloosa. Two or three times later he moved, settling for a while at Columbus, Mississippi, but finally ending his days in Texas. There were countless immigrants like Lincecum who delighted in freeing themselves from the restraints of civilization and settling in the "wildest, least trodden and tomahawk marked country."[48]

came into the Southwest in search of a mythical land which existed in their imaginations only. Timothy Flint, *Recollections of the Last Ten Years Passed in Occasional Residence and Journeys in the Valley of the Mississippi* (Boston, 1826), 240-241.

[47] Hall, *Travels in North America*, III, 131-132.

[48] Franklin L. Riley (comp.), "Autobiography of Gideon Lincecum," in Mississippi Historical Society, *Publications*, VIII (1904), 443 ff.

The permanent settler, like the nonpermanent, might be found moving into the Territory throughout the period under consideration. As a general rule, however, he was less inclined than the nonpermanent settler to migrate before the land was put up for sale. Such a move involved too many risks. The land on which he chose to settle might, when finally put up at public auction, bring more than he was willing or could afford to pay, thus forcing him into another costly move. The danger was especially great in the boom period following the War of 1812, when speculators and overly optimistic planters drove the price of choice of lands up to completely unrealistic levels—in some exceptional instances to more than one hundred dollars per acre.[49] It was far more practical for the individual who intended to settle permanently to purchase the land first and then move his family.

Within the category of permanent settler were to be found persons of every social rank. This category was not dominated, even in the latter stage of the migration, by the wealthy, established planter from the Upper South. The notion that the final phase of the settlement of the Southwest was dominated by wealthy planters from the older regions who quickly absorbed adjacent lands and drove the small planters and subsistence farmers either to some new frontier, or to the less fertile lands, is erroneous. The yeoman and poorer classes of farmers did not cease to exist in the Southwest when a handful of planters settled among them. Throughout the antebellum period the yeoman class constituted a very substantial portion of the population both of Alabama and Mississippi. Moreover, the great majority of wealthy planters in these states were self-

[49] Alabama *Courier*, March 19, 1819; *Niles' Register*, XV, 198.

made men who achieved their planter status in a single lifetime. Comparatively few planters from the older areas migrated to the Mississippi Territory with plantations full-blown.[50]

Finally, attention should be called to the fact that the society resulting from the advance of population into the Southwest bore many of the democratic traits generally ascribed to the frontier. While it is true that the plantation system emerged quickly in Alabama and Mississippi, the plantation was not an anachronism in frontier development nor did it upset the normal frontier process. The democratic spirit of Alabama-Mississippi society was pronounced. Individualism, self-reliance, and a sense of rough equality were no less evident in early Alabama and Mississippi than in Kentucky, Tennessee or Indiana.[51]

The man who assumed a superior attitude or considered himself better than his humblest neighbor was censured. One wealthy Mississippi planter, who employed his slaves to help rescue a sick neighbor's crops, met with sharp criticism because he did not work alongside his neighbors and slaves. He was told that if he "had taken hold of the plow and worked" along with the others all would have been well, but to sit on his horse and direct the Negroes was offensive even to those whose fields were benefitted.[52]

Even as late as 1849 the politician Albert Gallatin Brown could win wide approval in Mississippi by his boast of being "entirely a self-made man" and by accepting "every respectable man as his equal."[53]

[50] Craven, "The 'Turner Theories' and the South," 311.
[51] *Ibid.*, 308-311.
[52] Susan Dabney Smedes, *A Southern Planter: Social Life in the Old South* (London, 1889), 67.
[53] Craven, "The 'Turner Theories' and the South," 308-309.

65

A democratic spirit was reflected also in the constitutions of the states. The Alabama constitution granted to all white male citizens the right to vote and hold office without property or religious qualifications. And while the Mississippi constitution of 1817 required militia service or the payment of taxes for voting, and the possession of property and religious faith for officeholding, a second constitution of 1832 removed these qualifications and added a very emphatic statement about the equality of all men.[54]

[54] *Ibid.*, 308.

By Frank Otto Gatell

Secretary Taney and the Baltimore Pets: A Study in Banking and Politics*

Jacksonian Era scholars, in chronicling and interpreting the Bank War, the key American political issue of the 1830's, have properly concentrated upon the struggle between Andrew Jackson and Nicholas Biddle. Yet this "national level" view of the Bank War, while obviously important, has produced a tendency to slight or treat tangentially a significant segment of the problem — the role of the state-chartered banks. And in particular, the pet banks selected in 1833 to hold federal deposits remain relatively unexplored by political and economic historians. Glittering and unsubstantiated generalities abound concerning the role of these banks and their "not-pet" sister institutions in the political process, although details of what banks and which bankers produced which political results are singularly hard to come by.[1] That we lack a comprehensive history of deposit banking is understandable in view of the fact that case studies of nearly all pet banks have yet to be written.[2] The purpose of this article is to make a start in that

---

* Several grants from the General Research Board, University of Maryland, made possible the research for this article.

[1] For New York, for example, Jabez D. Hammond's overly relied upon *History of Political Parties in the State of New York* (2 vols., Albany, 1842), vol. II, p. 424, notes a Jackson majority in New York City in 1832, and asks: "Would this have been done had not the city banks been opposed to the existence of a branch of the Bank of the United States in that city?"

Bray Hammond labels the New York banks (presumably all the banks in the state) "congenial to the Albany Regency," which is not accurate even for the city of Albany. Hammond, *Banks and Politics in America from the Revolution to the Civil War* (Princeton, 1957), p. 352.

[2] Harry N. Scheiber has made some tentative remarks on pet bank politics: "The Pet Banks in Jacksonian Politics and Finance, 1833–1841," *Journal of Economic History*, vol. XXIII (June, 1963), pp. 196–214, but see especially p. 212. His article does not deal with the Baltimore pets. For part of the Detroit picture see Scheiber's "George Bancroft and the Bank of Michigan, 1837–41," *Michigan History*, vol. XLIV (March, 1960), pp. 82–90. The politics of selection of the thirty-five pets named from 1833 to 1836 is the subject of Frank Otto Gatell, "Spoils of the Bank War: Political Bias in the Selection of Pet Banks," *American Historical Review*, vol. LXX (Oct., 1964), pp. 35–58.

BUSINESS HISTORY REVIEW, 1965, vol. 39, pp. 205-227.

direction through a study of the politics of pet banking in Baltimore.

When federal funds were transferred to selected state banks in October, 1833, Secretary of the Treasury Roger B. Taney looked confidently to his native state and to the Baltimore bank selected. Everything pointed to a close, smoothly functioning relationship between the Treasury and the Union Bank of Maryland, a powerful, old institution under the direction of Thomas Ellicott, an imperious Quaker of domineering mien and character, as well as a strong opponent of Nicholas Biddle's Bank of the United States.[3] Taney knew the Union Bank, and he knew Ellicott, both as businessman and adviser on general banking problems. Surely this connection could not help but benefit everyone concerned. Instead, it proved a nightmare.

Ellicott had taken over the shaken but by no means stricken Union Bank during the panic year of 1819,[4] and did well enough, so that within a few years he figured in speculation concerning a new president for the national bank. The outgoing head, Langdon Cheves, favored Ellicott to succeed him. But influential voices among the Baltimore Branch stockholders, such as Robert Gilmor, questioned Ellicott's abilities, and by election time, November, 1822, his name had faded. With strong support from President Monroe and his cabinet, young Nicholas Biddle became third president of the Bank of the United States.[5]

Ellicott had not always hated the B.U.S. While the head of an important state bank might naturally be suspicious of the powerful national institution, in fact many state banks and the B.U.S. coexisted peacefully. Ellicott's hostility to the B.U.S. grew slowly. In 1819 he maintained friendly relations with it concerning debtors who abused state insolvency laws, and next year his recommendations on B.U.S. counsel in Baltimore won approval from President Cheves.[6] But on the eve of the B.U.S. presidential election in Philadelphia, the ambitious Baltimorean antagonized powerful interests. Richard Smith, cashier of the Washington Branch, complained in May, 1822 that Ellicott and the Union Bank had joined with several smaller Baltimore banks to fight the circulation of paper from the always shaky District of Columbia banks. This attempt to "usurp" centralizing and regulatory banking functions did not sit well with

[3] John E. Semmes, *John H. B. Latrobe and His Times* (Baltimore, 1917), pp. 399-400.
[4] Alfred C. Bryan, *History of State Banking in Maryland* (Baltimore, 1899), pp. 22-24, 64.
[5] Thomas P. Govan, *Nicholas Biddle: Nationalist and Public Banker* (Chicago, 1959), p. 75; R. L. Colt to J. White, Nov. 27, 1822, John C. White MSS (Md. Historical Society).
[6] Ellicott to J. Donnell, Dec. 17, 1819, Cheves to J. White, Dec. 7, 1820, *ibid*. See also Ellicott to Baltimore Branch of B.U.S., Feb. 26, 1820, Bank of the United States (Baltimore Branch) MSS (Library of Congress).

B.U.S. managers. But Ellicott ignored the effects of his course, and continued to press specie demands in Washington throughout the summer.[7]

In seeking an enlarged role for his bank, Ellicott scuttled whatever chance he had for the prime job at Philadelphia.[8] From this point to the outbreak of the Bank War, Ellicott's relations with the B.U.S. deteriorated. The settlement of an estate suit led to one especially icy exchange. "From the nature of thy inquiries," he informed B.U.S. cashier White, "thee seems to have very much mistaken the object of my call at the Branch Bank. . . . My only object was to explain the situation of the trust estate. . . . It was not to receive instructions as to what were our duties." At one point Ellicott haughtily chided the B.U.S. men for not knowing their business; at another, the B.U.S. furiously protested the Union Bank's timing of specie demands.[9]

More important than the substance of this petty warfare, was its effect on Roger Taney. Almost from his arrival in Baltimore in 1823 to practice law, Taney had acted as counsel for the Union Bank. The connection ripened into intimacy with Ellicott, both personal and professional. Taney invested his profits from fees into Union Bank stock and advised his relatives to follow suit. Many conferences with Ellicott, at office and home, confirmed his suspicions as to the B.U.S. and its capacity for evil.[10] By the time Taney became Attorney General in 1831, he and Ellicott were of one mind regarding the outline of a proper banking system. Details remained to be filled in, but a strong national bank did not fit into their picture. While Taney's thinking hardened, most Baltimoreans involved in banking or commerce — including many prominent Jacksonians — thought the B.U.S. would survive. Such "silk stocking" Democrats as Senator Samuel Smith and Congressman Benjamin C. Howard were pro-Bank, and others of their party and station echoed the sentiment, in varying degrees of enthusiasm.[11]

The B.U.S. decision to press for recharter during the presidential

---

[7] J. White to W. McIlvaine, May 27, 1822 (draft), R. Smith to J. White, Aug. 22, 1822, White MSS.

[8] There is no sizable body of Ellicott MSS. He preserved letters received from Taney, and they comprise the bulk of the Taney MSS at the Library of Congress. The John C. White MSS at the Maryland Historical Society contain a sprinkling of early Ellicott business documents.

[9] Ellicott to J. White, Feb. 10, 1825, June 13, 1827, R. Smith to J. White, June 24, 1828, White MSS.

[10] Stuart Bruchey (ed.), "Roger Brooke Taney's Account of His Relations with Thomas Ellicott in the Bank War," *Maryland Historical Magazine*, vol. LIII (1958), p. 64; Carl Brent Swisher, *Roger B. Taney* (New York, 1935), p. 92.

[11] Howard to R. Gilmor, Jan. 22, 1830, Benjamin C. Howard MSS (Md. Historical Society); Biddle to Smith, Jan. 30, Feb. 28, 1831, Misc. Biddle MSS (N.Y. Historical Society); Smith to S. Spear Smith, Jan. 12, 1832, Samuel Smith MSS (Library of Congress).

campaign of 1832 confirmed Taney's negative assessment. Since the "monied monopoly" could not hope to survive an election, it was demanding a twenty-year renewal before the people could be heard. Taney considered the B.U.S. by its nature politically motivated; Biddle's recent policies had merely revealed the fact to all.[12] The B.U.S. would get no mercy from the Attorney General, who ridiculed its highly regarded ability to create a national circulating medium: "The Bank of the U. States is no more entitled to the credit of creating a sound currency than the Bank of Hagers Town." [13]

As 1833 began, and the administration moved inexorably toward removal of public funds in the B.U.S., Ellicott's support of the Jackson-Taney position assumed greater importance. Nearness to Washington and intimacy with the Attorney General enabled him to work directly on the President. Ellicott drew close to the administration, partly because of his isolation from the Baltimore banking community stemming from his personality and the fact that only his bank and the Bank of Maryland paid interest on special deposits. Early in April, Ellicott presented his views in a long brief assuring Jackson that removal would not endanger the state repositories. A resolute removal policy guaranteed that the B.U.S. would "glide out of existence" with scarcely a twitch at the expiration of its charter in 1836.[14]

At Taney's invitation, Ellicott went further. Citing "the general impression which seems to prevail" of a coming pet bank system, he offered the government his bank's services. Again he hammered at the theme of state bank capability, and, with an alert and cooperative government, their invulnerability. This remarkably skillful argument came close to politics in several passages, but each time phrased the issues as much in terms of public policy as partisan advantage. An unstated premise designed to appeal to Andrew Jackson permeated the document: the identity of public policy and Democratic politics.[15]

Late in July the administration acted. Jackson dispatched Amos Kendall northward to sound out bankers on a state-banks network of government deposit. Kendall stopped first at Baltimore, and

[12] Biddle, drawn into politics reluctantly, had hoped that "if there is to be a new division of parties, the Bank may be kept out of them all." Biddle to Smith, Feb. 3, 1830, Letterbook, Nicholas Biddle MSS (Library of Congress).
[13] Taney to Ellicott, Jan. 25, Feb. 20, 1832, Roger B. Taney MSS (Library of Congress).
[14] Taney to Ellicott, Aug. 28, 1832, *ibid.*; J. L. Hawkins to D. Sprigg, March 2, 6, 1832, Misc. Hawkins MSS (Md. Historical Society); Ellicott to Jackson, April 6, 1833, John S. Bassett (ed.), *Correspondence of Andrew Jackson* (7 vols., Washington, 1926-1935), vol. V, pp. 49-52.
[15] Taney to Ellicott, May 5, 1833, Taney MSS; Ellicott to Duane, June 14, 1833, Treasury Department, National Archives (hereafter, TD/NA), Letters from Banks.

Ellicott received advance notice.[16] Kendall soon learned that his first series of proposals to the banks, involving the personal security of officers and stockholders and a measure of collective responsibility, would have to be scrapped. Even then, the security contemplated by the second less onerous series presented difficulties. The Bank of Maryland was the only bank which under its charter terms could probably comply with the original security demands, but its capital of only $300,000 stood next to lowest of these circulated. In addition, President Evan Poultney demanded a monopoly of Baltimore deposits.[17] Other solid and unspectacular banks, including the well-capitalized Bank of Baltimore, suggested further negotiations. But the Union Bank was the most highly capitalized, and Kendall believed other Baltimore banks unfriendly to Jackson. Instead of qualifications, Ellicott immediately offered cooperation. On specific matters, Ellicott bargained as sharply as other bank presidents, but he was obviously bargaining seriously and not equivocating.[18] When he left Baltimore, Kendall reported doubts about choosing among candidates, while Ellicott exuded confidence. With the deposit banks bandwagon rolling, he told New York pet-banker-to-be George Newbold of the Bank of America, that no one should miss out since the benefits of holding public funds would run high.[19]

When the time came to choose a pet for Baltimore, Roger Taney headed the Treasury Department. Rushed into service to remove the deposits when William Duane proved perversely independent, Taney now wielded the effective power of selection. By mid-September, after conferences with Kendall and Ellicott, Taney set a target date for removal, the first of October. To no one's surprise, the federal district attorney solemnly opined that the contract between the Treasury and the Union Bank seemed in order. The opposition press retorted dejectedly that the "Kitchen Cabinet and the stock jobbers" had won the day.[20]

Who selected the Union Bank? Ellicott commented cryptically that Jackson had chosen his bank not because he "was considered as

---

[16] Kendall to Ellicott, July 26, 1833, Taney MSS.
[17] Poultney to Kendall, July 31, Aug. 5, 1833, 23rd Cong., 1st Sess., *Senate Doc. No. 17* (hereafter, 23:1, SD #17), pp. 26, 33. This senate document contains much of Kendall's investigatory correspondence.
[18] Kendall to Jackson, Aug. 2, 3, 1833, Bassett (ed.), *Correspondence of Jackson*, vol. V, pp. 145–46; Ellicott to Kendall, Aug. 1, 2, 1833, 24th Cong., 2nd Sess., *House Report No. 193* (hereafter, 24:2, HR #193), pp. 444–48.
[19] Ellicott to Newbold, Aug. 5, 1833, George Newbold MSS (N.Y. Historical Society); Baltimore *Republican*, Aug. 2, 14, 1833.
[20] Taney to Jackson, September 17, 1833, Bassett (ed.), *Correspondence of Jackson*, vol. V, pp. 191–92; N. Williams to J. McCulloch, Sept. 28, 1833, TD/NA, Letters from Collectors; Baltimore *Chronicle*, Sept. 20, 1833; Baltimore *Gazette*, Sept. 23, 1833.

approving of the general policy and measures of the administration, but for other reasons." Taney's account stressed the impropriety of himself, a stockholder "to a small amount," making the choice. Taney set all his assembled facts on Baltimore banking in front of Jackson, and Ellicott came to Washington himself to see Jackson and Taney. At the Treasury, Ellicott also suggested that the Bank of Maryland be made a depository — showing a concern for the smaller institution which puzzled Taney at the time.[21] The President, in effect, merely ratified Taney's selection. Jackson's knowledge of the Baltimore situation came from Ellicott and his confidence in Taney made the choice automatic. The Union Bank had the paper qualifications — experience, capital, forceful leadership — and Ellicott's role in influencing the Removal made it almost certain that the Union would become the original deposit or pet bank. Within a few days, Taney doubtless wished that he had considered the alternatives for selection more carefully.

The initial trouble stemmed from transfer drafts. As a defense against B.U.S. incursions into state bank specie reserves, the administration armed several pets with drafts on government money held by Biddle and his branches. The Union Bank's original allotment was $100,000.[22] Taney stressed the precautionary nature of the moves: if the B.U.S. acquiesced in a quiet burial, the transfer drafts were not to be used. Since bankers quickly learned each other's business, Taney hoped that knowledge of the drafts' existence would restrain Biddle. Taney's advice went out on October 3, but the day before, Ellicott had already indicated that he coveted the deposits still in B.U.S. hands. "The public," he argued, "can derive no facility from the use of the public money as long as this state of things continues," while immediate transfer would "have a great tendency to render popular the course pursued by the government."[23]

A few hours after he dispatched the first draft to the Union Bank, Secretary Taney received visitors, emissaries from Ellicott. Reverdy Johnson and David M. Perine had come principally to discuss the shaky status of Baltimore's small Susquehanna Bridge and Bank Company, and the chain reaction its failure could produce. Conceivably, they explained, its collapse might indirectly harm the

[21] Swisher, *Taney*, p. 241; Thomas Ellicott, *Bank of Maryland Conspiracy* (Philadelphia, 1839), p. 10; Bruchey, "Taney and Ellicott," pp. 53, 131; G. Steuart to Jackson, Sept. 23, 1833, TD/NA, Misc. Letters Received.
[22] Each of the three New York pets received $500,000, as did the Girard Bank, Philadelphia.
[23] Ellicott to Taney, Oct. 2, 1833, TD/NA, Letters from Banks; Ellicott to Taney, Oct. 2, 1833, David M. Perine MSS (Md. Historical Society); R. Colt to Biddle, Oct. 23, [1833], Biddle MSS.

Bank of Maryland, or even the Union Bank itself. Since Johnson was a counsel, and Perine a director of the Union Bank, Taney respected their opinions. Furthermore, any state bank failure would psychologically damage all the pets. Only B.U.S. mischief could injure the Bank of Maryland, but to assure its condition, they suggested naming it a pet and sending it a $500,000 transfer draft. Taney parried the requests, but agreed to an additional $200,000 in contingent funds for the Union Bank.

Taney's accounts of these events leave much unanswered. Writing in 1834 and 1839 he described himself as an innocent gulled by the speculators from the city.[24] According to him, the emissaries assured him that all was well with the Bank of Maryland, and that the B.U.S. would not harm other Baltimore banks friendly to it. If all was well, why did he agree to send $200,000 more? Why the need for a special mission from Baltimore? But, as Taney noted, "the Union Bank was near me," and he could have waited until the "chain of events" revealed itself, since use of the additional drafts was contingent upon B.U.S. moves. In retrospect, Taney exaggerated his lack of comprehension and knowledge of the Baltimore banks. His intimacy with Ellicott could not have produced the naivité Taney claims he displayed on the evening of October 3.

Two days later Taney's chief clerk returned from a visit to Baltimore with disquieting news. Perine had told him of the use of both drafts drawn on the Baltimore B.U.S. Branch, and their endorsement to Poultney's Bank of Maryland. Taney, infuriated, began to write for explanations, when Ellicott appeared. It was then, claims Taney, that he first heard of the Bank of Maryland's woes. Slowly, fact by fact, the Secretary and Kendall, the Treasury's Fourth Auditor, extracted Ellicott's tale of high finance and low security. The Bank of Maryland was indeed in trouble, largely, because of speculations in Tennessee state bonds and railroad stocks. As he surrendered the details, Ellicott lost his composure, stammering and stuttering to find words to cloak his own involvement.

Desperately audacious, Ellicott pressed for yet another transfer draft, this time of $500,000. Taney waited a day before refusing, but he allowed Ellicott to keep the remaining $100,000 draft on the

[24] Taney to R. Johnson, July 25, 1834, Ellicott, *Bank of Maryland Conspiracy*, pp. 85–86; Bruchey, "Taney and Ellicott," pp. 133–39. The speculators "Club" consisted of Johnson, Perine, John Glenn, Hugh McElderry, Evan Poultney, and Evan T. Ellicott. The last two ran a private bank, "Poultney, Ellicott & Co." Thomas Ellicott entered the picture by underwriting Tennessee bonds to be sold in England at well over par, for a guaranteed, large commission. He stayed home instead, and working through Poultney, sold the bonds to the Union Bank at 108, so that Ellicott earned his commission at his own bank's expense. (J. Gordon to S. Gordon, June 3, 1834, James M. Gordon MSS, Maryland Historical Society.) For other sections of this jigsaw puzzle see Bryan, *Banking in Maryland*, pp. 91–94.

Philadelphia B.U.S., with the same injunction of self-restraint. Taney's friendship had created claims, of course, but, as Kendall noted, Taney had invested all his political prestige in the pets. To allow any part to buckle would be damaging; to allow a collapse of Ellicott and the Union Bank was unthinkable.[25]

Both men proceeded to burden the Baltimore-Washington mail pouches. Still angry from the face-to-face exchange, Taney brusquely called for performance instead of promises, bluntly warning Ellicott that he had lost much ground with the administration. The banker's stylized replies, short on logic and accurate data, were long on rhetoric about public duty and the "spirit" of Treasury instructions. Invoking his "duty to the city and the government," or his responsibility "in my humble degree, for the success of the measures of the administration," Ellicott's apologia expanded with each letter — an unconvincing mixture of self-esteem and high principle.[26]

Time and necessity brought a cooling off on Taney's part. More optimistic reports filtered in from Baltimore, and even the pesky Susquehanna Bank seemed solid again. While not happy over recent events, Taney looked to the future with a reminder that deposits did not exist as emergency funds for faltering speculators; pet banks, where the public funds were concerned, should operate on public and patriotic principles.[27] The reconciliation proceeded so well that by the end of October, Taney himself raised the possibility of using the third transfer draft against the B.U.S., a suggestion Ellicott acted upon within ten days.

The Treasury did not protest this time; Biddle's Panic was on. For protection, Ellicott called for close cooperation among the principal pets. At first he suggested a confederation under the central direction of Reuben M. Whitney, formerly a Philadelphia merchant and sometime B.U.S. director, who had turned on Biddle and worked himself into administration counsels during the Removal debates, but Taney would not accept the idea.[28] The Baltimore banker turned to pleas for voluntary cooperation, stressing the need to undercut the strongest B.U.S. argument, that the pets could not provide a system of uniform domestic exchange. He feared that without unity among the pets of the four leading Atlantic

---

[25] *Ibid.*, pp. 139–46; William Stickney (ed.), *Autobiography of Amos Kendall* (Boston, 1872), pp. 389–90.
[26] Ellicott to Taney, Oct. 8, 1833, TD/NA, Letters from Banks; Ellicott to Taney, Oct. 8, 10, 1833, Perine MSS.
[27] Taney to Ellicott, Oct. 11, 1833, Ellicott, *Bank of Maryland Conspiracy*, pp. 91–92; Ellicott to Taney, Oct. 12, 1833, Perine MSS.
[28] Ellicott to Whitney, Oct. 24, 1833, 24:2, HR #193, p. 441.

cities — Boston, New York, Philadelphia, and Baltimore — the advent of heavy government drafts would drop their notes below par, "a circumstance that would be much deplored by the friends of the measures which have been adopted by the Treasury Department." [29]

As a corollary aspect of security, Ellicott envisaged a truly "state bank" pet system, involving state governments in the posting of security. This could be done, not by chartering official state-owned banks as existed in some states (an idea repugnant to private banker Ellicott) but through state patronage of "certain state banks." Legislatures could nurture these local pets, particularly by refusing to charter new banks. Apparently Ellicott's anti-monopoly banking principles applied only to Biddle. As the panic session of Congress closed, he reminded Taney that while "I am not politician enough to form an opinion," B.U.S. recharter would come during the next session unless the administration imposed a system upon the pet bank expedient.[30]

While these moves and suggestions had, of course, been made confidentially, Secretary Taney and the Union Bank inevitably attracted public attention. Taney's promotion to the Treasury had provoked a heavy political debate, especially among opposition editors who blasted his ignorance of finance. Democratic defenders remained prudently imprecise about their new man, vaguely proclaiming his unswerving "hostility to monopolies and aristocracy." This proved to be the Democratic press strategy throughout the Bank War. The Baltimore *Republican* said as little about the Union Bank and its president as possible, replying to opposition charges with generalities and irrelevance.[31] The Union Bank remained unmentioned until over two months after selection. Then, Washington's *National Intelligencer* twitted Taney for selecting a Baltimore bank whose charter would soon expire, while impending charter expiration had been a principal argument used against the B.U.S. Ellicott then entered the public prints, and the Union Bank and its politics became an open issue. The issue flared again when George McDuffie of South Carolina lashed out against the pets, claiming they had been selected for political reasons, that their officers and directors were Democratic partisans. These charges, and steady needling by Baltimore editors concerning details of pet

---

[29] Ellicott to Newbold, Oct. 29, 1833, Newbold MSS; Ellicott to J. Schott, Dec. 9, 1833, Girard Bank MSS [part of the Lewis-Neilson MSS] (Historical Society of Pa.).
[30] Ellicott to Taney, Nov. 5, 1833, 24:2, HR #193, pp. 501–503; Ellicott to Taney, March 14, 1834, Perine MSS.
[31] Swisher, *Taney*, p. 238; Baltimore *Republican*, Sept. 28, Dec. 6, 1833.

operations, forced Democrats into denial. The Union Bank was not a Democratic institution, argued the *Republican*'s correspondent "Justice." Some of its directors had supported Adams in 1828 and Clay in 1832, and this was their right as citizens.[32]

Despite the editorials no paper published an annotated list of Union Bank directors' politics. Of the Union's eighteen directors, only four can be definitely identified politically.[33] Luke Tiernan and Solomon Etting were opposition men. Both men had served on the Union board since its inception in 1804, and were probably as much fixtures at the bank as its strongboxes.[34] Identifying Democrats is equally difficult. John H. B. Latrobe, a friend of Ellicott's, and a strong Jackson man, served briefly, and by his account, ineffectually. David M. Perine also falls into this category. The cashier, Robert Mickle, was apolitical.[35] Ellicott's politics cannot be classified precisely as Democratic. He supported the administration banking policy, but not the administration, a situation which did not escape the notice of his Washington connections. And because he personally ruled the Union Bank, the general question of the directors' politics becomes a phantom. While administration policy conformed to the Union's interests, Ellicott gladly played the role of ardent administration man, as when he tried to silence bank counsel John P. Kennedy, then about to break with the Democrats, with a reminder that the deposits would earn the Union Bank as much as $30,000 a year.[36] As for Bank War "ideology," one suspects that he responded to Taney's anti-B.U.S. exhortations with tongue resting comfortably against cheek.

Biddle and the B.U.S. watched every motion of the pet bank foundation, and nowhere were they more observant than in Baltimore. In addition to his normal channel of communication with Baltimore, Biddle cultivated Reverdy Johnson, then playing a double, if not a triple, game. Fresh from his Washington mission to Taney in October, Johnson began feeding inside information to the B.U.S., lacing his accounts with declarations of fealty and pathetic appeals for favors for the Bank of Maryland and an in-

---

[32] Baltimore *Chronicle*, Dec. 23, 25, 1833; Baltimore *Republican*, Dec. 12, 23, 25, 1833.
[33] For the Union Bank board in 1833 see *Matchett's Baltimore Directory, 1833-1834*.
[34] *The Biographical Cyclopedia of Representative Men of Maryland and the District of Columbia* (Baltimore, 1879), p. 497; Baltimore *Chronicle*, Aug. 7, Oct. 11, 1833; Etting to Biddle, June 10, 1834, Biddle MSS.
[35] Semmes, *Latrobe*, p. 366; Clayton C. Hall (ed.), *Baltimore, Its History and Its People* (3 vols., New York, 1912), vol. II, pp. 393-96; George W. Howard, *Monumental City* (Baltimore, 1883), p. 547.
[36] R. Whitney to W. D. Lewis, Nov. 13, 1833, Lewis-Neilson MSS (Historical Society of Pa.); Kennedy, MS Journal, Dec. 11, 21, 1833, Kennedy MSS (Peabody Institute Library, Baltimore).

surance company he managed.[37] Biddle listened and waited, confident that the Bank of Maryland and the Union Bank could manufacture enough trouble for themselves without B.U.S. interference. "We shall soon have occasion to see," he predicted to his Baltimore cashier, "when the funds of the Govt. in the B.U.S. are exhausted," whether the Union Bank could meet government demands as comfortably as the B.U.S. Branch had. "The other Banks may perhaps hereafter regret the absence of so forebearing a creditor as the B.U.S."[38]

Meanwhile the Union Bank had learned that deposit banking involved disbursements as well as receipts. In the heady days following removal, the pet prospered. Government funds flowed in steadily, and transfer drafts added to the total, while public officers drew on the B.U.S. By the end of 1833, however, the honeymoon period ended. When Taney warned of coming government demands, Ellicott, overcautious, drastically curtailed loans with vague explanations of high interest rates in northern cities having lured away money "properly belonging to Baltimore."[39] When the drafts came during January (accompanied by apparently well-timed B.U.S. demands for payments), Ellicott reacted badly. Instead of paying promptly, he whined that the Washington pet (the Bank of the Metropolis), had pressed demands on him without adequate notice. All this, despite Taney's careful attempt to prepare the way for withdrawals. No longer was Ellicott the confident and knowledgeable financier who had buttressed Jackson in the spring and summer of 1833. And with metallic currency ideas winning over more and more Democrats, Ellicott hesitated there, too: "I am a good deal of a *non-committed* man. . . . Let us not attempt to move too fast," he warned Taney. The new, "non-committed" Ellicott told Supreme Court Justice Henry Baldwin that Jackson and Taney knew nothing about finance, and that perhaps Biddle had been justified in curtailing discounts.[40]

"No one has any doubt about their standing," soothed a Democratic editor in February, referring to the Baltimore banks and the "brilliant exhibit" their published balance sheets presented. Within a month three banks, including the Bank of Maryland, had failed,

---

[37] Johnson to Biddle, Oct. 9, 1833, Biddle MSS. Bernard C. Steiner's life of Reverdy Johnson (Baltimore, 1914), is, to put it charitably, inadequate for this period. Perhaps someone interested in Maryland history as well as calligraphy will produce this much-needed study.
[38] Biddle to J. White, Nov. 10, 1833, White MSS.
[39] R. Colt to Biddle, Dec. 27, 28, 1833, Biddle MSS; Ellicott to Taney, Jan. 6, 1834, Perine MSS.
[40] R. Smith to J. White, Jan. 7, 1834, White MSS; Ellicott to Taney, Jan. 23, Feb. 20, 1834, Perine MSS; Baldwin to J. Hopkinson, Feb. 3, 1834, Hopkinson MSS (Historical Society of Pa.).

rocking the entire financial structure of the city and sending shock waves far beyond the state. The street outside the Bank's offices filled rapidly, first with businessmen, then short-tempered working class depositors. It appeared momentarily that Baltimore would then witness another of the outbursts which had already made it well known in American riot history. Many persons, some hopefully, thought the Union Bank would follow the bankruptcy trail, and Ellicott sustained a run on his bank for several days.[41]

On March 21, one day before the Bank of Maryland closed, Ellicott had advised Taney of coming troubles, casually giving details almost as if he were not involved. Fearing the letter would not achieve results in time, Ellicott dispatched Perine and Hugh McElderry (another member of the Union Bank-Bank of Maryland "Club") to Washington that night. Taney must have blinked at this repetition of the scenes of the previous October. This time the talk centered on the Union Bank's peril. As Taney remembered it, examination of the bank statement and information extracted from the visitors clearly indicated mismanagement by Union Bank officials. When he returned the documents he said he could not throw good public money away after bad private ventures.[42]

Taney's memory again played comforting tricks on him. He did try to rescue the Union Bank. The Treasury would support Ellicott vigorously — even to the extent of transferring government funds from other cities if essential, despite the political risks involved. Learning that Ellicott had become trustee for the defunct Bank of Maryland, Taney protested, knowing that any public link between the two banks provided an easy target for opposition editors. But so reassuring were Ellicott and his board, that Taney's early alarm subsided. There was no need for additional security; he had entire confidence in the Baltimore pet.[43]

Events soon proved Taney's confidence misplaced. As banking troubles mounted in other cities, Ellicott submitted no reports for several weeks. Amos Kendall swelled the chorus of advice descending on Ellicott. Banks might fail wholesale, "but if the '*Pets*' begin to go," anything might happen. The Union Bank must look to its self-interest "so that you *can stand amidst ruin*." Pet status made survival imperative. When Taney learned particulars of Ellicott's

[41] Baltimore *Republican*, Feb. 21, 1834; R. Wilson to J. James, July 31, 1834, Urbana Bank MSS (Ohio Historical Society); Philadelphia *National Gazette*, March 26, 1834; G. M. Dallas to G. Wolf, March 26, 1834, Wolf MSS (Historical Society of Pa.).
[42] Ellicott to Taney, March 21, 1834, Ellicott, *Bank of Maryland Conspiracy*, p. 93; Taney to Poultney, March 21, 1834 (copy), Perine MSS.
[43] Taney to Ellicott, March 25, 27, 30, 1834, Taney MSS; J. White to J. Meredith, March 28, 1834 (draft), White MSS.

heavy stock purchases and speculations, he knew that he would have to bail him out or absorb a severe political setback.[44]

Thoroughly dejected, Taney nevertheless moved to save the bank. He agreed that its long range outlook was not hopeless, but $200,000 was needed at once. Apparently he had had second thoughts about sending good money to rescue bad. When the Bank of the Metropolis made a surprise demand on the Union, Taney exploded: "I cannot agree to have the whole administration disgraced and defeated by such unaccountable folly," threatening to switch Washington pets. In no mood to hear explanations from any bankers, he told Ellicott to put his banking house in order whatever the sacrifices.[45]

When the Bank of Maryland collapsed, opposition men ascribed events to the removal of the deposits. Senator Henry Clay demanded an inquiry into the condition of the Union Bank and into Taney's own stock holdings. In reply, Ellicott denied any special connection with the Bank of Maryland, and scoffed at rumors of a heavy run on his bank.[46] Secretary Taney had to make his own explanations about the conflict of interest,[47] and in response to Clay's call he revealed ownership of 73 shares (worth about $5,000) on Removal day, with a drop to 63 shares since then, and no new purchases since 1831; all in all, a disappointing harvest for Clay.[48] But Taney did not reveal his intimate connection with the Union Bank — the factor which had prompted all the innuendoes and attacks. He was not simply another investor; Ellicott granted him favors in handling his account, which produced what Taney called, in one of the recurrent moments of crisis, "the peculiar relation in which I stand to you and to the Union Bank." [49]

Some Whig editors regretted that the Union Bank had not failed,

---

[44] Kendall to Ellicott, April 15, 1834, Taney to Ellicott, April 7, 18, 1834, Taney MSS.
[45] Taney to Ellicott, April 21, 24, May 1, 1834, *ibid.*
[46] Whitney to W. D. Lewis, March 25, 1834, Lewis-Neilson MSS; Ellicott to Taney, March 26, 1834, TD/NA, Letters from Banks.
[47] Bray Hammond's anti-Jackson *tour de force majeure* never lets the reader forget Taney's Union Bank investment (*Banks and Politics*, pp. 335, 414, 419, 431). Taney does not shine in this affair, but he deserves better of historians than Hammond's treatment (p. 419). Hammond charges Taney with conflict of interest, cheap speculation, and profiteering during Removal. He tells of the Secretary buying a "little more" Union Bank stock "as nominee" for female relatives. Actually, early in May, Taney asked Ellicott to invest his holdings for the ladies as a non-speculative venture in the Union, *or any other bank*. The stock would be his only nominally, with safe investment, "not profit on a resale" the only criterion. (Taney to Ellicott, May 5, 1833, Taney MSS.) The purchase, if made, was *not* in Union Bank stock.

Mr. Hammond wrote me on Aug. 17, 1963: "I do remember hesitating as to whether the matter was important enough to mention. It seems to me now it was not . . . The fiduciary purchase . . . seems to have been made too long before the selection of depositories to be relevant . . . I hope you will . . . correct me, with my specific concurrence. . . ."
[48] 23:1, SD #238, pp. 20-21; Ellicott to Taney, Feb. 20, 1834, Perine MSS.
[49] Taney to Ellicott, Jan. 7, May 20, 1834, Taney MSS.

some simply linked the two banks together. From Philadelphia, Robert Walsh's *National Gazette* hinted at murky conspiracies the year before between treasury agent Kendall and the Bank of Maryland's president. If Amos had had his way, the public money "might have been even more directly than it has been entrusted to Mr. Evan Poultney's leaky canoe." Whatever the editors' exaggerations, the psychological state of financial chaos in Baltimore was real enough.[50]

How would Biddle react? Reports of Ellicott's desperate maneuvers streamed into B.U.S. headquarters, many of them false, but by this time all of them plausible. Roswell Colt, who kept Biddle posted on Baltimore affairs, asked for aid for the community. The city banks needed specie to meet demands approaching panic proportions. Biddle adopted a policy of selective rescue, and within two weeks the crisis had eased. By the end of April all the Baltimore banks were secure, yet the only paper generating much confidence was B.U.S. notes. Biddle had reason to be pleased with the outcome of this skirmish.[51]

Ellicott had become more than ever shy about details. Taney sputtered in frustration that he wanted facts not misty generalities about specie drains and machinations against the Union Bank. At one point Ellicott had the temerity to arrive in Washington with an ultimatum. The Baltimorean reported that he had just returned from New York where the money men denied all aid while he maintained his Treasury contract. Unless he got extensive aid from Taney he would surrender the deposits, come out for restoration to the B.U.S., and presumably wreck all the pet banks. While Taney took a day to mull this over, Kendall interposed to relieve the Secretary who was "on all sides harrassed almost out of his senses," and instruct Ellicott on his duties. Taney rejected the ultimatum, calling Ellicott's bluff. Colt guessed correctly when he described Taney as "sick-sick at heart" of Ellicott, his bank, and the whole affair.[52]

Heartsick though he might be, Taney retained enough composure to move to salvage what he could. First he dispatched his able

[50] Portsmouth (N.H.) *Journal*, March 29, 1834; Philadelphia, *National Gazette*, Aug. 22, 1834; R. Johnson to V. Maxcy, March 30, 1834, Howard to V. Maxcy, May 14, 1834, Virgil Maxcy MSS (Library of Congress).
[51] Colt to Biddle [late March, 1834] and April 13, 1834, Biddle MSS; S. Smith to W. Mangum, April 19, 1834, Henry T. Shanks (ed.), *Papers of Willie Person Mangum* (5 vols., Raleigh, 1950–1956), vol. II, pp. 148–50; J. Atkinson to N. Wright, April 26, 1834, Nathaniel C. Wright MSS (Library of Congress); Joseph Cushing to Baltimore Branch of B.U.S., Nov. 10, 1834, B.U.S. (Baltimore Branch) MSS.
[52] Taney to Ellicott, May 15, 23, 1834, Taney MSS; Taney to D. M. Perine, June 2, 1834, Robert Oliver MSS (Md. Historical Society); Kendall to Ellicott, May 28, 1834, Taney MSS; Colt to Biddle, May 14, 17, 1834, Biddle MSS.

chief clerk, McClintock Young, to scout the Baltimore scene. Young's unhappy reports induced Taney to seek a new *confidante*, David M. Perine, who would replace Ellicott as Taney's man in Baltimore, and help displace Ellicott from the Union Bank. Perine and his fellow board members, argued Taney, must take control away from Ellicott, who had contended that his actions always gained the approval of his directors. Taney pleaded in personal terms as well: "I entreat you not to resign at this time, for I need the aid of my friends at the Board to carry me safe through my affairs there."

Taney's sense of betrayal and embarrassment naturally focused on Ellicott. "You know how much and how entirely I have confided in Mr. Ellicott," he reminded Perine. "I have thought he was obstinate and self-willed, but I have had the most entire confidence in his integrity." Ellicott's coyness in the face of frank inquiries, and his letters "so extraordinary and so ambiguous" could not be overlooked. Finally, the ultimatum ("such an extravagant and unreasonable proposition & such a breach of faith") shattered all trust in the man.

Ellicott's fall need not doom the bank. A complete investigation, ignoring Ellicott's objections and sophistries, would reveal the extent of the damage. Then, Taney promised, he could in good conscience do everything legally possible to help the bank. On evidence of a good report, Taney indicated his willingness to insure an average deposit of $400,000, and to return the Union Bank's security for the public money — the Tennessee bonds which Ellicott had obtained from the Bank of Maryland! Non-compliance, however, would bring a Carthaginian settlement, and the transfer of deposits to another Baltimore bank. "But if the Bank will do its duty every support and indulgence will be given." [53]

Taney's shift opened the way for Biddle as saviour of the Baltimore banking community. Reverdy Johnson proposed a "loan" of $500,000 for the Union Bank, exactly the amount of speculation in Tennessee bonds entered into by the club of investors, and the reverse side of the same coin rejected by the B.U.S. a year before, when McElderry sought $500,000 for the Bank of Maryland. Johnson, careful not to suggest direct B.U.S. aid this time, merely noted that perhaps "some of your capitalists" might like a good investment in Tennessee bonds. As bait he held out surrender of the deposits, which he labelled the first step towards B.U.S. recharter.[54]

[53] Taney to Perine, May 28, 29, June 2, 1834, Oliver MSS.
[54] McElderry to B.U.S., May 22, 1833 (copy), John Sergeant MSS (Historical Society of Pa.); Johnson to Biddle, May 29, June 4, 1834, Biddle MSS.

Johnson's maneuvering and sycophancy had not fooled Biddle. After a long talk with Johnson, Biddle calmly assessed the situation, financially and politically for Senator Daniel Webster, who had also seen Johnson. Short-term advantage might come out of the proposal. But Biddle, now waging a war, could not tarry over minor killings. Besides, B.U.S. involvement, given the state of public opinion, meant Biddle would be blamed. "I do not know Mr. Ellicott — but he has the reputation for considerable talents & energy. He will require it all now," Biddle observed drily. Disaster seemed assured if the Union Bank did not adopt certain measures, especially the surrender of government deposits. Financially, a Union Bank failure meant nothing to the B.U.S., "but I would rather see that institution prosper than suffer." Biddle promised nothing, but appeared to be offering a favorable hearing following cancellation of the Union Bank's Treasury contract.[55] He considered his position in Baltimore, far stronger than the pet's, and decided to allow time to work for him. The historic unpopularity of the Baltimore Branch had lessened, as Union Bank director Solomon Etting assured him.[56]

By June it was clear that while the Union Bank would survive, Ellicott could not remain as president. Everything he touched generated controversy. When his fellow trustees of the Bank of Maryland issued a report, Ellicott for murky reasons refused to join them, further diminishing confidence in any substantial settlement of accounts. Johnson, with no sympathy for his former associate, dismissed the possibility of Ellicott's hanging on. Only his resignation could save the bank, but that was too much to expect from his sort.[57] Baltimore had had enough of the bankers Ellicott, and all their cousins. The search for a scapegoat converged on Thomas Ellicott. But he stubbornly refused to resign, despite his isolation.[58]

Anti-Ellicott feeling among the stockholders soon crystalized into an organized force for change, as the election of directors approached. Initial disagreements over lists of candidates gave some trouble, but the aroused holders soon had a compromise ticket ready. Never a popular man, even the anti-reform slate of candi-

---

[55] Biddle to Webster, May 28, 1834, Biddle to Johnson, June 2, 1834, Letterbook, Biddle MSS.
[56] Etting to Biddle, June 10, 1834, *ibid*. Two months later, the new Union Bank managers applied for a B.U.S. loan. Samuel Jaudon, Biddle's cashier, chuckled as he informed his vacationing chief that he wanted to delay action on the request until Biddle's return: "I should like the answer to the Pet to come from you . . . It is worth something to have had such an application." Jaudon to Biddle, Aug. 11, 1834, *ibid*.
[57] Baltimore *Republican*, May 30, 1834; Baltimore *American*, June 12, 1834; Johnson to Biddle, June 7, 1834, Biddle MSS.
[58] Semmes, *Latrobe*, pp. 408–409; Taney to Perine, June 7, 1834, Oliver MSS.

dates did not include Ellicott's name.⁵⁹ Talk of the bank giving up the deposits once the new board took over prompted the *Republican* to wonder if the newcomers would foolishly weaken the bank further by trying to settle the Treasury account at once. Although the proposed board had a Whig majority (like the old, claimed the paper), the bank's interests ruled out any vindictive politics. Shortly thereafter, "Reform" assured *Republican* readers that politics played no part in the move to oust Ellicott. His "Janus-like" negotiations with Poultney were sufficient and compelling reasons.⁶⁰

Ellicott had lost control. With no board meetings for several weeks, directors Latrobe and Mayhew handled many day-to-day decisions on their own authority. As a final stratagem, Poultney, Ellicott & Co., which held 2,000 Union shares, "distributed" them to individuals who returned proxies (the charter limited voting rights on stock to the first sixty shares held per person). The reformers went to court charging fraud, and got an injunction which the Maryland Court of Appeals upheld. The legal moves gave Ellicott time to issue a public blast at the "organized and systematic" attempt to vilify him. He promised stockholders that whatever they decided he would subsequently make disclosures which would more than repay his tormentors. Taney dismissed the threat: "Rely upon it Mr. Ellicott will attack no body — except by dark insinuations intended to intimidate. He is conscious of what would in such a case await himself." ⁶¹ On July 14, the reformers easily carried the day, as Ellicott stood virtually alone. Reverdy Johnson had thrown his professional talent behind the ouster, and Taney signed over his shares to a proxy pledged to change the bank's direction.⁶²

The incoming regime had a decided political cast. The leading individual, cotton manufacturer Hugh W. Evans, had been chairman of the Baltimore National Republican Central Committee; Whigs Solomon Etting and Luke Tiernan, the perennials, remained on the board, and a number of the newcomers, if not rabid partisans, were professed Whigs. An important exception to this Opposition ascendancy was attorney Charles Howard. Brother of Democrat Benjamin C. Howard, and himself active in party affairs, Howard had just declined to serve as a government director of the B.U.S.⁶³

⁵⁹ James M. Gordon to S. Gordon, June 3, 1834, Gordon MSS; Baltimore *Republican*, June 7, 14, 17, 1834.
⁶⁰ Baltimore *Republican*, June 14, 21, 1834.
⁶¹ Semmes, *Latrobe*, p. 401; W. Block to Biddle, July 3, 1834, Biddle MSS; Baltimore *Republican*, July 3, 14, 1834; Taney to Perine, June 20, July 10, 1834, Oliver MSS.
⁶² Johnson to Biddle, Aug. 3, 1834, Biddle MSS; Bruchey, "Taney and Ellicott," p. 62; J. Gordon to S. Gordon, July 14, 1834, Gordon MSS.
⁶³ Baltimore *Chronicle*, July 26, 1833; Howard, *Monumental City*, p. 549; *Biographical Cyclopedia of Maryland and the District of Columbia*, p. 417; J. Forsythe to Jackson, July 12, 1834, Andrew Jackson MSS (Library of Congress).

That this political imbalance did not produce trouble with the administration is attributable to Evans' skillful salvage job and to Washington's overwhelming desire for peace in Baltimore banking. The new president had everyone's confidence. Roswell Colt, Biddle's man, thought Evans "a very intelligent and clever man, both in the English and Yankee acceptation of the word," who would give the B.U.S. no trouble, and might work with the national bank.[64] Evans also cultivated Democrats. Isaac McKim recommended transfers of specie from other pets to meet a shortage at the Union, and during a particularly serious crisis Taney interceded in the bank's behalf, praising Evans and the reconstruction he had accomplished. Furthermore, Taney had learned that Evans had curbed his Whiggery since assuming control of the bank.[65] As he wrote these lines, Taney's thoughts must have turned to Thomas Ellicott's betrayal. Forbearance had become a form of expiation for the bad situation the administration had aggravated by placing too much trust in Ellicott. Taney, rejected by the Senate as Secretary of the Treasury at the height of the Ellicott affair, could conveniently subordinate his own role and concentrate on the future struggles he and other "enemies of corruption" had yet to fight. For as the Whig *National Intelligencer* observed ominously of the complex series of frauds, there were "some things in it, which can be readily understood by any one; and they are ugly enough to look upon." [66]

Relations between Evans and the new Secretary, Levi Woodbury, proceeded smoothly. Evans avoided politics, and Whig plans to launch further investigations of Baltimore pet banking produced no appreciable results. When the Treasury finally settled the question of the Union Bank's solvency, and the Tennessee bonds, Evans gushed thanks for the government's "kindness, confidence, and liberality." He assured Woodbury of his bank's "earnest and continued desire and exertion, to promote the public welfare as well as to give the Government every support we are capable of." [67]

When word leaked out in April, 1835 that the Treasury might appoint a second Baltimore pet, a strong recommendation came in from New York for a bank about to be chartered, the Merchants

---

[64] L. Woodbury to Jackson, Aug. 2, 1834, Jackson MSS; Colt to Biddle, Oct. 2, Dec. 4, 1834, Biddle MSS.

[65] Louis McLane to Roswell Colt, Aug. 17, 1829, Colt MSS (Historical Society of Pa.); McKim to Woodbury, Sept. 22, 1834, Taney to Woodbury, Nov. 5, 6, 1834, Levi Woodbury MSS (Library of Congress).

[66] Taney to R. Vaux, July 19, 1834, Vaux MSS (Historical Society of Pa.); Washington, *National Intelligencer*, July 21, 1834.

[67] D. Webster to W. Mangum, Nov. 4, 1834, Shanks (ed.), *Papers of Mangum*, vol. II, p. 223; Evans to Woodbury, Sept. 9, 1835, Feb. 29, 1836, TD/NA, Letters from Banks.

Bank of Baltimore. Campbell P. White, originally from Baltimore, then a Jacksonian Member of Congress and director of a New York City pet, the Manhattan Company, confided that every effort would be made to place the new bank "in *good hands* in every sense of the *term*." The notorious political unfriendliness of all Baltimore banks made such a political selection necessary, claimed White: "It would be a great matter to establish *one* that might assist in counterbalancing that powerful interest." When the Merchants Bank came into existence, efforts to ensure "good" control failed. White and his brother John, the former cashier of the Baltimore B.U.S. branch who had turned against Biddle, failed in their plans to monopolize the stock subscription. The bank became Baltimore agent for the B.U.S. that summer, handling its accounts as Biddle began closing branches in preparation for the expiration of his charter. This certainly did nothing to raise its standing with the Treasury.[68]

Despite White's assurances, Woodbury watched patiently, and when results were not to his liking, he turned elsewhere. The Jacksonians' gaze focused upon the older Franklin Bank to supplement the services of Evans' institution. In April, 1835, Democrat Isaac McKim sent Franklin Bank president James Howard to call on Woodbury, with assurances that the selection would satisfy all concerned. Several months later McKim and his fellow Democratic Congressman, Benjamin Howard, again boosted the Franklin Bank.[69] The basis for this joint recommendation is not clear, unless it be nepotism. Both Howard and McKim had relatives involved: James Howard, the president, was Benjamin's brother, and McKim may have had the interests of his Whig nephew, William, not politics, partially in mind.[70] Only one other director can be identified as a Democrat, while Whigs abounded, including cashier James L. Hawkins.[71]

By December, 1835, Woodbury reasoned that the Union Bank stood firm, and would not totter with company. While the city's importance warranted an additional selection, he still hesitated before naming the second Baltimore pet. The Union Bank imbroglio had created the need for caution and consultation with his pre-

---

[68] White to Woodbury, April 25, 1835, Woodbury MSS; Woodbury to C. P. White, April 28, 1835, J. C. White MSS; S. Jaudon to J. White, Sept. 25, 1835, White MSS; R. Colt to Biddle, May 16, 1835, Biddle MSS; Baltimore *Republican*, Sept. 29, 1835.
[69] McKim to Woodbury, April 27, 1835, McKim and Howard to Woodbury, Nov. 25, 1835, TD/NA, Letters from Banks.
[70] *Biographical Cyclopedia of Maryland and the District of Columbia*, p. 335; F. A. Richardson, *Baltimore, Past and Present* (Baltimore, 1871), pp. 391–95.
[71] Most of them signed petitions in 1834 favoring restoration of the deposits. Other identifications are from Baltimore newspapers.

decessor, Roger Taney, now on the threshold of the chief justiceship of the United States, and still a key adviser on Maryland patronage. The Franklin Bank would "be more generally acceptable to our friends than any other," Taney affirmed without giving particulars.[72] Thus even out of office, Taney occupied an important role in Baltimore pet bank affairs. The opinions of Democratic city leaders McKim and Howard counted for much, but the ex-Secretary's approval produced action within two days. It can fairly be said that Taney chose both Baltimore pets.

The Franklin Bank joined the pet ranks without fanfare in December, 1835. A tiny notice in the *Republican* informed readers that Baltimore deposits would thereafter be shared by the Union and Franklin banks.[73] This time editor Samuel Harker left happy speculations about the new pet to the future.

Peace with the pets did not mean the end of banking acrimony, however. Late in July, 1834, George W. Gibbs, representing the Union Bank of Tennessee, had warned publicly against purchases of Tennessee bonds illegally held by the Union Bank of Maryland. Evans replied with a $500,000 libel suit, and Gibbs spent a night in jail until he could obtain bond. Similar legal snarls involving the Bank of Maryland trusteeship dragged through the courts, pulling the bank's paper down to twenty-five cents on the dollar and below.[74] In 1835 a new negotiator arrived from Tennessee, Anthony Van Wyck, who shortly described Ellicott as a man "who cannot be brought to accomplish a design or transaction but by a devious or indirect course." Finally, late in 1835, the Nashville bank sold its claim against the Bank of Maryland to a Baltimore businessman at a substantial loss. The Tennessee agents were not the only ones to lose their tempers. For months the Baltimoreans themselves issued public blasts and counterblasts — a pamphlet war which only inflamed public opinion against them all when the conspirators went on trial in nearby Bel-Air.[75]

The fury exploded on August 7, 1835. Disgruntled investors, fearful that their claims would never be settled, turned a protest

---

[72] Woodbury to Taney, Dec. 10, 1835, Taney to Woodbury, Dec. 14, 1835, Woodbury MSS.
[73] Baltimore *Republican*, Dec. 18, 1835.
[74] *Ibid.*, July 21, 23, 1834; J. McMahon to W. Rienhard, Feb. 6, 1835, Urbana Bank MSS; W. Frick to W. Carroll, April 1, 1835, Smith Collection (Washington's Headquarters National Park, Morristown, N. J.); Ellicott to George W. Andrews, May 7, 1835, B.U.S. (Baltimore Branch) MSS.
[75] Van Wyck to W. D. Lewis, Sept. 20, 26, 1835, Girard Bank MSS; J. M. Bass to J. Meredith, July 30, 1835, Van Wyck to J. Meredith, Nov. 6, 1835, Jonathan Meredith MSS (Library of Congress). The incredible tangle among the Union Bank of Tennessee, the Bank of Maryland, and the Union Bank of Maryland (the Girard Bank of Philadelphia also appears) is partly explained in correspondence in the Meredith MSS, and in the William Taylor MSS, also at the Library of Congress.

meeting into a three-day riot. The mob adopted a favorite sport, the destruction of homes of men connected with the Bank of Maryland. The inept mayor, Jesse Hunt, did too little too late, and old General Samuel Smith had to restore order with a makeshift volunteer force.[76]

Taney's reaction to the riots and their aftermath hardly fits the image of radicalism which his opponents tried to impose upon him, and which some historians have imposed oh the Jackson party. His no-nonsense approach called for a "whiff of grapeshot" immediately. Law officers should have fired on the rioters as soon as the first stone arched through the air, since "such a contest ought always to be treated as one in which the existence of free government is put to hazard."[77] In March, 1836, the legislature heard claims for property destroyed by the rioters, including Reverdy Johnson's. The conduct of some Baltimore Democrats, particularly Harker of the *Republican*, alarmed Taney. They soft-pedaled the mob outrage by emphasizing grievances and attacking the speculators: "the turpitude of the acts of the mob . . . must be measured and viewed with reference to the CAUSES which produced them." Taney would not coddle mob-ites, nor would he countenance any association of "the name of the party with any mob for the destruction of property." In defiance of the mob and its sympathizers, Taney sat beside Johnson the day John V. L. McMahon argued and won the claim.[78]

From spectacular beginnings, Baltimore pet banking settled down to the routine of other depositories. With the onset of the Panic of 1837 and the general suspension of specie payments, both Baltimore pets responded similarly, deploring the necessity for suspension, but declining to specify a resumption date. The suspension broke the contracts between pet banks and the Treasury, and in October, 1837 the Franklin balked at security requirements prescribed by Congress that year. Rather than agree, it gave up the public money, although Woodbury continued to use the bank for occasional special deposits.[79] Despite an enormous increase in the

---

[76] W. Bartlett to E. Stabler, Aug. 8, 1835, *Maryland Historical Magazine*, vol. IX, pp. 160–61; H. Wilkins to Mrs. J. Glenn, Aug. 11, 1835, John Glenn MSS (Md. Historical Society).

[77] Taney to J. Campbell, Aug. 19, 24, 1835, Howard Family MSS (Md. Historical Society). For Taney's Bank Riot correspondence see Frank Otto Gatell (ed.), "Roger B. Taney, the Bank of Maryland Rioters, and a Whiff of Grapeshot," *Maryland Historical Magazine*, vol. LIX (Sept., 1964), pp. 262–67.

[78] Taney to M. Van Buren, March 7, 1836, Martin Van Buren MSS (Library of Congress); Taney to J. Campbell, March 6, 17, 1836, Howard Family MSS; Baltimore *Republican*, March 5, 22, 1836; John T. Mason, *Life of John Van Lear McMahon* (Baltimore, 1879), pp. 62–65.

[79] R. Mickle to Woodbury, May 22, 1837, J. Howard to Woodbury, May 26, Oct. 24,

number of pets after 1836, the Department created no new ones in Baltimore and the Union Bank continued to receive the greatly reduced deposits.[80]

The Baltimore pet banking experience had sobered the Democracy. The party in Maryland, from Secretary Taney and Harker of the *Republican* to the ward captains, twisted uncomfortably as they tried to disengage from Ellicott. The administration needed an efficient primary pet bank in Baltimore, and Evans, a Whig but an apolitical banker, met this need. To have switched pets while Ellicott fell would have been too risky, financially and politically. The brief use of the Franklin Bank despite a majority of opposition directors shows the power of the established leadership of Howard, McKim, and, of course, Taney. With James Howard as president, nothing untoward could occur. Democrats had clearly won the Bank War, and Whigs like Hugh Evans saw no major inconsistency in their role as pet bankers. As long as Democrats remained in power, the B.U.S. was dead.

Whigs naturally stigmatized deposit banks as "pets," charging that the system existed to foster Democratic party interests, and in a large sense they were right. Functioning state depositories, as a replacement for the B.U.S., obviously represented a political triumph for Jackson. But bankers, however they vote, mean to make money. Some Whigs may have believed that pet bank vaults became the coffers of Democratic machines, but there is no evidence that this happened in Maryland, and business logic makes it highly unlikely. The word "pet" took on a different meaning in Baltimore. With regard to selection, the administration men (especially Taney), had special confidence in, or personal connections with individuals that earned them government favor. Personal commitments became as important as broad party considerations, if not more so, given the power of the single individual to sway decisions. Whatever the makeup of the Union Bank's board, or the details of its loan policies, or the amount of its circulation, Ellicott had been Taney's "pet." That was enough.

At Baltimore, the problems of pet banking touched administra-

---

1837, TD/NA, Letters from Banks; W. Frick to J. Campbell, Nov. 21, 1837, TD/NA, Letters from Collectors.

[80] The deposit act of 1836 allowed a bank to hold government funds amounting to no more than three-quarters of its paid in capital. The Union Bank ($1,845,000) and the Franklin Bank ($624,000) had sufficient capital to hold over $1,800,000. In May, 1837 the Union Bank held $839,000, the Franklin Bank $378,000, both amounts well below the statutory maxima. (Statement of the Condition of the Several Deposit Banks . . . On or Near the First of May, 1837; a printed table, distributed by Reuben M. Whitney, copy in the Girard Bank MSS). Thus in Baltimore, unlike other cities such as New York or Boston, pet bank capitalization proved ample, obviating the necessity for further selections.

tion counsels most often and most intimately, especially in the critical first year. The politics of removal and selection, the shift in power at the Union Bank, the choice of a second pet, the role of the local B.U.S. branch, all illuminate the larger question of the national pet bank apparatus, which, while not identical in detail, in many ways duplicated the Baltimore story in other commercial centers.

Coal-burning Locomotives: A Technological Development of the 1850's

BY

EDWARD F. KEUCHEL

HISTORIANS frequently write about technological developments by dealing with specific inventors and inventions. While the historian must recognize the stroke of genius in the individual inventor, he must also recognize that significant discoveries often have not been the result of a specific inventor but more of what Siegfried Giedion, the Swiss historian of mechanization, called anonymous history.[1] The conversion from wood to coal as a locomotive fuel in the United States fits this concept. No one inventor solved the problems concerning the burning of coal as a locomotive fuel, yet during the decade of the 1850's these difficulties were mastered and a significant technological advancement was achieved.

By 1840 there was little doubt that the United States was becoming conscious of the importance of railroads as a means of transportation and communication. Malin points out that by 1840 the railroad system of the United States had assumed the greatest magnitude in the world with 2,270 miles in operation and 2,346 miles under construction.[2]

Wood was the principal fuel used by locomotives during the 1830's and 1840's, and, although the supply was still plentiful, concern was expressed for future needs. In 1837 the *American Railroad Journal* reprinted an article from the *Geneva* (New York) *Gazette* on the importance of fuel, stating that:

In this climate a supply of fuel is of the first necessity; without it, the country could not have been inhabited. Hitherto, the forests have afforded this supply, and will continue to do so, for years to come. But the time will arrive, when we must look to other sources for this indispensable article.[3]

---

[1] Siegfried Giedion, *Mechanization Takes Command* (New York, 1948), 2–4.
[2] James C. Malin, *The Contriving Brain and the Skillful Hand in the United States* (Lawrence, 1955), 122.
[3] *Geneva Gazette*, as cited in *American Railroad Journal*, VI (Dec. 23, 1837), 684–685.

Coal was the "other source" that the *Journal* proposed to replace wood, and, in 1845, M. Carey Lea predicted in the *Merchants' Magazine* that "coal is evidently destined at some future period to entirely supersede wood as fuel."[4] After making a study of England's fuel supply, John Griscom, a chemist at Rutgers College, strongly recommended coal as a fuel in American mills and factories.[5]

Congress also was aware of the problems of providing fuel for the nation's new steam navy and commissioned Professor Walter R. Johnson of Philadelphia to analyze coal as a substitute for wood. In 1844 Johnson reported favorably on the use of coal as a fuel but was cognizant of the difficulties which would accompany its use in steam boilers. In the preface to his "Report to the Navy Department of the United States on American Coal," he remarked that:

The heat generated from burning coal warns us what to expect in regard to the durability of grate bars, and the adhesion of scoriae to those important appendages of the furnace. All subjects must necessarily engage the attention of engineers and furnace managers, and no little portion of the good or bad character in coal may be considered to depend on these circumstances.[6]

Most of the coal burned during the early years was anthracite. It was found mainly in the northeastern part of Pennsylvania and was readily accessible to the large centers of population by means of navigable rivers, canals, and railroads. Bituminous coal was distributed over a larger geographical area but did not come into general use until after the Civil War. Anthracite coal, a comparatively clean burning fuel, was preferred to bituminous because it was also easier to handle. The heat yield per pound was found to be about the same for either fuel.[7]

Comparing the heating value of coal with wood, the equation frequently employed was that one ton of bituminous or anthracite coal equaled at least one and three-fourths cords of wood commonly

---

[4] M. Carey Lea, "Coal of Pennsylvania and other States," *Merchants' Magazine*, XIII (July, 1845), 67–72.

[5] John W. Oliver, *History of American Technology* (New York, 1956), 149.

[6] Walter R. Johnson, "A Report to the Navy Department of the United States on American Coal," *Senate Docs.*, 28th Cong., 1 Sess. No. 386 (Serial 436), vi, vii.

[7] Jules I. Bogen, *The Anthracite Railroads* (New York, 1927), 3; Alfred W. Bruce, *The Steam Locomotive in America* (New York, 1952), 36; Courtney Robert Hall, *History of American Industrial Science* (New York, 1954), 48.

used for fuel purposes.⁸ This is a convenient figure by which to judge the relative costs of wood and coal for locomotive fuel purposes. Of course, costs of both wood and coal for fuel purposes varied according to nearness to source, but during the 1830's and 1840's George W. Whistler, Jr., reported in the *Journal of the Franklin Institute* that for railroads in the Northeast a figure of $8.00 a ton was general for coal. A common price for firewood in the Northeast during the 1830's was listed by the *American Railroad Journal* at $2.50 to $3.00 per cord.⁹ As more and more coal was brought to market, ¹⁰ however, the situation changed. In 1849 the *Journal* reported that anthracite coal could be purchased in Philadelphia for $3.25 to $3.50 per ton.¹¹ At such prices coal was competitive with wood and railroad managers looked forward to greatly reduced operating expenses since fuel wood constituted fifty-five per cent of the operating costs of a wood-burning locomotive.¹²

In addition to lower fuel costs there were other advantages to be gained from burning coal. For one thing, frequent stoppages were necessary on wood-burners to "wood-up," and it was particularly costly to stop heavy through-trains for long periods of time. As a train could not carry both freight and sufficient wood for long hauls, wood sheds had to be maintained at regular intervals and serviced by special wood trains. This meant, in most cases, that the wood was carried twice over most of the line: first by the wood train and then by the locomotive tender of the train that consumed it. This extra hauling also contributed to increased wear on both the rolling stock and the railway. In addition, coal burns at a more even temperature than wood, hence, a more regular supply of steam could be maintained. Finally, there were problems of theft from unguarded wood sheds. One railroad superintendent estimated that

---

8 Alexander L. Holley, *American and European Railway Practice in the Economical Generation of Steam* (New York, 1861), 71.

9 George W. Whistler, Jr., "Report upon the Use of Anthracite Coal in Locomotive Engines, made to the President of the Reading Railroad Company, April 20, 1849," *Journal of the Franklin Institute*, XLVIII (August, 1849), 82; *American Railroad Journal*, II (Jan. 12, 1833), 23.

10 The yearly coal production in the United States increased from 174,764 tons in 1830 to 805,414 tons in 1840. By 1850 the output was up to 3,332,614. Figures taken from *Debows Commercial Review*, XXV (August, 1858), 239.

11 *American Railroad Journal*, XXIV (Jan. 11, 1851), 32.

12 Zerah Colburn, *The Locomotive Engine* (Boston, 1851), 66.

one-fourth of the population within a half-mile of a wood shed was supplied with railroad wood.[13]

Lower coal prices during the 1840's encouraged experimentation with the use of this fuel in locomotives. In June, 1847, at the monthly meeting of the Franklin Institute, Professor Johnson reviewed the attempts of various railroads to burn anthracite coal. He observed that whereas anthracite coal was being successfully burned in the fireboxes of stationary steam engines and steam vessels, many difficulties had been encountered in its use in locomotives. Some short-haul lines, such as the Beaver Meadow and Hazleton railroads, were successfully burning anthracite on round trips of thirty to forty miles. Other lines, such as the Reading Railroad, however, were having a great deal of difficulty in burning the fuel on trips of approximately two hundred miles. Johnson noted that the Reading Railroad spent $202,061 for wood during 1846, and that savings as high as $125,000 per annum could have been realized by the use of anthracite coal. The difficulties in burning anthracite for locomotive purposes were summed up by Johnson when he listed the following impediments to its use:

1. The want of rapid ignition, and free, lively combustion.
2. The intense, concentrated, local heat, which is said to destroy the grate bars, to attack the rivets and laps of the fire-box and even to cause blisters to rise in the plates of which it is composed; and, finally, to fuse the ashes into a troublesome clinker.
3. The sharp, angular particles of coal projected by the violent, fitful blast of the escape-steam, obliquely into the ends of copper tubes, cuts them away within a few inches of the fire end. . . .
4. The difficulties of fitting in iron tubes, so as to make perfect joints. . . .[14]

In addition to the impediments listed by Johnson, certain other phenomena associated with burning coal should be mentioned. First of all, when a lump of coal was placed directly upon an open fire, it tended to disengage small particles, some possessing sufficient velocity to injure the sides of the firebox.[15] This was particularly destructive to the copper sheets that were introduced after the poor

---

[13] Charles B. George, *Forty Years on the Railroad* (Chicago, 1887), 31; Holley, 72.

[14] Walter R. Johnson, "Use of Anthracite Coal in Locomotives," *Journal of the Franklin Institute*, XLIV (August, 1847), 110–114.

[15] William M. Barr, *A Practical Treatise on the Combustion of Coal* (Indianapolis, 1879), 105.

quality iron sheets of the day proved unsatisfactory. These copper sheets became very ductile at high temperatures and were soon cut away by the mechanical action of the sharp particles of coal impinging upon them.[16] Also, the boiler sections, away from the fire, were attacked by sulphurous acid formed when combustion gases combined with moisture.[17]

Bituminous coal presented additional problems. When bituminous coal was first tossed onto a burning fire the bituminous portion had to be volatilized before the lump would burn. The process of volatilization is very cooling as anyone who has ever poured alcohol or ether on his hand can testify. Considerable heat was necessary to volatilize the bituminous constituents and a corresponding temperature drop was effected in the firebox. The gases, as emitted from the lump of coal, would not burn unless additional oxygen was supplied. The result was that they usually escaped, unburned, through the smokestack along with the combustion gases.[18]

In April, 1849, railroad engineer George W. Whistler, Jr., reported on the use of anthracite coal to the president of the Reading Railroad. He first listed comparative costs of various wood and coal-burning locomotives.[19]

Name or Class of Engine	Cost of Fuel (per 100 tons transported)
Baltimore Engine	$ 5.50 (coal-burner)
Novelty	7.95 (coal-burner)
Baldwin Large	13.09 (wood-burner)
Champlain	10.93 (wood-burner)
Reading Engine	14.40 (wood-burner)

In terms of fuel costs per unit weight of freight hauled, coal was far preferable to wood. But, as Whistler pointed out, coal-burners

---

[16] Whistler, 79.
[17] Barr, 162–168.
[18] Charles W. Williams, *The Combustion of Coal and the Prevention of Smoke* (London, 1854), 7–8.
[19] Whistler, 9–11, 79–85. This Whistler was the father of the celebrated American artist. In addition to his work with American railroads he was also actively involved in the construction of the Russian railway system in the 1840's. Indeed, the present five-foot gauge used by all Russian railways at the present time was based upon Whistler's recommendations in the building of the St. Petersburg-Moscow line. See J.N. Westwood, *A History of Russian Railways* (London, 1964), 30–31.

became more expensive to run because of the costs of repairing their burned-out fireboxes. The expense of renewing the firebox on the Baltimore engine of the Reading Railroad was listed at $486. The old copper taken out could be sold for $108, making the replacement price $378. The firebox needed replacement after fourteen months' service, making the yearly expense $324. Whistler deducted $75 as the depreciation of a wood-burning firebox during the same period, leaving $249 per annum as the extra cost of maintaining the firebox of the anthracite-burning locomotive. The total cost per year over wood-burning engines for each Baltimore coal engine on the Reading Railroad was listed at $456. This included grate replacement, boiler tube repairs, and various small repairs. The repairs took seven months, during which time the engine was inoperative. The extra yearly cost over wood-burners of an engine burning bituminous coal on the Baltimore and Ohio Railroad was given at $250.[20]

Whistler made some recommendations as to how repairs could be reduced, but even after considering these improvements the total yearly cost of an anthracite-burning engine over that of a wood-burning engine was estimated at $379.50[21] Clearly this was a cost the railroads did not want to incur, yet the increasing scarcity of wood presented problems they could not overlook.

Experiments continued throughout the late 1840's and early 1850's. In 1848 the Mine Hill and Schuylkill Haven Railroad announced that their experiments with coal-burners had been sufficiently encouraging to warrant the purchase of three new coal-burners for trial.[22] A. L. Roumfort reported in 1850 on the experiments conducted with the locomotive "Henry A. Muhlenberg" of the Columbia and Philadelphia Railroad. The "Muhlenberg" was fitted with a detachable firebox which did not solve the problems of burning coal without its destructive effects, but merely reduced the repair time for firebox replacement. The replacement cost was from $500 to $1,000, but it could be accomplished in twenty-four hours.[23]

---

[20] *Ibid.*, 79-85.
[21] *Ibid.*, 179.
[22] "Report of the Board of Managers to the Stockholders of the Mine Hill and Schuylkill Haven Railroad Company, at their Annual Meeting, January 10th, 1848," *Journal of the Franklin Institute*, XLV (March, 1848), 153.
[23] A. L. Roumfort, "Report of A. L. Roumfort, Superintendent Columbia and Philadelphia R.R., on the Experiments made with the Coal Burner, 'Henry A. Muhlenberg,'" *Journal of the Franklin Institute*, XLIX (February, 1851), 138-139.

In 1852, Richard E. Dibble, a mechanical engineer from New York, wrote to the editor of the *American Railroad Journal*, stating that:

The economy of coal over wood as a fuel for steam purposes is demonstrated beyond a question, as applied to steamboats and stationary engines, and there is no good reason why there may not be the same economy when applied to our railroads. . . . In any locality where coal can be delivered at $5, or less, per ton, it is a cheaper fuel than wood over $2 per cord.[24]

From the standpoint of fuel costs Dibble was correct, but the problems of burning coal effectively still remained. Indeed, by 1853, so little had been achieved in the use of coal as a locomotive fuel that the 1849 report of Whistler was still regarded by the *American Railroad Journal* as the standard work on the subject.[25]

As the problems of burning raw coal seemed almost insurmountable, railroad men turned to coke. Coke was used almost universally in England, and many of the railroads, such as the Manchester and Liverpool, were required by Act of Parliament to use coke exclusively for reasons of smoke abatement.[26] Professor Johnson's report of 1844 stated that coke evaporated more water per pound than free-burning coal, and he recommended the use of coke in locomotive boilers "in preference to any fuel where the price does not interfere to prevent it."[27] The Baltimore and Ohio and other railroads experimented with coke during the early 1850's and found it especially useful for passenger trains as it made few sparks and cinders, and was practically smokeless. By July, 1853, the *American Railroad Journal* had become enthusiastic about the use of coke, proclaiming:

It is evident that a great revolution is about to take place in the fuel employed in the propulsion of locomotives. Coke made from coal of the Cumberland region will, in a short time, be substituted for wood on all railroads in the Atlantic States that can obtain the requisite supplies.[28]

Coke, however, did not receive sufficient support in the United States. Although the fuel possessed advantageous burning properties,

---

[24] *American Railroad Journal*, XXV (Feb. 14, 1852), 106.
[25] *Ibid.*, XXVI (July 2, 1853), 435.
[26] John Holland, *Fossil Fuel, the Collieries and Coal Trade* (London, 1841), 424.
[27] Johnson, 307.
[28] *American Railroad Journal*, XXVI (June 4, 1853), 360; (July 23, 1853), 475.

it was expensive. Coke had increased locomotive fuel costs in England forty per cent.[29] Moreover, the United States did not have good coking coal accessible to the majority of its railroads.[30]

The impracticability of burning coke as a locomotive fuel in the United States turned attention to the problems connected with the burning of raw coal. Many reports of coal-burning locomotives were made during the period 1853-1857, and, although they gave promise that progress had been made, the major difficulties still remained. For example, in November, 1854, the Philadelphia *Public Ledger* reported that a coal-burning locomotive, built on the design of Leonard Phleger of Tamaqua, was operating on the Wilmington and Delaware Railroad between Philadelphia and Havre de Grace.[31] In July, 1855, the *American Railroad Journal* reported that the locomotive "Taunton," using a boiler constructed upon the patent plan of F. P. Dimphel, was successfully burning anthracite coal.[32] Both the Phleger and Dimphel designs are good examples of the various ways that railroad men were attempting to solve the problems of burning coal. The Phleger design used water-cooled grates to prevent them from burning, and the Dimphel plan called for water tubes around the firebox to lessen the injurious effects of intense heat. While they both helped to reduce firebox deterioration, they were too complex to gain wide acceptance. The Dimphel plan, in particular, was inadequate in that the maze of water tubes in the firebox region made repairs extremely difficult and costly.

What was needed was a solution that would be both simple and effective; a solution that would not result in any greatly increased price for new coal-burning locomotives and would allow the current wood-burners to be converted to coal with a minimum of expense. During the late fifties the solution was found—a solution which was partly the result of a continuation of the previous trial-and-error experimentation which had dominated the past and partly the result of increased knowledge about the nature of coal combustion.

One major break-through that established coal-burning locomotives on a firm basis was the introduction of the Delano Grate in

---

[29] Holland, 424.
[30] *American Railroad Journal*, XXX (Dec. 5, 1857), 777.
[31] *Public Ledger*, as cited in *Merchants' Magazine*, XXXIII (November, 1854), 635.
[32] *American Railroad Journal*, XXVIII (July 7, 1855), 424.

1856. Howard Delano, a locomotive mechanic of Syracuse, New York, designed grates which enabled bituminous coal to be forced from the bottom up through the bed of fire. The *American Railroad Journal* described these grates as follows:

> In an old engine, a section is cut out of the bottom of the grate corresponding to the size of the feeding box—say 13 or 14 inches square. Fitted to the space, cut out, is a movable grate to which is attached the feeding box. On drawing back the grate, the box filled with coal occupies its place. By very simple contrivance, the bottom of the box is thrown up to a level with the grate, discharging its contents directly into the furnace. The box is then drawn forward to receive another charge, the bottom of it remaining up until the movable grate gets into place. The bottom is then dropped down, and the box made ready to receive another charge.[33]

By this process the grates were shielded from the burning coal, and, more important, the bituminous constituents of the lumps were volatilized slowly, which enabled them to mix with air and burn. The lumps themselves, when finally forced to the top of the fire, had discharged all of their bituminous components and were free to burn. In addition, this gradual heating helped to prevent small coal particles from being projected against the sides of the firebox.

One of the earliest mentions of the Delano Grate in operation is to be found in the September 13, 1856, edition of the Pottsville, Pennsylvania, *Miners' Journal*. The *Journal* reported that the Boston and Worchester Railroad was operating a coal-burning engine with the Delano Grate "which seems so well to meet the wants of the road that all the engines of the Company, used in drawing freight, are to be altered to the new style."[34] With reference to the cost of installation, economy, and performance, the *Journal* stated that:

> Careful estimations of the precise cost of running this engine have been made, and it appears that with it, for 12 cents per mile, a common freight train can be run and make the usual speed. A wood engine to run the same train costs 30 cents. . . . The cost of altering a common wood engine to fit it for burning coal is but $150.[35]

---

33 *Ibid.*, XXX (Dec. 5, 1857), 778.
34 Pottsville, *Miners' Journal and Pottsville General Advertiser* (hereinafter cited as *Miners' Journal*), Sept. 13, 1856.
35 *Ibid.*

The Delano Grate was well received, and by June, 1858, the *American Railroad Journal* was able to report that they were in "successful use on several roads."[36]

The real significance of the Delano Grate was not its mechanical construction, but rather that it solved the problem of burning coal by applying correct methods of combustion. The Delano Grate enabled the most unskilled of firemen to burn coal efficiently and without injurious effects to the firebox. This grate did not continue in use because it was found that educating firemen in the correct methods of burning coal achieved the same end. As the mechanical engineer Zerah Colburn later remarked: "the aim of the stoker must be to have a proportionate mixture of coal in all stages of combustion spread over the grate. . . ."[37] Indeed, British firemen had found a way to burn bituminous coal by merely placing the fresh coal under the fire door upon the hind part of the grate and moving it forward as the heat of the fire volatilized its bituminous elements.[38]

The Delano Grate was based upon sound principles of coal combustion—principles which were successfully applied by all railroad firemen. Indeed, by 1858, it was recognized that simply to burn coal successfully in a wood burner, proper stoking and an arch of firebrick, costing about ten dollars, were all that were necessary.[39] Greater expense than this was necessary, however, to convert a wood burner to burn coal with maximum efficiency.

Now that the problem had been resolved into a technique of burning the coal, other successful engines began appearing. One approved locomotive, built according to Boardman's patent, was reported by the *Merchant's Magazine* and the *Miners' Journal* as being well received and running at costs ranging from 10.64 to 12 cents per mile, as compared with 15.14 to 18.2 cents per mile for the best available wood-burners.[40] By the summer of 1857 the *American Railroad Journal* was able to report that:

[36] *American Railroad Journal*, XXXI (June 19, 1858), 393.

[37] Zerah Colburn, *Locomotive Engineering and the Mechanism of Railways* (London, 1871), 217.

[38] *Ibid.*, 296

[39] Zerah Colburn and Alexander L. Holley, *The Permanent Way and Coal-Burning Locomotive Boilers of European Railways* (New York, 1858), 161.

[40] *Miners' Journal*, Aug. 15, 1857; *Merchants' Magazine*, XXXVII (September, 1857), 379.

The final success claimed for the experiment with coal-burning locomotives at different points perhaps renders it unnecessary further to illustrate the subject.... It can now be demonstrated that coal can be used at one-half the cost of wood, and that better time can be made, because of the facility afforded of keeping up a uniformity in generating steam.[41]

The reports from the *Merchants' Magazine*, the *American Railroad Journal* and the *Miner's Journal* after 1858 no longer mention problems connected with burning coal, but rather the large number of railroad companies adopting coal and the savings afforded. The Illinois Central Railroad introduced coal-burning locomotives in 1858. The Hudson River Railroad had eight coal-burning locomotives in use during 1858 and reported costs of little more than one-fourth as compared with wood-burners.[42] Superintendent Watson of the Great Western Railroad was very satisfied with coal-burners, stating that "all that we hoped for them is being realized." He believed that the general adoption of coal as a locomotive fuel would result in yearly savings of "more than $10,000,000, or one per cent of the entire cost of all the railroads in the United States."[43] Early in 1859 the *Miners' Journal* looked to the past year and cast an eye to the future when it reported that:

Sufficient experiments have been made in the last year, to demonstrate the great superiority of Coal as a fuel in Locomotives, producing a saving of fully *one-third* ... the expenses of wood. Its use for this purpose will cause a large demand for Coal, because all the New Locomotives will be built for its use as fuel, and the old ones altered as rapidly as circumstances will permit, wherever Coal can be procured.[44]

By 1860, the conversion from wood to coal was an accomplished fact. Alexander L. Holley, who was better known for his role in introducing the Bessemer process into the United States but who was also interested in coal-burning locomotives, reported that "since so large a proportion of American railways are committed to the use of coal, the more important question is, not the economy of using coal in place of wood, but *how to burn coal economically....*"[45]

---

[41] *American Railroad Journal*, XXX (Aug. 29, 1857), 557.
[42] *Merchants' Magazine*, XXXIX (August, 1858), 250; *American Railroad Journal*, XXXI (June 5, 1858), 364.
[43] *Ibid.*, XXXI (June 19, 1858), 393.
[44] *Miners' Journal*, Jan. 15, 1859.
[45] Holley, 73.

From 1860 on through the nineteenth century coal bec principal fuel for locomotives, although the use of wood was common in areas where it was abundant: chiefly the South and parts of New England, or in the West where coal was scarce. In a period of about two decades the problems of burning coal had been met and solved. During that time the railroads had tried solutions in the form of fireboxes made of metals other than iron, other forms of coal (coke), various mechanical contrivances; and finally succeeded by applying correct theories of combustion. As Colburn and Holley stated:

No other reform, so great as that of the fuel bills of our railways, rests upon so few, so simple and so entirely available conditions as those of burning coal correctly. . . . While we have observed the simple laws which science has indicated for our guide, PRACTICE, so omnipotent with practical minds—a practice more intelligent and successful than our own—has proven their absolute correctness.[46]

A great deal of work still remained to be done before the American railroads became the later efficient transporters that revolutionized American transportation. Various technological advances, steel rails and bridges, standardization of track gauges, automatic couplers, air brakes, and other improvements followed. In the meantime, the conversion to coal had its immediate effects. More uniform steam meant that more rigid schedules could be maintained. The use of the cheaper fuel resulted in more economical operating expenses. Most important, however, was the fact that heavy freight trains could now carry sufficient fuel for long hauls without frequent stops.

---

[46] Colburn and Holley, 161.

From 1860 on through the nineteenth century coal became the principal fuel for locomotives, although the use of wood was still common in areas where it was abundant: chiefly the South and parts of New England, or in the West where coal was scarce. In a period of about two decades the problems of burning coal had been met and solved. During that time the railroads had tried solutions in the form of fireboxes made of metals other than iron, other forms of coal (coke), various mechanical contrivances; and finally succeeded by applying correct theories of combustion. As Colburn and Holley stated:

No other reform, so great as that of the fuel bills of our railways, rests upon so few, so simple and so entirely available conditions as those of burning coal correctly. . . . While we have observed the simple laws which science has indicated for our guide, PRACTICE, so omnipotent with practical minds—a practice more intelligent and successful than our own—has proven their absolute correctness.[46]

A great deal of work still remained to be done before the American railroads became the later efficient transporters that revolutionized American transportation. Various technological advances, steel rails and bridges, standardization of track gauges, automatic couplers, air brakes, and other improvements followed. In the meantime, the conversion to coal had its immediate effects. More uniform steam meant that more rigid schedules could be maintained. The use of the cheaper fuel resulted in more economical operating expenses. Most important, however, was the fact that heavy freight trains could now carry sufficient fuel for long hauls without frequent stops.

---

[46] Colburn and Holley, 161.

By *Robert S. Starobin*

The Economics of Industrial Slavery
In the Old South

From the pre-Civil War period until the present day, historians and economists have offered theories and evidence regarding the extent and profitability of southern plantation slavery.[1] After all, if slavery was not economically viable, would not slaveholders have abandoned their "peculiar institution?" Was a bloody civil war necessary if slavery was dying of its own weight? Obviously, the moral question of chattel slavery far transcends economics. But since economic arguments were so prominent in the slavery controversy, then and now, objective analyses of their form, substance, and verity are worthwhile.

Curiously, and despite its political and social significance, scholars have devoted little attention to either the extent to which slave labor was employed in industries in the Old South or to the economic feasibility of such employment. This article is the core of a larger study intended to pose and explore these questions.[2]

THE EXTENT OF INDUSTRIAL SLAVERY

Southern industry's most distinctive aspect was its wide and intensive use of slave labor. In the 1850's, for example, 160,000 to

---

[1] H. D. Woodman, "The Profitability of Slavery," *Journal of Southern History*, XXIX (1963), 303–325; S. L. Engerman, "The Effects of Slavery Upon the Southern Economy," *Explorations in Entrepreneurial History*, second series, IV (1967), 71–97.

[2] See *Industrial Slavery in the Old South* (New York, Oxford University Press, 1970), which is based on my doctoral dissertation of the same title, done at the University of California, Berkeley, in 1968. These works contain a full discussion of the use of slaves in southern industries, 1790–1861; working and living conditions; black resistance to bondage and work routines; means of control of slave workers; relations between black and white laborers; and the political relationship of slave-based industrialization to the coming of the Civil War.

200,000 — about 5 per cent of the total slave population — worked in industry. About four-fifths of these industrial slaves were directly owned by industrial entrepreneurs; the rest were rented by employers from their masters by the month or year. Most were men, but many were women and children. They lived in rural, small-town or plantation settings, where most southern industry was located, not in large cities, where only about 20 per cent of the urban slaves were industrially employed.[3]

Many southern textile mills employed either slave labor or combined both bondsmen and free workers in the same mill, contradicting the myth that southern textile manufacturing was the sole domain of native poor whites. The manufacture of iron was also heavily dependent on slave labor, and southern tobacco factories employed slave labor almost exclusively. Slave labor was crucial to hemp manufacturing, and most secondary manufacturing — shoe factories, tanneries, bakeries, paper-makers, printing establishments, and brickmakers — used bondsmen extensively. Sugar refining, rice milling, and grist-milling together employed about 30,000 slaves. At ports and river towns, slaves operated mammoth cotton presses to recompress cotton bales for overseas shipment.

The southern coal and iron mining industry was greatly dependent upon slave labor, gold was mined throughout the Piedmont and Appalachian regions largely with slaves, and lead mining employed many bondsmen. Salt was produced with slave labor; slaves were used to log the pine, cypress, and live-oak in the swamps and forests from Texas to Virginia; and the turpentine extraction and distillation industry was entirely dependent upon slave labor. Southern internal improvements enterprises were so dependent upon slave labor that virtually all southern railroads, except for a few border-state lines, were built either by slave-employing contractors or by company-owned or hired bondsmen. Most southern canals

---

[3] This figure has been computed from tables on urban slave population in 1850 in Richard Wade, *Slavery in the Cities* (New York, 1964), 30 and appendix; J.D.B. De Bow, *Statistical View of the United States . . . Being a Compendium of the Seventh Census* (Washington, 1850), 94, estimated that about 400,000 slaves lived in southern cities and towns in 1850, while Kenneth Stampp, *The Peculiar Institution* (New York, 1956), 60, estimates that the *total* city, town, and non-agricultural slave population was about 500,000 in 1860. If two-thirds of the *male* population of towns *and* cities was engaged in industrial enterprises and occupations, then the 160,000 to 200,000, or 5 per cent, figure seems approximately correct for the years 1850 to 1860. See tables in my dissertation for some large slaveholdings by deep-South industries. For further estimates, see J. C. Sitterson, *Sugar Country* (Lexington, 1953); J. F. Hopkins, *A History of the Hemp Industry in Kentucky* (Lexington, 1951); J. C. Robert, *Tobacco Kingdom* (Durham, 1938); C. Eaton, *Growth of Southern Civilization* (New York, 1961), 64–65, 134–35, 99–101, and ch. 10; K. Bruce, *Virginia Iron Manufacture in the Slave Era* (New York, 1930); and S. Lebergott, "Labor Force and Employment, 1800–1860," National Bureau of Economic Research, *Output, Employment, and Productivity in the United States After 1800* (New York, 1966), XXX, 117–210.

and navigation improvements were excavated by slave labor, and most other southern transportation facilities — turnpikes, plank roads, ferries, keelboats, and steamboats — were dependent upon slave work forces.

Though little studied, the reliance of municipalities, states, and even the federal government upon slave labor to build public works throughout the South was important. Slaves commonly cleaned and repaired the streets of towns and cities, and some states required the annual service of slaves to build roads and levees and to clear rivers and harbors. Other states purchased or hired slaves to construct state-controlled levees, canals, and railroads. Agencies and departments of the federal government were deeply involved in the industrial use of hundreds of slave hirelings, and only reluctantly disassociated themselves from this practice. Slave stonequarriers and common laborers helped erect the first national capitol at Washington and participated in its reconstruction after the War of 1812. Slaves manned government dredge boats, and federal fortifications, naval installations, and arsenals also frequently used slaves.[4]

The wide use of industrial slaves by state and federal agencies suggests not only the centrality of industrial slavery to the southern economy, but also the extent of southern control of the national political structure. That private enterprise also relied greatly on slave labor suggests the dependence of the private economic sector upon slavery as well as the determination of Southerners to use that labor system most advantageous to their program of economic development.

The Economics of Industrial Slavery

To study the economics of industrial slavery requires the consideration of several questions. First, could slave-employing industries expect to earn reasonably profitable rates of return on their capital investments? In this analysis, "profit rate" means either the annual dividend paid on common stock or the annual net income expressed as a percentage of the net worth of the industrial enterprise. A "reasonably profitable investment" means at least a 6 per cent annual return on capital — the average rate of return on other forms of investment.[5]

Second, was industrial slavery *generally* as efficient and as eco-

---

[4] Starobin, *Industrial Slavery*, ch. 1.
[5] H. O. Stekler, *Profitability and Size of Firm* (Berkeley, 1963), ch. 1 and 2; A. H. Conrad and J. R. Meyer, *The Economics of Slavery* (Chicago, 1964), ch. 3.

nomical as an alternative labor system? Were slaves as efficient as free whites? Was slave labor — directly owned or hired — less expensive to employ than free labor? Did slave labor entail higher capital and maintenance costs than free labor?

Third, were there *specific* competitive advantages to industrial slavery — that is, did the use of slaves enable southerners to compete with the North and with Britain, where industrialization had progressed further? Did the exploitation of slave women and children, the training of slave managers, and the coupling of common slaves with skilled foreign technicians enable southern industries to reduce their costs and to raise their quality in order to become competitive in national market places?

Fourth, what were the sources of capital for slave-based industries? Did southerners have sufficient investment capital to support industries? Did the funding of industries with slave capital have a detrimental effect on financial structures by reducing the flexibility of capital and the mobility of labor? [6]

The Profitability of Industrial Slavery

Under normal operating conditions, slave-employing industries and transportation projects could expect to earn reasonable profits

[6] U. B. Phillips, "The Economic Cost of Slave-holding in the Cotton Belt," *Political Science Quarterly*, XX (1905), 257–75; U. B. Phillips, *American Negro Slavery* (New York, 1918), especially chs. 18 and 19; C. W. Ramsdell, "The Natural Limits of Slavery Expansion," *Mississippi Valley Historical Review*, XVI (1929), 151–71; E. D. Genovese, "The Significance of the Plantation for Southern Economic Development," *Journal of Southern History*, XXVIII (1962), 422–37; E. D. Genovese, *The Political Economy of Slavery* (New York, 1965); D. North, *The Economic Growth of the United States, 1790–1860* (Englewood Cliffs, 1961), esp. 132. Cf. R. R. Russel, "The General Effects of Slavery Upon Southern Economic Progress," *Journal of Southern History*, IV (1938), 34–54, and "The Effects of Slavery Upon Non-Slaveholders in the Ante-Bellum South," *Agricultural History*, XV (1941), 112–26, and *Economic Aspects of Southern Sectionalism* (Urbana, 1924); D. Dowd, "A Comparative Analysis of Economic Development in the American West and South," *Journal of Economic History*, XVI (1956), 558–74; Wade, *Slavery in the Cities*, esp. chs. 1 and 9, argues that slavery was dying in the largest southern cities and thus tends to lend weight to the old Ramsdell hypothesis.

Certain theoretical and methodological problems also should be noted. While the above questions are obviously interrelated, an affirmative answer to one of them does not necessarily imply an affirmative answer to the others. Much confusion has resulted from a failure to distinguish the differences between the questions. Precise analysis of the economics of industrial slavery is also difficult, since information on the sources of finance, the capital cost and maintenance of labor, and the profits of enterprise is scarce. Available statistics are unsatisfactory because not all businesses kept records and only a few fragmentary accounts have survived. Those that have do not necessarily constitute representative samples of non-agricultural enterprises, since most records pertain to large establishments, and it is not certain whether small industrial operators were as successful as large ones. Moreover, the data on costs of labor and rates of profit is often unclear because of the peculiarities of antebellum accounting and the difficulty of finding long-term statistical series. Company reports tended to underestimate expenses and to exaggerate earnings to promote southern enterprise, while official censuses were haphazardly taken and must be used cautiously. Prices varied, while business cycles caused fluctuating profit rates and frequent bankruptcies. Variables such as location, luck, competition, and caliber of management also make computations of the profitability of industrial slavery difficult for those slave-employing industries whose records survive.

on their capital investments. Some enterprises failed, of course, but most industrial entrepreneurs employing slave labor enjoyed highly satisfactory rates of return on their investments. Most slave-employing enterprises whose records are available matched or exceeded an annual rate of return of about 6 per cent.

The records of southern textile mills employing slave labor indicate that they usually earned annual profits on capital ranging from 10 to 65 per cent and averaging about 16 per cent. The DeKalb, Martin & Weekly, Roswell, and Tuscaloosa textile companies, to give but four examples, annually paid between 10 and 20 per cent. The Woodville mill, which went bankrupt with free labor, annually paid 10 to 15 per cent dividends after switching to slave labor. "The Saluda Manufacturing Company . . . is doing a flourishing business . . . [and] pays large dividends," ran a report of one slave-employing cotton mill (see Table 1).

The available records of southern iron works employing slaves suggest further that substantial profits could be made in this industry. As early as 1813, one slaveowning iron manufacturer reportedly could "afford to work as cheap as others, and always do so but not at an under rate." From 1835 to 1845, a Mobile iron foundry made 25 per cent annually; during the 1850's, a South Carolina iron works earned 7 per cent yearly. The famous Tredegar Iron Company averaged annually better than 20 per cent returns from 1844 to 1861.[7]

Other kinds of manufacturing and processing enterprises employing slave labor evidently earned similar profit rates. One hemp manufacturer testified that he realized more than 42 per cent profits per annum in the 1840's. A tannery reported 10 per cent yearly between 1831 and 1845. A gas works also earned a 10 per cent return in 1854.[8] According to official reports, most Louisiana sugar mills earned better than 7 per cent returns in 1830 and almost 11 per cent in 1845. During the 1850's, a cotton press made 10 per cent; the Haxall Flour Mills of Richmond reportedly "made large fortunes for their owners for over half a century."[9]

Similarly, slave-employing enterprises in the extractive industries

[7] D. Ross to J. Staples, Sept. 16, 1813, Ross Letterbook (Virginia Historical Society, hereafter cited as VHS); De Bow's Review, VI (1848), 295; Charleston Mercury, February 18, 1859; Tredegar Stockholders' Minutebook, reports for 1838–48, and Tredegar Corporate Holdings, 1866, 7–9 (Virginia State Library, hereafter cited as VSL).
[8] Report of Court of Claims, #81, 34 C., 3 s., 1857, 62–64; L. McLane, Documents of Manufactures, House Doc. #308, 22 C., 1 s., 1832, 676–77; New Orleans Picayune, January 2, 1853.
[9] Niles' Register, XXXIX (1830), 271–72; F. L. Olmsted, A Journey in the Seaboard Slave States (New York, 1861), 686–88; Report of the Secretary of the Treasury, House Exec. Doc. #6, 29 C., 1 s., 1845, 708–09, 748; J. G. Taylor, Slavery in Louisiana (Baton Rouge, 1963), 96–101; Sitterson, Sugar Country, 157–66, 178–84, 197, and passim.

generally made handsome profits. Though one turpentine manufacturer "believed sincerely that no money can be made at the business while labour is so extremely high," in the 1850's, turpentine enterprises in North Carolina and Georgia did achieve satisfactory returns. In 1850, *De Bow's Review* proclaimed that "compared to other labor, this [turpentining] has, for the last ten years, been deemed the most profitable of all." The profitability of lumbering is suggested by one Louisiana woodyard that annually earned 12.5 to 25 per cent returns between 1846 and 1850. In addition, the Dismal Swamp Land Company reportedly "realized almost fabulous proceeds from the timber," while a Carolinian maintained that "I have no doubt from all I have heard . . . that more money can be made in this business [West Florida lumbering] than any other when [slave] manual labor is used." Fisheries usually earned at "a level with the ordinary industrial pursuits of the country," though "enormous profits" were "sometimes realized." [10]

Most southern mining enterprises employing bondsmen also earned substantial profits. As early as 1807, the Missouri lead-smelter Frederick Bates declared that "few labors or pursuits in the U. States, yield such *ample*, such *vast* returns — A slave, with a *Pick* and *Shovel* is supposed to do nothing, if the nett proceeds of his labor, do not amount, annually, to the sum of 400 dollars — the price which his master probably paid for him." Later, Bates added: "You will see [in my letter to Albert Gallatin] the vast profits arising from the prosecution of this lucrative business." Official records indicate that between 1834 and 1845, several Key West salt works earned 8 per cent annually. Many southern gold seekers failed, to be sure, but scores of mines were as profitable as, for example, John C. Calhoun's which yielded nearly $1,000,000, and Samuel J. Tilden's which earned $4,000,000.[11]

[10] J. E. Metts to J. R. Grist, December 5, 1858, Grist Papers (Duke University Library, hereafter cited as Duke); letters dated 1854–60, Williams Papers (University of North Carolina Library, hereafter cited as UNC); *De Bow's Review*, VII (1849), 560–62, and XI (1851), 303–05; W. H. Stephenson (ed.), *Isaac Franklin* (Baton Rouge), 1938), 114, 177–80, 213–17; *Harper's Monthly*, XIII (1856), 451, and XIV (1857), 441; J. M. Cheney to E. Bellinger, February 16, 1855, Misc. Mss. (South Carolina Historical Society, hereafter cited as SCHC); C. H. Ambler (ed.), *Correspondence of R.M.T. Hunter*, AHA *Annual Report* (1916), 176–77.

[11] T. M. Marshall (ed.), *Papers of Frederick Bates* (St. Louis, 1926), I, 111–12, 244. For lead mining, see inventory of January 1, 1851, and Will of May 1, 1856, Desloge Papers (Missouri Historical Society, hereafter cited as MoHS); Report on Salt Springs, and Lead and Copper Mines, *House Doc.* #128, 18 C., 1 s., 1824, 20–22, 130–33; Report of Secretary of Treasury, *House Exec. Doc.* #6, 29, C., 1 s., 1845, 660, 664–65; and *Mining Magazine*, 1 (1853), 164–66. For gold mining, see 1828 memo and undated account sheets, Fisher Papers (UNC); High Shoal Gold Mine Records (UNC); letters of 1830–33 and account book, vol. X, 1843, Brown Papers (UNC); Expense Book of S. Burwell and J. Y. Taylor, 1832–39 (UNC); T. G. Clemson to J. C. Calhoun, January 23 and 24, 1843, and T. G. Clemson to P. Calhoun, October 12, 1856, Clemson Papers (Clemson University Library, hereafter cited as Clemson); *Mining Magazine*, XI (1858), 211, and XII (1859–

From the 1790's to 1861, the majority of transportation enterprises employing slaves realized profitable returns. Some southern railroads paid annual dividends as high as 20 per cent, and most other lines averaged about 8 per cent. Some canal companies, such as the Roanoke, did not do as well as most railroads, but others, such as the Louisville & Portland and the Dismal Swamp, paid nearly as well. Plank roads and turnpikes, however, generally did not earn returns greater than 4 per cent on the capital invested (see Tables 2 and 3).

A few unusually complete statistical series for such slave-employing enterprises as sawmilling, steamboating, and gold mining do survive to permit the further computation of the profitability of industrial slavery. As early as 1794, Alexander Telfair's sawmills made him one of Georgia's wealthiest citizens. The Hart Gold Mining Company yielded a similar fortune for another Georgian. The earnings of the *Thomas Jefferson* permitted a Virginia steamboat company to average acceptable dividends between 1833 and 1849.[12]

The surviving records of two rice mills are complete enough so that some idea of the profitability of this industry can be determined. Though it is impossible to separate the profits of rice planting from rice milling, James Hamilton Couper's Georgia rice estate annually averaged 4.1 per cent return on capital between 1833 and 1852, despite his financial losses from natural disasters and from long agricultural experimentation. However, Couper's 4.1 per cent return does not take into account personal expenditures to support his sumptuous living standard and the appreciation of his lands and slaves. Between 1827 and 1841, for example, the plantation appreciated in value as much as 26 per cent; between 1827 and 1845, the slaves multiplied from 380 to about 500 — almost a 20 per cent increase on their original valuation.[13] Couper's average total annual return on capital was therefore greater than 6 per cent. Similarly,

---

60), 365–66; *De Bow's Review*, XII (1852), 542–43; *American Farmer*, series 1, XII (1830), 230; *Hunt's Magazine*, XXXI (1854), 517.

[12] Telfair Account Books, 1794–1861, #s 90, 87, 88, 89, 152, 153, 155, and 156 (Georgia Historical Society, hereafter cited as GHS); Hart Gold Mine Company Accounts, 1855–57, Latimer Plantation Book (University of Georgia Library, hereafter cited as UG); Journal of James River Steamboat Company, 1833–49 (VSL); the Palfrey Account Books, 1842–61 (Louisiana State University Library, hereafter cited as LSU) reveal that Louisiana sugar plantations and mills earned substantial profits.

[13] J. H. Couper Accounts, 1827–52 (University of North Carolina Library, hereafter cited as UNC); the financial statements and reports for Hopeton and Hamilton plantations, 1830's to 1843, 1849, and 1853–65, and J. H. Couper to F. P. Corbin, March 28, 1859, Corbin Papers (New York Public Library, hereafter cited as NYPL) permit the computation of Couper's profits from rice milling beyond the year 1852; T. P. Govan, "Was Plantation Slavery Profitable?" *Journal of Southern History*, VIII (1942), 513–35, demonstrates the profitability of Couper's rice milling enterprise from 1827 to 1852.

the records of the Manigault family's Savannah River rice mills reveal average annual returns of 12 per cent between 1833 and 1839, and 12.2 per cent from 1856 to 1861. The natural increase in the number and value of the Manigaults' bondsmen compensated for losses from three cholera epidemics, the absence of an experienced overseer between 1855 and 1859, a destructive freshet in 1852, and a devastating hurricane in 1854.[14]

The records of those industrial enterprises which hired bondsmen instead of purchasing them outright further reveal that reasonably profitable returns on invested capital could be earned. In such cases, of course, slave hirers computed only the cost of labor against their net income to estimate their profit rate, while slaveowners computed the amount of rent against their investment to estimate their profits for the year. In 1817, Ebenezer Pettigrew noted the expenses and earnings from a hired slave lumberman as follows:

Hire	$80.00
Clothing	17.00
Victuals	27.40
	$124.40

Net proceeds of said fellow geting Juniper Shingles is found to be $250.00

Moreover, from 1830 to 1860, the annual rates of return from slave hiring ranged, according to one study, from 9.5 to 14.3 per cent in the upper South, and from 10.3 to 18.5 per cent in the lower South.[15] Such earnings suggest that slave hiring was at least as profitable as direct slave ownership for industries.

Finally, it should be recalled that industrial entrepreneurs, like most other slaveowners, profited from slavery's intermediate product — marketable and productive slave offspring. Many industrial establishments owned slave women whose progeny could easily be sold, and both women and children could be employed in light and heavy industries. Slave women and children therefore gave competitive advantages to employers of industrial slaves.[16] It may therefore be

---

[14] Account Book, vol. IV, 1833–39, 1856–61, Manigault Papers (UNC); Govan, "Was Plantation Slavery Profitable?" 513–35, demonstrates the profitable returns on Manigault's rice milling investment. For the earnings of other rice mills, see J. B. Irving's Windsor and Kensington Plantations Record Books, 1840–52, and T. Pinckney's estate appraisals and account books, 1842–63 and 1827–64 (Charleston Library Society, hereafter cited as CLS).

[15] R. Evans, Jr., "The Economics of American Negro Slavery," *Aspects of Labor Economics* (Princeton, 1962), 217; B. Wall, "Ebenezer Pettigrew" (Ph.D. dissertation, University of North Carolina, 1946), 308.

[16] See below on the use of women, children, and superannuates in southern industries; cf. Conrad and Meyer, *The Economics of Slavery*, ch. 3.

concluded that industrial enterprises, which either owned or hired slave labor, earned profitable returns on their investments.

THE GENERAL EFFICIENCY OF INDUSTRIAL SLAVERY

It is possible that industrial slavery was an inefficient or uneconomical labor system, even though it was simultaneously profitable to most industrial enterprises. Slaves were so troublesome and so unwilling, according to some historians, that they were less efficient than free workers. After all, did not slaves have to be coerced, while free workers responded eagerly to wage incentives? Industrial slave labor may also have been so expensive compared to free labor that it was, objectively, an unviable labor system. Given these questions, it is necessary to examine further the general efficiency of industrial slave workers and the costs arising from their ownership.

The available evidence indicates that slave labor was not less efficient than the free labor available in the Old South. To be sure, the slave's indifference to his work and his resistance to bondage tended to diminish his productivity somewhat. But this does not necessarily mean that competent managers could not make industrial slaves work or would have found free labor more efficient to employ. Physical coercion, or the threat of it, was an effective slave incentive, and masters often gave bondsmen material rewards for satisfactory production. In addition, industrial slaveowners could exploit women and children more fully than could employers of free labor. The average industrial bondsman was disciplined more rigorously than the typical free worker. Slaveholders were not troubled by labor organizations and were not obliged to bargain openly with their employees. "These advantages," concludes one authority, "more than compensated for whatever superiority free labor had in efficiency." [17]

In theory, slave labor may be less efficient than free labor over the long run, but for this study the practical comparison is between southern Negro slaves and the alternative free labor — poor whites, yeomen, and immigrants — available to the Old South. If this comparison is made, then it may be seen that the available free labor

---

[17] Stampp, *Peculiar Institution*, ch. 9, and Starobin, *Industrial Slavery*, chs. 2, 3, and 4, for a discussion of slave resistance and discipline. According to R. W. Fogel and S. Engerman, *The Reinterpretation of American Economic History* (New York, 1968), part 7, the manuscript census schedules reveal that about one-half of the slave population was in the labor force — a figure which is close to, if not at, the maximum possible participation rate. Since 44 per cent of the slaves were children under fourteen years old and 4 per cent were adults over sixty, virtually every able-bodied adult slave and most teen-agers were compelled to work. The slave participation-rate in the labor force was, moreover, 60 per cent greater than that of white workers.

— particularly the poor whites and immigrants — was less efficient than slave labor, since these whites were less tractable than slaves.[18]

Testimony from southern manufacturers who employed free labor supports the conclusion that it was not very efficient. White "hands had to be trained," admitted an associate of Daniel Pratt, the well-known Alabama businessman. "These [whites] were brought up from the piney woods, many of them with no sort of training to any kind of labor; in fact, they had to learn everything, and in learning, many mistakes and blunders were made fatal to success." Southern poor whites were not disciplined to sustained industrial labor, conceded the treasurer of William Gregg's Graniteville, South Carolina, cotton mill — another southern showpiece employing southern white workers.[19] Moreover, such testimony has been confirmed even by those scholars who argue that the level of productivity (that is, output per man) of slave labor was "low." "When white labor was used in Southern factories, it was not always superior to slave labor," admits one historian. ". . . [Southern white] productivity was much lower than in the North. . . . The use of whites did not guarantee a better work force than did the use of Negroes, for the South lacked an adequate pool of disciplined free workers." [20]

The efficiency, or total output, of slave labor compared to free labor can also be estimated by comparing the prices paid for slave hirelings with the wages paid southern free labor. From 1800 to 1861, white wages did not increase substantially; they remained fairly constant at about $300 per annum.[21] On the other hand, between 1800 and 1833, slave rents increased by about 50 per cent. Then, in the 1840's and the 1850's, slave hires again increased by another 50 per cent. At the same time, the value of slaves was increasing proportionately.[22] This suggests that both the productivity of and the demand for slave labor were increasing substantially

[18] For the unreliability of white workers in the South, see Starobin, *Industrial Slavery*, ch. 4.
[19] Genovese, *Political Economy of Slavery*, 226–27; for the politics of industrial slavery, see Starobin, *Industrial Slavery*, ch. 6.
[20] Genovese, *Political Economy of Slavery*, 37 fn. 13.
[21] For evidence on this point, see my dissertation, "Industrial Slavery," ch. 5, fn. 20. The wages of white southern textile workers were somewhat lower.
[22] See appendix on the cost of slave hiring in my dissertation:

Summary of Cost of Slave Hiring at Southern Industries
(in dollars)

Period	Daily	Monthly	Annually
1799–1833	.76	13.14	66.39
1833–52	.77	16.51	100.55
1853–61	1.44	19.68	150.00

during the first half of the nineteenth century. Thus, no matter how inefficient slave labor may have been, it was not less efficient than the free labor available to Southerners at the time.

It is often argued that the use of slaves entailed expenditures that were avoided by the employers of wage labor. The initial investment in blacks, the interest and depreciation on slave capital, the constant risk of financial losses from death, injury, disease, and escape, and the expense of maintaining slaves were all special expenses supposedly peculiar to slave ownership. These extra costs, according to some scholars, made slave labor more expensive and less economical than free labor.

It is clear, however that these special costs did not make slave ownership more expensive than free labor. Many industrialists did not bear the cost of initial slave capitalization, since they had inherited their bondsmen or had shifted them from agriculture to industry. Interest on capital was a current operating expense only if bondsmen were purchased on credit rather than with cash. Depreciation of slave capital was not a cost for most slaveowners, since slaves were appreciating in value and were producing saleable offspring. The prospect of financial disaster from losses of bondsmen was beginning to be alleviated in the 1840's and 1850's as many owners began to insure the lives of their Negroes. Finally, industries that hired slaves rather than purchasing them did not bear directly the cost of initial capitalization.[23]

Yet, when industries did purchase bondsmen, considerable expenditure of capital was involved, which should be compared to the costs of wage labor. The purchase of slaves entailed a different sort of expense than wages of free labor, since it was capitalization of future expenditures on labor and the payment all at once of a portion of what an employer of free labor would pay over a period of years. The cost of Negroes and their maintenance were, as one historian has argued, part of the wages an employer of free labor would expect to pay, and what masters were willing to pay for the right to fully control the time and movements of their workmen.[24] Slavery thus involved long-term capitalization of labor, while free labor involved the current expense of wages.

The surviving evidence also demonstrates that maintaining industrial slave labor cost much less than paying wages to available free

---

[23] Govan, "Was Plantation Slavery Profitable?" 513–35; Engerman, "The Effects of Slavery," 71–79; Stampp, *Peculiar Institution*, ch. 9; and Starobin, "Industrial Slavery," ch. 3.

[24] Conrad and Meyer, *The Economics of Slavery*, ch. 3; Stampp, *Peculiar Institution*, ch. 9.

labor. For directly-owned industrial slaves the largest annual expenditures were for maintenance and supervision — specifically for food, clothing, shelter, medical care, and management, as well as such incidental expenses as taxes, insurance, and incentive payments (see Table 4). The records of typical slave-employing enterprises reveal that the cost of important maintenance items and of supervision varied considerably. Suits of clothing, for example, ranged in price from $4 to $7, while shoes cost between $1 and $1.50, and boots from $1.50 to $2.50 a pair. Hats and caps sold for 50 or 75 cents, while blankets cost $1 or $2 each. Doctors ordinarily charged from $1 to $3 per visit; treatment of diseases such as syphilis cost from $5 to $15; medicine cost between 50 cents and $1 per illness. Life insurance ranged between $1.66 and $5 per hundred dollar valuation but averaged about $2 per hundred, or 2 per cent of valuation.[25] Depending on self-sufficiency and locale, the annual per capita cost of food varied between $10 and $125; clothing varied from about $3 to $30 annually per capita, housing cost between $5 and $10, and management ranged from about $200 to $3,000 a year.[26]

Despite such wide variations, industrial records indicate that between 1820 and 1860 food annually averaged about $50 per slave and clothing about $15.[27] Medical attention annually averaged about $3 per slave, housing probably cost about $7, and supervision amounted to about $800 per thirty hands, or about $27 per annum per slave. Incidental expenses annually cost little more than $5 per slave.[28] The annual average maintenance cost per industrial slave therefore amounted to about $100. Obviously, this was higher than the maintenance of slaves on plantations, which were much more self-sufficient. But how did these expenses compare with the cost of free labor in the Old South?

In the antebellum South, the daily wages of white common laborers ranged from 75 cents to $2 and averaged about $1 a day, while skilled whites earned daily from $2 to $5 and averaged about $3. The wages of common white workers did not increase appreciably between 1800 and 1861.[29] Thus, for a 310-day working year, and depending on skill, white wages ranged from $225 to $1,500 annually. But the bulk of unskilled white workers who figure in this study averaged only about $310 per year. Like slaves, wage laborers required supervision, but they ordinarily fed, clothed, and

[25] These prices are taken from the records of slave-employing industrial enterprises.
[26] See table 4 on maintenance costs.
[27] Ibid.
[28] Ibid.
[29] For white wages, see Starobin, "Industrial Slavery," ch. 5, fn. 20.

housed themselves, unless their board was furnished for them or they lived in company towns where their maintenance costs were automatically subtracted from their wages. The cost of free labor thus totaled about $335 per annum, including supervision. The annual average maintenance cost per industrial slave was therefore less than one-third the annual cost of wages and supervision of free common laborers.

The surviving reports from those "integrated" companies which used both slave labor and free labor simultaneously (or in succession) in the same workplace also reveal that slave labor was much less expensive than free labor. At the Cape Fear & Deep River Navigation Works, white workers cost 40 cents per day to board, while slaves cost 30 cents. In 1849, the Jackson *Mississippian* reported that whites cost 30 cents per day to board, while slaves cost 20 cents. In the late 1830's and 1840's, the Graham Cotton Mill in Kentucky listed white board at from $65 to $71 per year, while slave board ranged from $35 to $50. The accounts of the Roanoke Valley Railroad for 1852–1853 indicate that slaves were boarded more cheaply than whites, and the records of the Jordan & Davis iron works in Virginia for 1857–1858 demonstrate that whites were boarded for $8 per month, while slaves cost $7.[30]

Similarly, in the 1820's, the proprietors of the Maramec Iron Works in Missouri (another such integrated enterprise) reported that slaves were cheaper than free workers. Whites cost on the average about $15 per month, *excluding* supervision and free housing. Slaves hired for $100 per annum; their supervision and maintenance ran no more than $80 per year. Maramec's proprietors also testified that the cost of labor per cord of wood chopped by slaves compared favorably with the cost when whites performed the task.[31] A Kentucky hemp manufacturer, who converted from free labor to slave labor, claimed that slaves reduced his costs by 33 per cent. In 1854, it was reported that Kanawha River, Virginia, slave miners produced $2 a day more than free miners at Pittsburgh, Pennsylvania, pits. The next year, the Virginia & Tennessee Railroad reported that slave labor cost only about $11 monthly while free labor cost $40 to $50 monthly. The manager of one South Carolina cotton

---

[30] Boarding bills for October and December 31, 1859, Treasurers' Papers: Internal Improvements: Cape Fear & Deep River Navigation Works (North Carolina Department of Archives and History, hereafter cited as NCA); Jackson *Mississippian*, May 4, 1849; Graham Cotton Mill Daybook and Inventory, 1837–41 (University of Kentucky Library, hereafter cited as UK); vol. XXVI, Hawkins Papers (UNC); account sheet, 1857–58, Jordan and Davis Papers (State Historical Society of Wisconsin, hereafter cited as WSHS). For further information on "integrated" companies, see Starobin, *Industrial Slavery*, ch. 4.
[31] J. Norris, *Frontier Iron* (Madison, 1964), 40–41.

mill estimated that in 1851 slaves cost less than half as much as whites.[32] Therefore, at such integrated industrial enterprises, where the only variable was the nature of the labor force, slave labor was very much less expensive to employ than free labor.

Unusually complete records of several other integrated enterprises provide additional evidence that industrial slave labor was much cheaper than free labor. The labor rolls of the Gosport Navy Yard reveal that in the 1830's slaves produced as much as white workers for two-thirds the cost — that is, the use of industrial slaves was, in this case, almost twice as efficient as the use of whites. This was partly because the daily rent of slave hammerers ranged only from 72 to 83 cents, averaging close to 72 cents. The daily wages of white hammerers ranged from $1.68 to $1.73. Of course, the cost of maintaining the slaves probably amounted to about 30 cents daily, which increased the cost of slave hammerers to about $1 per day. Even so, it was less expensive to employ slaves than whites.[33]

The account sheets for Robert Jemison, Jr.'s Alabama construction projects further indicate that in 1858 bondsmen were 26 per cent cheaper to employ than free laborers. In 1859, slaves were 46 per cent less expensive than whites. The accounts of the Graham textile mill in Kentucky reveal that from 1837 to 1843 unskilled slaves annually cost 26 per cent less than unskilled whites, while skilled slaves cost between 15 and 22 per cent less than skilled whites. As late as 1851, slave carders, weavers, and spinners still cost less than comparable whites. The records of the Woolley textile mill in Kentucky also indicate that, between 1856 and 1861, most skilled slaves annually cost 57 per cent less to employ than skilled whites.[34]

Another integrated industrial enterprise, Richmond's Tredegar Iron Works, offers an interesting example of the cheapness of slave

---

[32] D. Myerle's testimony in *Report of Court of Claims*, #81, 34 C., 3 s., 1857; memorial of F. G. Hansford, *et al.*, Virginia *Board of Public Works Report, 1854*, 403; Report of the Virginia and Tennessee Railroad, Virginia *Board of Public Works Report, 1855*; J. S. Buckingham, *The Slave States* (London, 1842), I, 264-65 and II, 112-13; *De Bow's Review*, XI (1851), 319-20; E. M. Lander, "Slave Labor in South Carolina Cotton Mills," *Journal of Negro History*, XXXVIII (1953), 170-71.

[33] Rolls of Labor, 1831-32, and Memorandum of Work, 1831, Baldwin Papers and Selekman Notes (Baker Library, Harvard University Graduate School of Business Administration, hereafter cited as Baker). However, the daily rent of slave *common* laborers averaged about 72 cents, while the daily wages of white common laborers averaged $1.01. Therefore, savings came in the cost of the more skilled hammerers, where slaves were cheaper than whites.

[34] Cost of Hands for 1858, 1859, and undated, loose inserts in Jemison and Sloan Company contract account book, 1856-59, Jemison Papers (University of Alabama Library, hereafter cited as UA); Daybook and Inventory, 1837-41, Ledger and Inventory Book, 1832-45, Factory Time Book, 1847-52, Ledger, 1846-47, and Account Book, 1847-50, Graham Cotton Mill Papers (UK); Wages Ledger, 1856-61, Daybook, 1856-59, and G. Woolley to W. Peck & Sons, June 2, 1861, Lettercopybook, Woolley Papers (UK). However, the Woolley Papers reveal that common slave weavers cost almost as much to hire and to maintain as common white weavers. Again, savings came with skilled labor, where slaves were cheaper than whites.

labor. After commencing to hire slaves in 1848, Tredegar's proprietor, Joseph Reid Anderson, stated that slave labor "enables me, of course, to compete with other manufacturers." Competitiveness was achieved by combining slaves with white iron workers, which reduced the average cost of labor per ton of rolled iron. Between 1844 and 1846, *before* slaves were employed, for example, labor cost more than $12 per ton; from 1850 to 1852, *after* slaves were fully at work, labor averaged $10.59 per ton. The introduction of slaves thus enabled Anderson to reduce his labor costs by 12 per cent.[35]

A confidential report by the chief engineer of the South Carolina Railroad, which employed free labor at its Charleston terminal but used slave labor for its upcountry stations, offers additional evidence on the comparative cost of bondsmen and free workers. "It is a subject well worthy of enquiry whether the labor at the Charleston Depot could not be performed by slaves more economically than by whites," confided the official to the president of the line in 1849. "What cannot fail to strike you in the abstract of Depot expenses for August last is the fact that 1570 days [of] *white* labor at Charleston Depot cost $1,206, or 77 cts per day, while 1033 days [of] *slave* labor cost at the three *upper terminii* only $524 or 51 cts per day," he continued. "This statement also shows that it took 50 per cent more labor to load merchandise and unload cotton [at Charleston by white labor] than to load cotton and unload merchandise [in the upcountry by slave labor], or the cost of the former was two & a third (2⅓) times the latter."[36] Similarly, an 1855 report by the State Engineer of Louisiana also reveals that slave labor was much less expensive than free labor.[37]

Whatever the capital costs of slave ownership, these hardly concerned the employers of slave hirelings. Slave hirers bore only the expense of rent, maintenance, and supervision, even though other costs might be hidden in the slave rent. Slave hiring was thus similar to paying wages to free labor. Moreover, industrial slave hirelings, like directly-owned Negroes, were also more economical to employ than the free labor available. This is confirmed by comparing the total cost of hiring slaves with the cost of free labor. Throughout the slave states during the period from 1833 to 1852,

[35] C. Dew, *Ironmaker to the Confederacy* (New Haven, 1966), 18–20, 29–32, tables, and notes for prices of Tredegar iron products; J. R. Anderson to H. Row, January 3, 1848, Tredegar Letterbook (VSL).
[36] J. McRae to J. Gadsden, November 4, 1849, McRae Lettercopybooks (WSHS).
[37] This report is particularly valuable since it was based on detailed accounts and considerable experience with both slave and free labor, *De Bow's Review*, XIX (1855), 193–195.

the average annual rent of slave hirelings was $100; from 1853 to 1861, it was $150. During the same spans, per capita slave maintenance annually averaged about $100. The total cost of employing slave hirelings thus ranged from $200 to $250 per annum from 1833 to 1861. However, between 1800 and 1861, the annual average cost of employing free common laborers remained at about $310, not including supervision. By comparing these figures, it can be seen that slave hirelings remained between 25 and 40 per cent cheaper to employ than wage laborers.[38] Therefore, industrial slaves — whether hired or owned — were apparently more efficient and economical than the free labor available in the Old South.

Specific Competitive Advantages of Industrial Slavery

It is well known that southern industrialization lagged behind that of the North and of Great Britain. At least by the 1830's, northern and British industrialists had longer experience, more efficient management, larger markets, superior technology, and the ability to ship directly to the South. Northern products were of a better quality; Pennsylvania's iron and coal ores, for example, were superior to Virginia's and Kentucky's.[39] The earlier development of internal improvements in the North reduced transportation costs, which in turn reduced the prices of northern products generally. The availability of cheap labor — native and immigrant — in the North lowered prices further; the immigration of skilled Europeans increased the quality of northern products even more. The abundance of commercial capital for industrial investment enabled northern manufacturers to expand production, absorb business losses, withstand depressions, and, most important, to engage in cutthroat competition with southern producers. Thus, whatever the long-range causes and consequences of southern industrial backwardness,[40] the immediate question facing southern businessmen — especially manufacturers — was how best to compete with outside producers.

Southerners attempted to overcome their competitive disadvantages in various ways. They tried to foster direct trade with consumers of cotton, to promote internal improvements, and to recapture western markets.[41] But the most interesting means by which south-

---

[38] For the cost of slave hiring and white wage rates, see my dissertation, "Industrial Slavery," appendices, and ch. 5, fn. 20. In addition, R. Mills, *Statistics of South Carolina* (Charleston, 1826), 427–28, calculated that in Charleston in 1826 black common laborers cost half as much as whites; skilled blacks averaged 82¾ cents per day, while skilled whites averaged $1.37½.
[39] Dew, *Ironmaker to the Confederacy*, 32; North, *Economic Growth, passim.*
[40] See works cited above in fn. 1 and 6.
[41] See *De Bow's Review*, 1846 to 1862, for such programs.

erners attempted to raise the quality and reduce the cost of their products was the use of industrial slave labor in several specific ways. First, southern businessmen extensively exploited slave women and children (and sometimes superannuates). Second, they trained a Negro slave managerial group to complement white overseers. Finally, they "coupled" inexpensive slave workers with highly skilled white technicians — northern and foreign. In short, Southerners attempted to take advantage of the efficiency and inexpensiveness of slave labor to improve their competitive position in national market places.

Slave women and children comprised large proportions of the work forces in most slave-employing textile, hemp, and tobacco factories. They sometimes worked at "heavy" industries such as sugar refining and rice milling, and industries such as transportation and lumbering used slave women and children to a considerable extent. Iron works and mines also employed them to lug trams and to push lumps of ore into crushers and furnaces.[42]

Slaveowners used women and children in industries in several ways in order to increase the competitiveness of southern products. First, slave women and children cost less to capitalize and to maintain than prime males. John Ewing Colhoun, a South Carolina textile manufacturer, estimated that slave children cost two-thirds as much to maintain as adult slave cotton millers. Another Carolinian estimated that the difference in cost between female and male slave labor was even greater than that between slave and free labor.[43] Evidence from businesses using slave women and children supports the conclusion that they could reduce labor costs substantially.[44]

Second, in certain light industries, such as manufacturing, slave women and children could be as productive as prime males, and sometimes they could perform certain industrial tasks even more efficiently. This was especially true in tobacco, hemp, and cotton manufacturing, where efficiency depended more upon sprightliness and nimbleness than upon strength and endurance. The smaller hands and agile fingers of women and children could splice cotton or hempen threads more easily than the clumsy fingers of males. Delicate palms and dexterous digits processed tobacco more carefully. "Indeed it is well known that children are better adapted to some branches of manufacturing labor than a grown person," edi-

---

[42] See Starobin, *Industrial Slavery*, fn. 42–48.
[43] J. E. Colhoun Commonplace Book (Clemson); J. D. B. De Bow, *Industrial Resources* (New Orleans, 1853), II, 178.
[44] See, for example, Graham Cotton Mill Papers (UK); Woolley Papers (UK); and King Papers (Georgia Department of Archives and History, hereafter cited as GA).

torialized the Jackson *Mississippian*.⁴⁵ In addition, some industrialists believed that slave women could do as much work in some heavy occupations as males. "In ditching, particularly in canals . . . a woman can do nearly as much work as a man," concluded a Carolinian.⁴⁶

Third, industrialists used slave women and children in order to utilize surplus slaves fully. "Negro children from ten to fourteen years of age are now a heavy tax upon the rest of the planter's force," editorialized the Jackson *Mississippian*. "Slaves not sufficiently strong to work in the cotton fields can attend to the looms and spindles in the cotton mills," concluded a visitor to a cotton mill where 30 of 128 slaves were children, "and most of the girls in this establishment would not be suited for plantation work." Placing Negroes in cotton mills "render[s] many of our slaves who are generally idle in youth profitable at an early age," observed a textile promoter. "Feeble hands and children can perform this work," concluded a rice miller, "leaving the effective force for improvements or to prepare for another crop." ⁴⁷

The intention of industrialists to utilize slave capital fully by employing women and children extensively is confirmed by an analysis of the manuscript census schedules. This study reveals that almost one-half of the slave population was in the labor force — a figure which is close to, if not at, the maximum possible participation rate. Since 44 per cent of the slaves were under fourteen years of age and 4 per cent were adults over sixty, then most slave women, most teen-age slaves, many slave children, as well as most adult males seemed to be at work. Moreover, the slave participation rate in the labor force was 60 per cent greater than the white participation rate.⁴⁸ This suggests that slaves of all age groups were forced to labor more extensively than whites.

It has already been seen that one of the greatest costs at southern industries was supervision. Since the cost of management contributed to the price of industrial products, Southerners sought to reduce its expensiveness and to increase its competence. Each of the types of free white management available — personal supervision, native white technicians, and imported directors — had serious

---

⁴⁵ Jackson *Mississippian*, March 19, 1845; cf. *De Bow's Review*, XXV (1858), 114.
⁴⁶ *Southern Agriculturist*, VI (1833), 587; *De Bow's Review*, XVIII (1855), 350–51; cf. Northrup, *Narrative*, 155–56; K. W. Skinner to C. Manigault, February 8, 1852, Manigault Papers (Duke).
⁴⁷ Jackson *Mississippian*, March 19, 1845; *De Bow's Review*, XI (1851), 319–20, and XXII (1857), 394, 397; Richmond *Enquirer*, October 7, 1827; *Southern Agriculturist*, VII (1834), 582, and IV (1831), 368; New Orleans *Picayune*, October 16, 1858.
⁴⁸ Fogel and Engerman, *Reinterpretation*, part 7.

limitations. When more than thirty slaves were employed, personal supervision was difficult, since sales, supplies, and bookkeeping occupied the owner's time. Native white managers were scarce, and they were often technically incompetent. Imported directors — northern and foreign — commanded high salaries for their superior abilities and to compensate for the rigors of the southern climate. No matter what the source, therefore, free white industrial management was expensive, ranging from $200 to $3,000 per annum and averaging about $800.[49] Given these circumstances, industrial enterprises often trained their own Negro slave managers.

Black slave managers were used by many southern industries. Simon Gray and Jim Matthews, slave hirelings of the Andrew Brown Lumber Company of Natchez, were responsible for rafting lumber and sand down the Mississippi River to customers along the way and to a New Orleans depot. Simon Gray directed as many as twenty raftsmen — both free whites and slaves — either owned or hired by the company. He disciplined the crewmen, distributed the wages of the white workers and the overtime payments to the slaves, and he paid the expenses of both. Gray bargained with planters and sawmillers, kept accurate accounts, and collected and disbursed large sums of money. He once delivered $800 to a creditor; on another occasion he escorted a newly purchased bondsman from a slave market to the industrial site — a responsibility ordinarily entrusted only to white men. He had his own pass, and he could charge goods to his personal account at the company store.[50]

Simon Gray had many counterparts in southern industry. As early as the 1790's, Andrew, a slave, rafted lumber down Georgia rivers, directed other slave raftsmen, and responsibly delivered bills of lading as well as valuable lumber for saw-miller Alexander Telfair.[51] Other slave managers handled large sums of money with fidelity. One slave ferryboat operator faithfully collected company tolls, controlled disbursements, and seemed to manage the entire business without difficulty. One railroad company hired Phocian, a slave, who served as a business agent, delivered company correspondence, faithfully handled sums of money ranging up to $200, and received many privileges, including a pass to visit his wife. Harry, a slave, delivered iron and procured supplies for an iron works during the 1830's and 1840's.[52] Other industrial slave managers were

---

[49] See appendix on the cost of management in my dissertation, "Industrial Slavery."
[50] J. H. Moore, "Simon Gray," *Mississippi Valley Historical Review*, XLIX (1962), 472–84.
[51] Account Books, 1794–1863, Telfair Papers (GHS); Northrup, *Narrative*, 89–99.
[52] G. Rogers, *Memoranda of Travels* (Cincinnati, 1845), 196, 310; D. B. McLaurin to

also trained as business agents. From as early as 1857 until 1862, Nathan, a fifty-seven-year-old bondsman, responsibly transacted much of the affairs of a North Carolina tannery. Without much supervision, Nathan made week-long business trips to sell leather at markets within a fifty-mile radius of the company.[53]

Many slave engineers skillfully operated complicated industrial machinery. Two slave rice millers, Frank the "headman" and Ned the engineer, capably ran the steam engine and the milling machinery at one establishment. Sandy Maybank was the slave head carpenter at another Georgia rice mill. A "full-blooded" black man superintended a Carolina cotton mill; a slave machinist attended the machine shop of a Virginia railroad; and Emanuel, a locomotive engineer owned by a Louisiana line, had an admirable record during ten years' service. One master's coal pits were, according to Edmund Ruffin, "superintended and directed entirely by a confidential slave of his own (whom he afterwards emancipated, and then paid $200 a year wages), and the laborers were also slaves; and they only knew anything of the condition of the coal." [54]

Some slave managers were quite talented. Horace, a slave architect and civil engineer, and Napoleon, his slave assistant, designed and executed Black Belt bridges for Robert Jemison, Jr., a wealthy Alabama planter-industrialist. Horace's most notable achievement for the year 1845 was the erection of a bridge in Columbus, Mississippi, for which he served as "chief architect." This project won Horace his employer's praise as "the most extensive and successful Bridge Builder in the South." Upon the completion of Horace's next project, a bridge in Lowndes County, Mississippi, Jemison wrote: "I am pleased to add another testimony to the style and despatch with which he [Horace] has done his work as well as the manner in which he has conducted himself." [55]

There can be little doubt that industrial slave managers were less expensive to employ than white managers, and that by reducing

---

W. H. Richardson, April 19 and January 27, 1855, and pass dated April 20, 1855, Richardson Papers (Duke); M. Bryan to W. W. Davis, June 30, 1846, Jordan and Davis Papers (WSHS).

[53] Vols. 39, 44, 17, 30, 45, and 46, Hawkins Papers (UNC).

[54] F. Kemble, *Journal* (New York, 1961), 113, 116–17, 168, 176, and 187–88; C. C. Jones to Sandy, August 15, 1853, and C. C. Jones to T. J. Shepard, March 30, 1850, Jones Papers (Tulane University Library, hereafter cited as Tulane); H. T. Cook, *David R. Williams* (New York, 1916), 140; J. B. Mordecai, *A Brief History of the Richmond, Fredericksburg and Potomac Railroad* (Richmond, 1941), 17; Richmond *Times-Dispatch*, January 31, 1943 (VHS); C. Sydnor, *Slavery in Mississippi* (New York, 1933), 7; *Farmers' Register*, V (1837), 315; Duke de la Rochefoucault-Liancourt, *Travels Through the United States* (London, 1799), III, 122–23.

[55] R. Jemison, Jr., Letterbooks, 1844–54 (UA); cf. Olmsted, *Seaboard Slave States*, 553.

the costs of supervision, they increased the competitiveness of southern industries. Simon Gray, the riverman, clearly reduced the management costs for the Andrew Brown Lumber Company. As a head raftsman, Gray at first received twelve dollars monthly; this was about one-fourth the wages of a white head raftsman. Even when Gray's incentive was raised to twenty dollars monthly, the same wages as ordinary white raftsmen, it was still only *half* that of white head raftsmen. A white manager with Gray's skills and responsibilities would have cost the lumber company annually almost as much as Gray's total market value.

Similarly, Nathan, the tannery business agent, cost much less than a comparable white manager. Nathan received for his services only a dollar or two per trip, for about ten trips per year. He incurred in addition only his maintenance, which amounted to several cents per day. A white business agent with Nathan's responsibilities would have cost at least $2.50 daily in wages alone and might have been less trustworthy than the slave. Sandy Maybank, the slave head carpenter at the Georgia rice mill, was as skillful as, yet less expensive than, a comparable white manager. Moreover, his master reaped extra financial benefits from Maybank's ability to hire himself out in the slack season. Horace and Napoleon, the slave bridge builders, cost only five dollars daily plus board; two comparable white managers probably would have cost twice as much. Even at these rates, Jemison considered Horace's services so indispensable and profitable that he continued to engage Horace for many years.[56]

While some industries employed slave managers, others used highly skilled white technicians — imported from the North or Europe — to improve the quality and the competitiveness of industrial products. Of course, imported managers were more expensive than native ones — free or slave; but businessmen discovered that the use of inexpensive slave common laborers made possible the employment of expensive skilled foreign technicians. By "coupling" common slaves with these skilled white managers, industries could raise the quality of products without increasing overall labor and management costs. By engaging the best foreign technicians available, southerners thus attempted to compete with northern and British manufacturers.[57]

Among the many southern industries which coupled cheap slaves

---

[56] For further information and sources on slave managers in industry, see my dissertation, "Industrial Slavery," ch. 3 and 5, fn. 64.

[57] See earlier discussion on the costs of slave ownership.

with expensive white engineers was textile manufacturing, where competitiveness depended greatly on quality. As early as 1815, cotton millers realized the advantages of skilled management, when one Carolinian who hired three northern superintendents "thought it best so to do — for to depend upon our hands to learn would take a considerable time before we could cleverly get underway." Similarly, a Tennessee textile mill employed a Providence, Rhode Island, foreman; John Ewing Colhoun, whose products were so widely praised, also employed a northern superintendent; and an experienced "Loweller" managed a Mississippi mill.[58]

Combining inexpensive slaves with skilled technicians was also common in extractive industries. Mining companies often hired experienced Welsh, English, Cornish, and other foreign supervisors to direct the blasting, tunneling, seam tracking, and other work performed by common slave miners.[59] Lumbering enterprises often engaged skilled sawyers from Maine or northwest forests to supervise unskilled slave lumbermen. "Those who would engage in a scheme of this kind," advised an early shipbuilding promoter, "would however find it their interest to instruct negroes in the art of working on ships under two or three master-builders."[60]

Experienced foreign civil engineers likewise directed many heavy construction projects, since native southern technicians were scarce. Architect B. H. Latrobe designed the New Orleans Water Works, Loammi Baldwin administered the Gosport Navy Yard, while his brother, James, executed the Brunswick & Altamaha Canal. After 1819, Hamilton Fulton, an Englishman, supervised North Carolina's and then Georgia's river improvement programs. European-trained J. Edgar Thompson planned the Georgia and the Southern Pacific railroads. Charles Crozet, a French engineer, served the Virginia Board of Internal Improvements.[61]

[58] E. M. Lander (ed.), "Two Letters by William Mayrant on His Cotton Factory, 1815," *South Carolina Historical Magazine*, LIV (1953), 3–4; *American Farmer*, series 1, IX (October 12, 1827), 235–314; E. M. Lander, "Development of Textiles in South Carolina Piedmont," *Cotton History Review*, I (1960), 92; *Jackson Mississippian*, December 4, 1844; Charleston *Courier*, February 19, 1845. Another industry which often used white managers was sugar milling.
[59] Report upon . . . the N.C. Gold Mining Co., September 5, 1832, Fisher Papers (UNC); *Richmond Enquirer*, March 23, 1839, and January 9, 1840; B. Broomhead to B. Smith, September 7, 1857, Smith Papers (Duke); Olmsted, *Seaboard Slave States*, 47–48; *Harper's Monthly*, XV (1857), 297; *De Bow's Review*, VII (1849), 546–47, and XXIX (1860), 378; S. Ashmore to C. Thomas, March 26, 1860, Silver Hill Mining Company Papers (NCA).
[60] N. Pendleton, "Short Account, 1796," *Georgia Historical Quarterly*, XLI (1957); J. Baker to S. Plaisted, November 23, 1839, Plaisted Papers (LSU); *Niles' Register*, XLVII (1834), 55; account book, 1812–17, Telfair Papers (GHS); entry for June 6, 1857, D. C. Barrow Diaries (UG); C. L. Benson to M. Grist, October 1, 1855, Grist Papers (Duke).
[61] Latrobe Papers (Tulane); Baldwin Papers (Baker); C. K. Brown, *A State Movement in Railroad Development* (Chapel Hill, 1928), 12; M. S. Heath, *Constructive Liberalism*

The coupling of inexpensive bondsmen with skilled white artisans was also important to the iron industry which attempted to compete with northern and foreign producers. South Carolina's Nesbitt Manufacturing Company imported several New York founders. Four experienced Connecticut Yankees managed the Hecla Iron Works in Virginia. Another iron company employed a "Jersey founder;" William Weaver's hiring agent tried to engage one of Virginia's most famous colliers, while another iron monger sought the services of James Obrian, Weaver's skilled hammerer.[62]

Many southern businessmen clearly understood the competitive advantages of combining skilled white technicians with inexpensive slaves. Textile manufacturers and promoters, such as E. Steadman, who advocated paying cotton mill superintendents well enough to attract the "best talent and skill" to the South, seemed especially aware of these advantages. If the Saluda cotton mill had only hired "a carder, spinner, dresser, weaver, and an active and skillful young man as overseer, taking the best talents that Massachusetts could afford . . . and offered inducements that would have commanded the very best," editorialized the Columbia *South Carolinian* in 1844, the company would have been more successful. "If it is desirable to establish cotton factories in the South," agreed a "practical" English manufacturer who visited South Carolina, "let the proprietors select the proper man to make out the plans, select the machinery, manage the manufacturing details, and let them pay such men sufficient remuneration for their services, and I venture to affirm that there will be no difficulty in building up a manufacturing business, equally as successful, and much more profitable, than the majority of Northern factories."[63]

Other manufacturers were also aware of the advantages of coupling slaves with skilled managers. Manufacturing was less expensive in the South, according to one promoter, mainly because "the manual labor, costing even now as little as northern labor, may be and will be, under a . . . skilful and eminently practical management, made, by the judicious intermingling of slave male and female

(Cambridge, 1954), 241; 261; C. Goodrich, *Government Promotion of American Canals and Railroads* (New York, 1960), 98; *De Bow's Review*, XXVII (1859), 725.

[62] J. M. Taylor to F. H. Elmore, June 25, 1840, and agreements of September 30 and October 5, 1837, Elmore Papers (Library of Congress, hereafter cited as LC); Richmond *Enquirer*, June 3, 1851; Jordan, Davis & Co. to W. Weaver, November 24, 1830, W. Ross to W. Weaver, November 27, 1831, and G. P. Taylor to W. Weaver, October 7, 1831 and May 2, 1832, Weaver Papers (Duke); W. Rex to W. Weaver, January 8, 1859, Weaver Papers (University of Virginia Library, hereafter cited as UV).

[63] E. Steadman, *Brief Treatise on Manufacturing in the South* (Clarksville, Tenn., 1851), 108; Columbia *South Carolinian*, December 18, 1844; *De Bow's Review*, XXVI (1859), 95-96.

labor with that of native whites, and their imported tutors, cheaper than it can possibly be had for in any northern locality. Here then, with all the elements of cost at the lowest rate," he concluded, "the wares of this factory would contend successfully, even for a foreign market, with the keenest Yankee competition." [64]

Transportation companies also comprehended the advantages of skilled management. As early as 1822, the Upper Appomattox Company of Virginia purchased slave laborers and placed them under "as industrious and enterprising an overseer as we could obtain" with excellent results,[65] and during the 1830's and 1840's, other transportation projects also engaged skilled engineers. By the 1850's, as railroad construction forged ahead, the advantages of "coupling" had become widely known.[66] When the directors of the Southern Pacific Railroad pondered the merits of various labor forces, for example, promoter Thomas Jefferson Green proposed to combine common bondsmen with skilled engineers: "It may be safely estimated that the natural increase of negroes upon the healthy line of our road together with the increased value of turning field labour into railroad mechanicks will aqual 15 pr. cent per annum, whilst the interest upon their cost would be 6 pr. cent — leaving a difference of 9 pr. ct. in favor of the company which would go far toward covering Engineering expenses & head mechanics — and other incidental charges." [67] Green thus understood that the use of slaves would save the company enough money to permit the employment of high-salaried civil engineers.

It remained, however, for Joseph Reid Anderson of the Tredegar Iron Works of Richmond, Virginia, consciously to systematize the coupling of common slaves with expensive technicians in order to increase competitiveness. In 1842, Anderson contracted skilled white "puddlers" to train common slave apprentices. Then, in 1847, some of these bondsmen, now more skillful, were promoted to the position of puddler. The next year Anderson explained the theory behind the practice: [68]

---

[64] *De Bow's Review*, XIV (1853), 622–23.
[65] Report of the Upper Appomattox Company, Virginia *Board of Public Works Report*, 1822, 33.
[66] Richmond *Dispatch*, December 9, 1853, for example.
[67] T. J. Green to the Executive Committee of the Southern Pacific R.R. Co., October 14, 1856, Green Papers (UNC).
[68] J. R. Anderson to H. Row, January 3, 1848, Tredegar Letterbook (VSL). However, employing slave labor could not solve Tredegar's problems of competing entirely, since American labor and transportation were costly and Virginia coal and iron were of a low quality. Tredegar's "most glaring weakness" lay not in its use of slave labor, according to Dew, *Ironmaker to the Confederacy*, 32, but in its "pitifully inadequate raw materials base" and in the southern transportation system.

> ... I am employing in this establishment [Tredegar] as well as at the Armory works, adjoining, of which I am President, almost exclusively slave labor except as to Boss men. This enables me, of course, to compete with other manufacturers and at the same time to put it in the power of my men to do better for themselves.

Throughout the 1850's, Anderson continued these arrangements, and he was soon able to reduce his labor cost per ton of rolled iron by 12 per cent.[69]

The Character of Industrial-Slave Capitalization

It is possible that the capitalization of the slave labor force crippled the finances of industries, even though industrial slavery was both profitable to investors and an efficient labor system. In this respect, industrial slavery may have been unviable in the long run because it reduced the flexibility of capital and the mobility of labor. Slave capital was so frozen, according to some scholars, that it could not easily be converted into cash. To transfer slaves from one place to another or to use them in different kinds of employment was allegedly difficult. "Negro slave labor was expensive," argued one historian, "because it was overcapitalized and inelastic. . . . Circulating capital was at once converted into fixed capital. . . . The capitalization of labor lessened its elasticity and its versatility; it tended to fix labor rigidly in one line of employment" — namely, in agriculture.[70]

Contrary to this view, the available evidence suggests that slave ownership did not seriously lessen the mobility of labor nor did slavery inhibit investment in industrial enterprises. Indeed, the funding of slave-based industries was primarily an internal process, intimately linked to slave-based agriculture.[71] Many industries were actually capitalized by transferring bondsmen from farming or planting to manufacturing, milling, mining, and transportation. And slaveowners themselves, not merchants or bankers, were the chief source of capital for industrial investment.

---

[69] See earlier discussion on the efficiency of slave labor and Starobin, *Industrial Slavery*, ch. 4 for further information on Tredegar. For sources on the use of white managers at other industrial enterprises, see my dissertation, "Industrial Slavery," ch. 5, fn. 66 and 67.

[70] Phillips, "The Economic Cost of Slaveholding," *Political Science Quarterly*, XX (1905), 257–75; Phillips, *A History of Transportation* (New York, 1908), 388–89; F. Linden, "Repercussions of Manufacturing in the Ante-Bellum South," *North Carolina Historical Review*, XVIII (1940), 328.

[71] As a source of industrial capital, the money derived from mercantile activity — mentioned in L. Atherton, *The Southern Country Store* (Baton Rouge, 1949), ch. 6, 194, 204 — seems less important than agricultural-based accumulation.

Slaveowning planters capitalized many manufacturing enterprises, such as cotton mills and hemp factories, by shifting some less-than-prime field hands or house servants to weaving and spinning. In such cases slave labor itself contributed to capitalization, while profits from planting or slavetrading provided additional funds. "The staples of the lower country require moderate labour, and that at particular seasons of the year," reported a Virginian to Alexander Hamilton, as early as 1791. "The consequence is, that they have much leisure and can apply their hands to Manufacturing so far as to supply, not only the cloathing of the Whites, but of the Blacks also." A visitor to Kentucky calculated: "The surplus [farm] labor is chiefly absorbed by the rope and bagging factories, which employ a vast number of slaves."[72]

To finance larger textile factories, slaveowning planters often pooled their slaves and cash and sold stock to neighboring agriculturists. David Rogerson Williams's South Carolina Union Factory was but one example of a textile mill where close financial relationships developed between investors and the company. James Chesnut, the prominent planter, bought company stock and arranged to rent to Williams's factory several of his surplus slaves. The company credited Chesnut for the amount of the hirelings' rent, against which he drew cotton and woolen goods manufactured at the mill. The factory purchased Chesnut's raw cotton, paying him in cash or credits which he used to buy finished textile goods for his plantation hands. Of course, Chesnut also received a share of the company's earnings.[73] To the company, Chesnut was a welcome source not only of capital, but of labor and raw material at comparatively low prices, while Chesnut's plantation served as a market for its manufactured goods. To Chesnut, the mill absorbed surplus slaves, cash, and cotton, while the company provided comparatively cheap manufactured goods and yielded profitable returns on his investment. Such financial relationships were mutually beneficial to planters and manufacturers alike.

Slaveowning planters also financed many iron works — the Nesbitt Manufacturing Company, a large South Carolina concern, being an

---

[72] D. Standard and R. Griffin, "Textiles in North Carolina," *North Carolina Historical Review*, XXXIV (1957), 15–16; E. Coulter, "Scull Shoals," *Georgia Historical Quarterly*, XXVIII (1964), 41–43; G. A. Henry to wife, December 3 and November 28, 1846, Henry Papers (UNC); Spinning Book, 1806–7, Taylor Papers (VHS); E. Carrington to A. Hamilton, October 4, 1791, A. H. Cole (ed.), *Correspondence of Alexander Hamilton* (Chicago, 1928), 94, 145; E. G. Abdy, *Journal of a Residence* (London, 1935), II, 349.

[73] Bill, May 9, 1829, account sheet, May 14, 1831, and J. N. Williams to J. Chesnut, May 17, 1831, Chesnut-Miller-Manning Papers (SCHS); same to same, ca. October, 1831, Chesnut Papers (WSHS); account sheets, 1829, 1830, June 19, 1830, and February 14, 1835, Williams Papers (South Caroliniana Library, Columbia, hereafter cited as USC).

interesting case. Like other Nesbitt investors, its president, Franklin Harper Elmore, a leading slaveholding and landowning banker, had strong personal, political, and financial ties in South Carolina and neighboring states. To raise capital, the company's founders agreed to permit investors to purchase stock with an equivalent value of blacks. Financial records reveal that several planters, including Wade Hampton, Pierce Mason Butler, and the Elmore Brothers, each invested thousands of dollars' worth of bondsmen in return for company certificates. The iron works thereby accumulated about 140 Negroes, worth about $75,000. Though two nearby banks loaned cash, a large portion of the company's capital consisted of slave labor.[74]

Similarly, slaveowning planters capitalized many extractive enterprises. As early as 1804, Moses Austin observed Missouri farmers sending or accompanying their slaves to the lead diggings after harvest to supplement their incomes. In the 1840's, John C. Calhoun periodically worked some of his cotton plantation slaves at his Dahlonega, Georgia, gold mines. In 1849, the *American Farmer* reported that Alabama cotton planters were shifting their bondsmen into turpentine extraction and distillation.[75]

Slaveowning planters and farmers also financed the majority of southern railroads, canals, and turnpikes. Some planters bought company stock with cash; others purchased or received shares for the labor of their slaves. "The cleaning, grubbing, grading, and bridging of the road," reported the Mississippi Central Railroad, "have been undertaken by planters residing near the line, who, almost without exception, are shareholders in the company. They execute the work with their own laborers, whose services they can at all times command."[76] Some slave-employing railroad contractors were paid company stock instead of cash, while some planters exchanged their slaves' labor for the privilege of having a railroad pass nearby their plantation.[77] The advantages of such financial relationships were clearly understood by many southern railroad officials, including the president of the Charlotte and South Carolina line, who reported in 1849:[78]

[74] Various financial documents, Elmore Papers (LC and USC); account sheet, ca. 1840, P. M. Butler Papers (USC).
[75] M. Austin, Description of the Lead Mines, 1804, *American State Papers: Public Lands*, I, 207; various letters between T. G. Clemson and J. C. Calhoun, 1842–43, Calhoun and Clemson Papers (USC and Clemson); *American Farmer*, series 4, IV (1849), 252.
[76] *American Railroad Journal*, XXVIII (1855), 577.
[77] For further information on the financing of southern transportation enterprises, see *De Bow's Review, American Railroad Journal, Western Journal and Civilian*, and the reports of boards of internal improvements and public works of the various slave states.
[78] *American Railroad Journal*, XXIII (1850), 9.

The practice of allowing stockholders to pay up their subscriptions in labor, is one of recent origin; is admirably calculated to increase the amount of stock subscribed, to facilitate its payment; and gives to the slave States great advantages over the free in the construction of railroads. . . .

While private investment by slaveowners predominated, public investment in industries and internal improvements by state and local authorities comprised only a small portion of the total capitalization of southern industries. Such public funds went almost entirely into slave-employing transportation projects rather than into other types of industry.[79] Moreover, federal [80] and foreign [81] funding of southern industries was also negligible. This situation contrasted with the process of capitalization in the North and West, where more industrial capital came from commercial surpluses, rather than agricultural, and where state, federal, and foreign funding of industries played an important role.[82] Indeed, the ratio of public to private investment, especially in transportation, seemed lower in the South than in the North. Thus, Southerners derived industrial capital from their own internal, private sources, specifically from the earnings of plantation agriculture. Southerners seemed to be developing industries in their region almost exclusively by their own efforts.

[79] For information on the public capitalization of southern internal improvements and the ratio of public to private investment, see C. Goodrich, "The Virginia System of Mixed Enterprise," *Political Science Quarterly*, LXIV (1949), 366-69 and tables; Goodrich, *Government Promotion of American Canals and Railroads*, passim; G. R. Taylor, *Transportation Revolution* (New York, 1951), passim; Heath, *Constructive Liberalism*, 287-89; A. G. Smith, *Economic Readjustment of an Old Cotton State* (Columbia, 1958); 179-83; J. W. Million, *State Aid to Railways in Missouri* (Chicago, 1896), 232-36; D. Jennings, "The Pacific Railroad Company," *Missouri Historical Society Collections*, VI (1931), 309; T. W. Allen, "The Turnpike System in Kentucky," *Filson Club History Quarterly*, XXVIII (1954); 248-58 note; C. B. Boyd, Jr., "Local Aid to Railroads in Central Kentucky," *Kentucky Historical Society Register*, LXII (1964), 9-16; S. J. Folmsbee, *Sectionalism and Internal Improvements in Tennessee* (Knoxville, 1939), 28, 122, 135, 265; S. G. Reed, *A History of Texas Railroads* (Houston, 1941), passim; M. E. Reed, "Government Investment and Economic Growth: Louisiana's Ante-Bellum Railroads," *Journal of Southern History*, XXVIII (1962), 184, 189; and railroad reports in *De Bow's Review*, XXVI (1859), 458-60, and XXVIII (1860), 473-77; *American Railroad Journal*, XXVIII (1855), 771; and the *Western Journal and Civilian*, XIV (1855), 292. Taylor, *Transportation Revolution*, 92, estimates that in 1860 about 55 per cent of the investment in railroads in the eleven Confederate states came from state and local authorities. This accounts for neither private investment in internal improvements other than railroads, nor investment generally in such states as Maryland, Kentucky, and Missouri.

[80] For federal funding of southern enterprises, see Goodrich, *Government Promotion*, 156-62; Taylor, *Transportation Revolution*, 49-50, 95-96, 67-68; Reed, "Government Investment . . . Louisiana's Railroads," 184, 189.

[81] For foreign funding of southern enterprises, see Kemble, *Journal*, 104-22; R. Hidy, *House of Baring* (Cambridge, 1949), 281, 330, 336; B. Ratchford, *American State Debts* (Durham, 1941), ch. 5; R. McGrane, *Foreign Bondholders and State Debts* (New York, 1935), 89, and ch. 9-13; *Niles' Register*, XXIX (1825), 178; XL (1831), 270; and XLII (1832), 91; Governor's Message, Milledgeville *Federal Union*, November 6, 1849, and December 2, 1845; Barclay, *Ducktown*, passim; correspondence for 1845-48, and 1855-61, Gorrell Papers (UNC); F. Green, "Gold Mining, Virginia," *Virginia Magazine of History and Biography*, XLV (1937), 357-65; F. Green, "Gold Mining, Georgia," *Georgia Historical Quarterly*, XIX (1935), 224-25; *American Journal of Science*, LVII (1849), 295-99.

[82] Taylor, *Transportation Revolution*, 97-102, and passim.

Regarding the flexibility of industrial slave capital, the records of several southern enterprises reveal that slave ownership did not cripple industrial finance. It is, of course, possible that larger industrial enterprises and wealthier businessmen were able to manipulate their slave investments more easily than smaller operators and less secure investors. But it is also true that industrial slavery reduced neither the flexibility of capital nor the mobility of labor to the extent that financial problems could not be solved. At the Nesbitt Manufacturing Company, a large South Carolina iron works, for example, finance remained quite flexible. In 1840, a planter-investor proposed to rent twelve blacks to the company rather than to invest them. The annual rate of hire would be $120 for each slave, the duration of hire four years, and the rent paid in company stock at the end of each year. The company accepted this proposal. The same year Pierce Mason Butler decided to withdraw some of his slave capital. Having transferred $12,315 worth of slaves to the company in 1837, Butler now withdrew eight bondsmen worth $4,850, including four whose skill and value had increased. Even when the company terminated operations and settled its obligations, the original stockholders were reimbursed merely by returning their slaves, whose offspring counted as a bonus.[83]

The Nesbitt Company's slave capital was sufficiently flexible so that in the first case the investor obtained shares by renting his slaves, utilized some of his surplus bondsmen, received company earnings, and withdrew his slaves when they had become more skillful and valuable. In the second and third instances, investors suffered little financial embarrassment and they retained appreciated slave capital when the enterprise was terminated. In each case, slave capital seemed sufficiently mobile to meet the company's needs.

Slave labor supposedly was less flexible than wage labor during commodity market fluctuations and business depressions when income dropped and labor costs had to be reduced. However, many slaveowning industries found that during such periods slave labor was as flexible as wage labor, even though whites could be dismissed and slaves could not. "The certainty of a regular and adequate supply of mining labor at reasonable prices is the surest avenue of success in coal mining," privately confided the slaveowning coal miner William Phineas Browne in 1847. "In this respect slave labor owned by the mining proprietors is greatly superior to free labor

---

[83] Financial documents and letters, Elmore Papers (LC and USC), and P. M. Butler Papers (USC).

even if the latter were as abundant as it is in Europe or in the mining districts of the North." To Browne, the purchase of slave coal miners was less expensive than paying wages to free laborers. To retain large stocks of coal during dull periods the capital required to sustain free-labor mining operations economically would amount to nearly enough to purchase Negroes, he argued. If enterprises owned slaves, on the other hand, sufficient funds could be realized from current sales to maintain full mining operations without financial embarrassments during periods of depressed market conditions. Browne also argued that slave ownership enabled enterprises to capitalize on market fluctuations. Mining companies should therefore depend mainly upon slave labor; free labor should be worked only as a "subordinate adjunct" to the regular slave force. "The employment of slave labor besides being more in harmoney with our institutions," concluded Browne after much experience, "ensures a successful business against all contingencies and will enable proprietors to pass through all disturbing crises without being sensibly affected by them." [84]

Browne's confidence in the flexibility of slave capital was confirmed by the experiences of many southern transportation enterprises. The Upper Appomattox Company of Virginia, which owned its black diggers, was able, in 1816 and again in 1835, to rent out twenty bondsmen to obtain funds to complete the work. The Roanoke Navigation Company of Virginia, which also owned Negroes, was able, in 1823, to obtain capital by either selling or renting out several slaves. During the panic of 1837, this company rented out some bondsmen for five months for $3,167; within a few months the company thereby recouped 23 per cent of its original $14,025 investment in thirty-three slaves. Of course, the company still operated the canal and owned its slaves. In 1839, the company sold half its blacks for $7,044, rented some of the remaining bondsmen to a nearby railroad, and thereafter earned additional income by hiring out slaves each winter and spring, while using them for repairs during summers.[85] Similarly, after 1827, the Slate River Company of Virginia, which owned five Negroes worth $1,900, rented out four of them at $235 per annum each. One Alabama railroad, which owned $9,575 worth of slaves, realized $2,503 annually

---

[84] W. P. Browne to G. Baker, January 30 and 31, 1857, Browne Papers (Alabama Department of Archives and History, hereafter cited AA).

[85] Reports of Upper Appomattox Company, Virginia *Board of Public Works Reports, 1816*, ii, and *1835*, 42; Reports of Roanoke Navigation Company, Virginia *Board of Public Works Reports, 1823*, 69; *1838*, 98, 100, *1839*, 125, and *1854*, 478–88.

(a 26 per cent return) by hiring them out in the 1830's. Upon the completion of the slaveowning Bayou Boeuf Navigation Works in Louisiana, the company totally reimbursed its original investors and continued to pay them dividends.[86]

Confidence in the flexibility of slave capital was also evident in the financial schemes of A. C. Caruthers, a Tennessee turnpike promoter. "We have a Charter for a Road to Trousdale's Ferry," confided Caruthers to a friend in 1838. "We will build our Road — the State takes half. The plan is devised — a few men — 8 or 10 — will take the stock — pay it all in at once — get the State Bonds — & with the fund build the Road." Proposing a clever plan of finance, the promoter concluded: [87]

> With this fund, they can buy say 300 negroes, who will do the work in one year. The interest of the $70,000 borrowed — the tools — support of hands, mechanicks & all cant cost more than $40,000. When the Road is done the $140,000 is theirs — the bond to the Directors is cancelled. The 300 Negroes are theirs — They can sell them for an advance of at least of $100 each = $30,000. The whole sale would be $140,000 original cost & $30,000 profit = in all $170,000. Out of this they must repay the $70,000 borrowed, & the $40,000 expenses in all $110,000 — leaving a clear profit of $60,000 & their road stock, which is $6,000 each partner & $10,000 in road stock. . . .

The experiences of slaveowning industries regarding the flexibility of slave capital have been confirmed by some recent studies. In South Carolina there seemed to be adequate sources of capital for industrial investment, while Texas masters converted slave capital into liquid capital, according to one historian, by selling, mortgaging, or renting out their Negroes. "At the same time that slave labor was being used as an instrument of production, that labor was also creating capital," he concludes. "It is difficult to understand how the notion became current that the slave became a frozen asset and a drain upon the capital resources of a region." [88] Of course, slave hiring was an even more flexible use of capital than slave ownership, and since demand for slaves remained high, slave capital tended to remain liquid.

---

[86] Report of Slate River Company, Virginia *Board of Public Works Report, January 24, 1828,* 23; *American Railroad Journal,* 5 (1836), 817; J. Andreassen, "Internal Improvements in Louisiana," *Louisiana Historical Quarterly,* XXX (1947), 46–47.

[87] A. C. Caruthers to W. B. Campbell, January 28, 1838, Campbell Papers (Duke), for the entire scheme.

[88] Smith, *Economic Readjustment of an Old Cotton State,* 115–34; G. R. Woolfolk, "Planter Capitalism and Slavery: the Labor Thesis," *Journal of Negro History,* XLI (1956), 103–16.

SLAVERY AND INDUSTRIAL BACKWARDNESS

Even if these findings — that slave labor in southern industries was profitable, efficient, and economically viable — are valid, it still should be explained why southern industry did not develop more rapidly. While the reasons for this are, of course, complex, an explanation seems to rest in the limitations of southern markets, the South's difficulty competing with northern and foreign producers, unfavorable balances of southern trade, and, perhaps most important, in the ability of southern agriculture to outbid industry for investment capital.

The slow development of southern industries stemmed partly from various restrictions on consumer demand. Slaveowners usually maintained their slaves at subsistence living standards, and some of the largest plantations were almost entirely self-sufficient. The poor whites lacked purchasing power because they did not produce for regional markets. Isolated from transportation facilities, yeoman farmers produced only for limited markets and had difficulty competing with more efficient planters. Moreover, the South lacked urban markets, since by 1860 only about 10 per cent of its population lived in cities, compared to the Northwest's 14 per cent and the Northeast's 36 per cent. Except for New Orleans and Baltimore, the South had only a handful of cities with populations over 15,000,[89] and many urban dwellers were slaves or free blacks whose purchasing power was minimal. Relatively few foreigners emigrated to the South, where economic opportunity was poorer and the climate sicklier than in the North. In addition, as late as 1861, the southern transportation network still primarily tied plantation districts to ports, rather than providing a well-knit system which might have increased internal consumption. Finally, the distribution of wealth, which helps determine consumption propensities, was less even in the South than in the North,[90] although the rate of growth and the level of southern income compared favorably with other sections.[91]

---

[89] Genovese, *Political Economy of Slavery*, 20, 24–25, 34, 37, fn. 13, 159–72, 185, 276–77; North, *Economic Growth*, 126, 130, 132, 166, 170, 172–76, 205–06.

[90] Genovese, *Political Economy of Slavery*, *passim*; North, *Economic Growth, passim*. For further discussion of southern income distribution, see Engerman, "Effects of Slavery," 71–97; F. L. Owsley, *Plain Folk of the Old South* (Baton Rouge, 1949), *passim*; F. Linden, "Economic Democracy in the Slave South," *Journal of Negro History*, XXXI (1946), 140–89.

[91] However uneven income distribution was *within* the South, recent comparisons of regional and national wealth for 1840 and 1860 suggest that southern income levels and rates of growth compared favorably with those of the free states. See Engerman, "Effects of Slavery," 71–97, especially fn. 35, and R. Easterlin, "Regional Income Trends, 1840–1950," in S. Harris (ed.), *American Economic History* (New York, 1961), 525–47. If one revises upward the maintenance cost of slaves (as Genovese, *Political Economy of*

Southern industries also lagged because southern manufacturers had difficulty competing in national market places. Compared to northern and foreign producers, southerners had less experience, less efficient management, smaller markets, inferior technology, poorer transportation, indirect trade routes, and, perhaps most important, smaller capital resources. Credit arrangements and unfavorable balances of trade drained plantation profits northward and permitted northern merchants increasingly to dominate the commerce in cotton, the leading export both of the South and of the nation. Imports came first through New York, rather than directly to the South, because ships were assured of more cargo on the westward passage from Europe to northern ports than to southern ones.[92] The South would have had to pay for loans and services obtained from the North in any event, but capital accumulated by northern merchants, bankers, and insurance brokers tended to be reinvested in northern industries and transportation enterprises rather than in southern ones.

Southern backwardness was not inevitable; rather, it was the result of human decisions which could have led in a different direction. After all, from the 1780's to about 1815, southern planters had been investing much of their surplus capital in industries and transportation projects. During these years, when the South sustained one-third of the nation's textile mills, southern industrial growth seemed to be paralleling that of the North.[93] After 1815, however, southern industries waned as the rapidly developing textile industry of Britain and New England demanded cotton, the invention of the cotton gin stimulated short-staple cotton cultivation, and fertile southwestern plantations yielded quick profits to investors. Southerners now began to invest more in new lands and in slave labor than in industry and internal improvements. This decision stemmed not only from the agrarian tradition and the prestige of owning real property, but also because the production of staples seemed to promise the easiest financial success. In the competition for capital, agriculture thus outbid industry.[94]

Slavery, 275–80, has done, and as I have also done above), then the size of the southern market and the demand for manufactured goods also increases.

[92] Dew, *Ironmaker to the Confederacy*, 32; North, *Economic Growth*, 122–26; Genovese, *Political Economy of Slavery*, 159–65; W. Miller, "A Note on the Importance of the Interstate Slave Trade of the Ante-Bellum South," *Journal of Political Economy*, LXXIII (1965), 181–87.

[93] Taylor, *Transportation Revolution*, chs. 1 and 10; R. Griffin, "Origins of Southern Cotton Manufacture, 1807–1816," *Cotton History Review*, I (1960), 5–12; R. Griffin, "South Carolina Homespun Company," *Business History Review*, XXXV (1961), 402–14; and Starobin, "Industrial Slavery," ch. 1 and tables.

[94] Taylor, *Transportation Revolution*, chs. 1 and 10; Eaton, *Southern Civilization*, chs. 9 and 10; Stampp, *Peculiar Institution*, 398–99.

As a result of this process, by the 1830's key slave-state industries were already a generation behind those of the free states, and they were having great difficulty competing against outsiders. By 1860, the South had only one-fifth of the nation's manufacturing establishments, and the capitalization of southern factories was well below the national coverage. Thus, as Eugene Genovese has pointed out, southerners could provide a market for goods manufactured by Northerners and foreigners, but that same market was too small to sustain southern industries on a scale large enough to be competitive.[95]

Though these factors helped inhibit southern industries, it is hard to demonstrate that slavery was the *sole* cause of industrial backwardness. Slavery was only partly to blame for the South's difficulty competing with outside manufacturers, for unfavorable patterns of trade, and for restricted consumer demand. Other factors, such as geography, topography, and climate, were at least as important as slavery in retarding southern industry. Can slavery be blamed, for example, for the natural attractiveness of farming in a fertile region? Was slavery responsible for the South's natural waterway system, which delayed railroad development? It therefore seems doubtful that slavery alone decisively retarded the industrialization of the South.[96]

However, it must also be understood that, in the long run, extensive industrialization would have been difficult, if not impossible, under a rigid slave system. To develop according to the British or northern pattern, the rural population of the South would have had to be released from the land to create a supply of factory workers and urban consumers. Greater investment in education for skills and greater steps toward a more flexible wage labor system would have been necessary than were possible in a slaveholding society. Changes in the southern political structure permitting industrialists, mechanics, and free workers greater participation in decision-making processes affecting economic development were prerequisite to any far-reaching program of modernization.

On the other hand, even if slavery is theoretically and practically incompatible in the long run with full industrialization, the point

---

[95] Genovese, *Political Economy of Slavery*, 165–66.
[96] Stampp, *Peculiar Institution*, 397. Though limited markets restricted southern industrialization, the extent of this phenomenon should not be exaggerated. Plantation self-sufficiency, slow urbanization, and other market factors did not restrict consumption entirely. Recent studies have also shown that plantation and slave consumption were higher than once thought, and that some southern businessmen found substantial markets outside of the slave states. See Genovese, *Political Economy of Slavery*, 25, 159–62, 170, 185, and 276–77; and North, *Economic Growth*, 130.

at which this inconsistency would manifest itself had, apparently, not yet been reached between 1790 and 1861. Tensions were present in southern society, to be sure, but southerners were not yet foundering upon their domestic contradictions. The time when slavery would be absolutely detrimental to southern industries remained quite far off, and the development of slave-based industries was still necessary and desirable, given the imperatives of the proslavery ideology and the political realities of the period leading up to the Civil War.

Facilities for the Construction of War Vessels in the Confederacy

By WILLIAM N. STILL, JR.

WHEN WAR BROKE OUT IN 1861 THE CONFEDERATE STATES DID NOT then have the facilities necessary to construct or equip vessels of war. Shipyards of adequate size were scarce as were plants for the manufacture of iron plate, marine machinery, and ordnance. Secretary of the Navy Stephen R. Mallory and his colleagues were well aware of the deficiencies but clung to the optimistic belief that adequate facilities could be established.[1]

Of the ten navy yards operated by the United States government in 1860, two were in the South: one at Norfolk, the other at Pensacola.[2] The smaller was the Pensacola yard which was primarily a coaling and refitting station. Nevertheless, two large sloops (the *Pensacola* of over 2,000 tons and the *Seminole* of 800 tons) were built there during the 1850's. At the Gosport Navy Yard in Norfolk, thirteen warships including four screw steamers had come off the ways before 1861.[3] Both yards were obviously important in Confederate plans for the construction of war vessels.

Of privately owned shipyards in the South, there were a few large ones and many small yards in operation. The exact number is difficult to determine because of frequent abandonment and new construction. Most of them were rather simple affairs—a

---

[1] *Official Records of the Union and Confederate Navies in the War of the Rebellion* (30 vols., Washington, 1894-1927), ser. 2, II, 77; Jefferson Davis to Congress, February 25, 1862, Dunbar Rowland (ed.), *Jefferson Davis, Constitutionalist: His Letters, Papers, and Speeches* (10 vols., Jackson, Miss., 1923), V, 205.

[2] In 1844, Congress voted $100,000 to equip a navy yard at Memphis. The yard, however, was never a success, and in the 1850's the buildings were sold and it ceased to exist. Walter Chandler, "The Memphis Navy Yard," *West Tennessee Historical Society Papers*, I (1947), 70-71. Both New Orleans and Charleston lobbied for years trying to obtain a yard but neither had been successful.

[3] Two frigates built at Norfolk, the *Roanoke* and *Colorado*, received their engines, boilers, and other machinery (excepting the shafts) from the Tredegar Iron Works. A revenue cutter, the *Polk*, was built from the keel up at the Richmond works. Kathleen Bruce, *Virginia Iron Manufacture in the Slave Era* (New York, 1931), 262.

THE JOURNAL OF SOUTHERN HISTORY, 1965, pp. 285-304.

small clear area on a beach, or the bank of a river, creek, or inlet. They required only water deep enough for launching, and timber and other materials nearby. According to the 1860 census on manufacturing, there were thirty-six yards in the states that were to form the Confederacy.[4] But the census figure is questionable. There were no yards listed in Mississippi, for example, yet there were small establishments at Pearlington and Gainesville on the Mississippi River.[5] There were none listed for Tennessee or Texas, although in the 1850's there were at least eight steam merchant vessels built in Tennessee and ten in Texas.[6] One writer in 1850 estimated that the South, including Maryland, Kentucky, and the area that became West Virginia, had 145 shipyards.[7] Although there was a decline in Southern shipbuilding in the decade of the 1850's, the total number of Southern yards lay somewhere between the estimated 1850 figure of 145 and the 1860 census figure of 36.[8]

In the United States as a whole between 1849 and 1858, slightly over 8,000 vessels, of wood or iron, were constructed. Of this number approximately 1,600 were built in the South.[9] Steam vessels were constructed in all the Southern states, and the localities which were construction centers before the war became the first centers of shipbuilding in the Confederacy. The most important centers were Norfolk, Charleston, Savannah, Mobile, and New Orleans—coastal towns which were also the most im-

[4] U. S. Census Office, Eighth Census, 1860, *Manufactures of the United States in 1860* (Washington, 1865), 716-18. This includes the area that became West Virginia. Wheeling was the most important boat-building center (in numbers built) in the South prior to the war.

[5] New Orleans *Picayune*, July 17, August 1, 1861.

[6] Compiled from William M. Lytle, *Merchant Steam Vessels of the United States, 1807-1868* . . . , edited by Forrest R. Holdcamper (Mystic, Conn., 1952). Lytle was deputy commissioner of navigation in the United States Commerce Department from 1927 to 1932. The list was compiled from abstracts of registers, licenses, and enrollments issued to various vessels and the copies of the actual documents in the Bureau of Navigation.

[7] Thomas P. Kettell, *Southern Wealth and Northern Profits* (New York, 1860), 85.

[8] The decline of shipbuilding in the South during the 1850's was offset, at least partly, by the action of several states, including Alabama and Louisiana, in granting state bounties for tonnage built within their borders. Victor S. Clark, *History of Manufactures in the United States* (3 vols., New York, 1929), I, 470.

[9] This number includes ships, brigs, schooners, sloops, canal boats, and steamers. An examination of the Lytle list indicates that approximately 370 merchant steam vessels were constructed before the war in the states that comprised the Confederacy.

portant shipping ports.¹⁰ Nevertheless other smaller towns, not only seaports but river towns, became important construction centers, particularly after the capture of Norfolk and New Orleans in the spring and summer of 1862. Mallory realized the unsuitability of some of the new locations for building large seagoing steamers, but by that time the strategy of defense was dominant. What he wanted and needed were shallow-draft vessels capable of navigating in the shoal waters of the South. These vessels, including ironclads, should be flat-bottomed, of slight draft and simple to construct—the type of vessels that could be built in small shipyards of limited means.

Related shipbuilding facilities—those for the manufacture of iron, marine machinery, and ordnance—were in short supply in the South when war broke out. There was little problem if ships were to be made of wood, since the South had an adequate supply of sawmills; nearly every community or large plantation possessed a mill for cutting timber. But facilities for the manufacture of iron were limited. In 1860 the South had ninety-six foundries and eighty-two rolling mills or other establishments that produced bar, sheet, and railroad iron.¹¹ Most of these establishments were small-scale. There were only eleven rolling mills of any size, five in Virginia, three in South Carolina, one in Georgia, and two in Tennessee. Furthermore, at the outbreak of war, none of these were able to roll two-inch plate. Virginia was by far the most important Southern state in the manufacture of iron products, but in contrast with the leading state in the United States, Pennsylvania, its production was quite small.¹²

The availability of adequate facilities for manufacturing ship machinery is difficult to determine. Shortly after the outbreak of

---

¹⁰ Henry Hall, "Report on the Ship-building Industry of the United States," U. S. Census Office, *Tenth Census, 1880* (Washington, 1882), VIII, 130; *Official Records of the Union and Confederate Navies*, ser. 2, I, 500, 507; Ernest M. Lander, Jr., "Charleston: Manufacturing Center of the Old South," *Journal of Southern History*, XXVI (August 1960), 341. For Mobile see J. H. Scruggs, Jr., *Alabama Steamboats, 1819-1869* (Birmingham, 1953).

¹¹ At the same time the nation as a whole had 1,412 foundries and 256 rolling mills. U. S. Census Office, *Manufactures in 1860*, 716-18.

¹² Two-inch rolled iron plate was standard on Confederate ironclads. In 1860 Virginia produced $1,666,885 worth of iron products, while Pennsylvania was producing over fifteen million dollars worth. For this and other comparisons see Clement Eaton, *A History of the Old South* (New York, 1949), 433, and Richard N. Current, "God and the Strongest Battalions," in David Donald (ed.), *Why the North Won the Civil War* (Baton Rouge, 1960), 15-23. See also J. P. Lesley, *The Iron Manufacturer's Guide to the Furnaces, Forges, and Rolling Mills of the United States* . . . (New York, 1859), 747.

hostilities, Mallory wrote that Tennessee was the only state that had factories capable of producing complete engines.[13] But he seems to have been mistaken. Of the steamships built in Norfolk before the war, two of them had engines (complete except for their shafts) made in the Tredegar Iron Works in Richmond. In addition, there were by 1860 nearly a hundred shops, large and small, which had built and repaired steam engines.[14] The census figures, unfortunately, do not indicate the type of machinery. Other sources make clear, however, that there were a number of large foundries in the South making marine engines, such as the Nobles Foundry of Rome, Georgia, the Leeds Company and the Clarke Foundry of New Orleans, Skates and Company of Mobile, and the Shockoe Foundry of Richmond.[15] There were also machine shops and foundries in the river towns and seaports where shipping was a major industry. Nevertheless the lack of suitable facilities for producing marine machinery proved to be a serious weakness in the Confederate shipbuilding program.

Ordnance facilities were nonexistent in Southern states. Prior to the war nearly all naval ordnance used by the United States navy came from the naval gun factory in Washington. There is no evidence to indicate that naval guns (or any cannon, for that matter) were manufactured in the states that were to form the Confederacy. There were two small powder mills, but no facilities for manufacturing shot, shell, and ordnance stores.

Thus at the outbreak of the Civil War, although there were a number of facilities in the Confederate states that could be used to provide a navy, the value or importance of the facilities was not uniform. For example, a potentially adequate number of shipyards were available, but there was a complete absence of certain essential facilities for the outfitting of warships such as ordnance and ropewalks. At the same time the magnitude of the problem of building war vessels in the South was increased be-

[13] *Official Records of the Union and Confederate Navies*, ser. 2, II, 77.
[14] The 1860 census lists 115 establishments in the South out of a nationwide total of 1,173 steam engine manufacturing firms. U. S. Census Office, *Manufactures in 1860*, 716-18. See also J. Leander Bishop, *A History of American Manufactures from 1680 to 1860* (3 vols., Philadelphia, 1868), II, 460.
[15] Ethel Armes, *The Story of Coal and Iron in Alabama* (Birmingham, 1910), 185; Papers of Talbott & Sons, Richmond, Virginia (Virginia Historical Society); Mobile *Advertiser and Register*, May 30, 1850; *Official Records of the Union and Confederate Navies*, ser. 2, I, 503-504. The making of marine machinery was a relatively recent industry in the South. Prior to 1850 the East and Midwest had a monopoly. Clark, *History of Manufactures*, I, 508; Louis C. Hunter, *Steamboats on the Western Rivers* (Cambridge, 1949), 105-109.

cause of the obvious interdependence of the naval industries. The Confederacy might develop adequate plants for casting guns, rolling plate, and manufacturing machinery, but they were useless unless sufficient iron was available. And the various naval establishments were to compete with each other as well as with other industries throughout the war for raw materials, transportation, and labor. This competition might not have proven so disastrous if time were available, but the Confederacy was born in war and with an immediate need to provide a navy to fight it.

When the Confederate Navy Department was created in February 1861, it had neither ships nor the necessary facilities to begin immediate construction and fitting out of war vessels. For this reason the naval secretary concentrated in the early months of the war on purchasing the type of vessels that could be quickly converted into warships. The acquisition of the huge yard at Norfolk gave the Confederate navy its first important establishment. Mallory's initial efforts to locate additional facilities were on the whole unsuccessful; and in fact, with the exception of a few wooden gunboats contracted for at New Orleans, he failed to obtain supplemental construction yards until August 1861.[16]

In the old United States navy all yards came under the Bureau of Yards and Docks, and ship construction was the responsibility of the Bureau of Construction, Equipment, and Repair. The senior naval officer in charge of the latter bureau was responsible for initiating the navy's shipbuilding program as laid out by the secretary of the navy and approved by Congress. When the Confederate Navy Department was organized, four bureaus were set up,[17] but there was no bureau of construction, equipment, and repair or its equivalent. In fact, the Confederate naval secretary retained direct control of the navy's shipbuilding program throughout the war. In 1862 semiautonomous positions were created to handle specific aspects of the program; William P. Williamson was appointed engineer-in-chief and John L. Porter was appointed chief naval constructor. These positions were basically administrative, however; Mallory continued to initiate ship construction. The details of construction and the supervision of the program in the various yards, both private and public, were left

[16] For the early development in the Confederate shipbuilding program see William N. Still, Jr., "The Construction and Fitting Out of Ironclad Vessels-of-War Within the Confederacy" (unpublished Ph.D. dissertation, University of Alabama, 1964), especially the first three chapters.

[17] The offices of Orders and Detail, Ordnance and Hydrography, Provisions and Clothing, and Medicine and Surgery.

to Porter and to the "acting constructors" under him. Porter and his assistants, William A. Graves and Joseph Pierce, made frequent trips to inspect construction in progress and to make recommendations.[18]

Mallory followed three different plans in his shipbuilding program: (1) ships were constructed in navy yards directly under the supervision of the department, (2) agents of the department directed the construction, or (3) ships were built in private yards under contract. A large majority of the vessels constructed (or at least laid down) in the Confederacy were built under contract.[19] The reason for this was not so much necessity, as some writers have emphasized, as it was a reflection of the prevailing economic philosophy of the Confederacy. The Confederate government was never interested in developing shipbuilding or other industries of its own. It tried many ways (bonuses, government subsidization of new industries, and advances) to encourage private industry. The government, nevertheless, did attempt to mobilize industry for wartime production by controlling profits, labor, and transportation.[20] Shipbuilding under contract was, generally speaking, controlled throughout the war. In a few instances ships under construction were taken over completely by the department, but, for the most part, naval officers were ordered to assist the contractors in fulfilling their obligations.

The shipbuilding program got under way early, with wooden vessels and ironclads under construction in both government and private shipyards. Most of the early contracts were with experi-

[18] Travel vouchers and copies of orders of Porter, Graves, and Pierce can be found in Section BZ, Personnel Records, Naval History Division, Department of the Navy (National Archives Building, Washington). See also *Official Records of the Union and Confederate Navies*, ser. 2, II, 272. Porter wrote that "the Secretary refers most of the matter concerning the building of vessels, buying materials, etc. to me." Porter to Reverend Moore, November 4, 1861, in John L. Porter Papers (Confederate Museum, Richmond).

[19] Records are inadequate, but apparently the *Virginia, Virginia II, Fredericksburg,* and *Richmond,* all of the James River squadron, were built either at Norfolk or at Rocketts, the navy yard at Richmond. In South Carolina, the *Pee Dee* was built at a navy yard (Mars Bluff), while the *Jackson* was constructed at the Columbus (Georgia) Naval Iron Works. The brothers Asa and Nelson Tift acted as agents of the department in building the *Mississippi* (New Orleans) and converting the *Fingal* into the *Atlanta* (Savannah); and Thomas Weldon and John McFarland were appointed agents of the department to contract for ironclads on the Yazoo River.

[20] Wilfred B. Yearns, *The Confederate Congress* (Athens, 1960), 127-28; Frank E. Vandiver, *Rebel Brass: The Confederate Command System* (Baton Rouge, 1956), 65; Charles W. Ramsdell, "The Control of Manufacturing by the Confederate Government," *Mississippi Valley Historical Review*, VIII (December 1921), 249.

enced builders who already had the necessary facilities. Arrangements were made to construct or convert vessels at New Orleans, Memphis, Nashville, Mobile, Jacksonville, Savannah, Charleston, Wilmington, and Norfolk. Ships were also laid down directly under the supervision of the Navy Department at yards in Norfolk, Pensacola, and New Orleans.[21] When Matthew Fontaine Maury's plan to build one hundred small wooden gunboats was adopted by the Confederate Congress, eighteen of these vessels were laid down at various localities, including yards on the Pamunkey, Rappahannock, and York rivers in Virginia. By the end of the first year of the war there were at least eighteen yards building vessels for the navy.

In the spring and summer of 1862 a significant change occurred in the location of Confederate naval establishments. Up to that time, most naval industries were concentrated in the large shipping ports such as Norfolk and New Orleans, but disaster overtook many of these localities in 1862. Nashville was lost in February, Jacksonville in March, New Orleans in April, Pensacola and Norfolk in May, and Memphis in June. As a result of these setbacks, the decision was made to locate naval facilities in the interior whenever possible.[22] To do so provided more security from attack, but it had a serious disadvantage—decentralization. There was no one point where everything needed to construct and outfit a vessel was located. Shipyards were in various localities, ordnance stores and laboratories in others, and foundries, machine shops, iron works, and ropewalks were in still other locations.[23] Transportation was obviously essential, but transportation within the Confederacy, particularly over railroads, was never adequate; and as the war progressed, chaos increasingly characterized railroad operations, seriously affecting

---

[21] An effort was also made to build gunboats at Elizabeth City, North Carolina, and Cerro Gordo, Tennessee, but the localities had to be evacuated before much could be accomplished.

[22] James D. Bulloch, *The Secret Service of the Confederate States in Europe* (2 vols., London, 1883), II, 209-10; Victor Ernst Von Sheliha, *A Treatise on Coast Defense* (London, 1868), 100.

[23] On October 29, 1862, the Shelby Iron Company was ordered to ship twenty-five tons of iron ore to the Columbus (Georgia) Naval Iron Works. The ore was taken to Columbiana, Alabama, by wagon where it was loaded in cars and sent by rail to Selma, transferred to riverboat and carried to Montgomery, loaded back on cars and sent to Columbus. At Columbus this ore was used in the building of machinery for two ironclads under construction at Selma. In this instance ore had been sent to one locality and the manufactured product returned to the vicinity of the original shipping point. Colin J. McRae to Jones, n.d., in Shelby Iron Company Papers (University of Alabama Library).

the naval building program. A majority of the railroads were small fragmented lines with different gauges; there was a chronic shortage of operating equipment and rails which continued to deteriorate because of no replacement; skilled labor was scarce; and the competition between the Confederate government and the various state governments for use of the roads all contributed to the chaos. Although the Confederate government gradually monopolized the railroads, the navy found its use curtailed because of army control. Mallory, naval officers, builders, and contractors constantly complained with some justification that their requirements were ignored by the army. For example, Captain William F. Lynch, CSN, reported to the secretary of the navy that "Fourteen car loads of plate iron arrived last evening, and for a week past we have had two car loads waiting transportation to Kinston and Halifax [North Carolina]. The whole rolling capacity of the road, except passenger trains, has been monopolized by the army, and I fear the completion of the gun boats at those places will be delayed." The note was passed on to the secretary of war who replied that "at present the food and forage necessary for our armies in the field demand our entire transportation."[24] This is one illustration of the inadequate logistical cooperation which continued to trouble the navy throughout the war; the destruction of a great many ironclads and other vessels while they were still on the ways was at least partially a result of this difficulty.[25]

The new interior sites (some of which were not connected by railroad lines) included Richmond; Edward's Ferry and Whitehall, North Carolina; Mars Bluff, South Carolina; Saffold and Columbus, Georgia; Yazoo City, Mississippi; Selma, Montgomery, and Oven Bluff, Alabama; and Shreveport, Louisiana. At Richmond the yard known as Rocketts was chosen as a navy yard. Here three ironclads were completed and others were under construction when the war ended. In North Carolina, Gilbert Elliott and William P. Martin received contracts to build two ironclads and a wooden gunboat; only one of these, the *Al-*

[24] Seddon to Mallory, March 18, 1864, in Secretary of War Letterbooks, Record Group 109 (National Archives).

[25] The significant role of railway transportation during the war is well described in two monographs: Robert C. Black, *The Railroads of the Confederacy* (Chapel Hill, 1952), and George E. Turner, *Victory Rode the Rails: The Strategic Place of the Railroads in the Civil War* (Indianapolis, 1953). There are a number of letters in the Shelby Iron Company Papers and in the official papers and letterbooks of Governor Zebulon B. Vance of North Carolina (North Carolina Department of Archives and History) about the navy and the transportation problem.

bemarle, constructed at Edward's Ferry, reached operational status. A sister ship, the *Neuse*, was completed at Whitehall. Wooden gunboats were laid down at Mars Bluff, South Carolina, and Saffold, Georgia, while the ironclad *Jackson* [*Muscogee*] was finished at Columbus, Georgia. Yazoo City became the site of a yard, not by design but because the unfinished *Arkansas* was towed there from a shipyard near Memphis. Here she was taken over by the department and finished under a naval officer's supervision. Later three other warships, including two ironclads, were laid down there, but they were never launched.[26] At Oven Bluff, Alabama, some sixty miles up the Tombigbee River from Mobile, the wooden hulls of three ironclads were built, but the incompetence of the contractors plus the location of the yard near a malaria-ridden swamp prevented the completion of the vessels. They were later towed to Mobile, but were still without armor and machinery when the city surrendered in the spring of 1865. The keels of four ironclads were laid down at Selma but only three (*Tuscaloosa, Huntsville,* and *Tennessee*) were completed. At these inland yards a total of nine ironclads and one wooden gunboat were commissioned; at least ten others, including eight ironclads, were destroyed before becoming operational.[27]

[26] They were destroyed in May 1863 when a Federal expedition pushed up the river as far as Yazoo City. Porter to Welles, May 24, 1863, in David D. Porter Papers (Library of Congress).

[27] There is very little information on most of these yards. Scattered information, however, can be found under the various categories (yards, ship construction, etc.) in the Subject File, Record Group 45 (National Archives). For the contracts of vessels built at private yards see the appendix to *Report of Evidence Taken Before a Joint Special Committee of Both Houses of the Confederate Congress to Investigate the Affairs of the Navy Department* (Richmond, 1863). For the yards in North Carolina see the Confederate Shipyards at Wilmington Marker File (North Carolina Department of Archives and History); Richard Iobst, "Ram Built Under Hardships," Kinston (N. C.) *News*, November 25, 1861; William F. Martin Papers (Southern Historical Collection, University of North Carolina); and Robert Minor to Mallory, February 16, 24, 1864, in Minor Family Papers (Virginia Historical Society). For Yazoo City see Porter to Welles, May 24, 1863, in David D. Porter Papers, and Robert M. Bowman, "Yazoo County in the Civil War," *Publications of the Mississippi Historical Society*, VII (1903), 65. On Selma see William N. Still, Jr., "Selma and the Confederate States Navy," *Alabama Review*, XV (January 1962), 19-37; and on Montgomery and Oven Bluff see Admiral Franklin Buchanan Letterbook, 1861-1863 (Southern Historical Collection, University of North Carolina). Information on the navy yard at Mars Bluff can be found in the Edward J. Means Letterbook (Louisiana State University Library). Apparently very few wooden gunboats were laid down in the inland yards. The primary reason was the decision of the Confederate Navy Department to concentrate on ironclad construction after the spring of 1862. See William N. Still, Jr., "Confederate Naval Policy and the Ironclad," *Civil War History*, IX (June 1963), 153.

The area which presented the most problems in establishing naval facilities was the trans-Mississippi West. With the exception of a yard at Shreveport, where the ironclad *Missouri* was built, and several small yards in Texas, where wooden gunboats were converted from merchant steamers, little was accomplished. President Davis, who usually showed little interest in naval matters, became concerned about the lack of naval vessels in the West. He wrote General Edmund Kirby Smith to investigate the possibility of constructing gunboats and marine machinery at Little Rock, Arkansas. In 1863 a naval constructor sent to Texas to ascertain whether ironclads could be built there reported that materials were not available.[28]

The availability of materials anywhere in the Confederacy, and in particular the availability of iron for armor plate, presented a problem that continuously affected the shipbuilding program. Four out of every five ships built in the Confederacy after the spring of 1862 were ironclads, and the acquisition of armor plate for these vessels was a chronic difficulty. At the outset of the war there were no rolling mills within the South capable of producing two-inch plate, and during the course of the war only three were developed. Tredegar Iron Works and the Scofield & Markham Iron Works converted their machinery in 1862, and the Shelby Iron Company of Columbiana, Alabama, followed in 1863. Attempts were made by the department to increase the number of rolling mills capable of producing two-inch plate, but without success.[29]

In July 1861, the president of Tredegar agreed to roll the armor plate for the *Virginia*. The first contract called for one-inch plate, but after the navy conducted experiments on Jamestown Island the rollers were changed to produce two-inch armor.[30] From then until hostilities ended, Tredegar rolled plate for the navy. The *Virginia, Virginia II, Fredericksburg, Richmond,* and several of the North Carolina ironclads received their armor from the Richmond plant.

The next largest rolling mill in the South was the Atlanta Roll-

---

[28] Carter to Mallory, April 4, 18, 1864, and January 18, 1865, in J. H. Carter Letterbook, Record Group 45 (National Archives).

[29] Bruce, *Virginia Iron Manufacture*, 383.

[30] Anderson to D. N. Ingraham, September 15, 1861, in Letterbook, 1861-1862, Tredegar Rolling Mill and Foundry Papers (Virginia State Library). See also Brooke to Jones, July 10, 1874, in Construction at Norfolk file, Subject File, Record Group 45 (National Archives), and *Official Records of the Union and Confederate Navies*, ser. 2, I, 786.

ing Mill, owned and operated by Scofield & Markham. It had been established originally to manufacture rails for railroads, but in November 1861 the Scofield & Markham works, under pressure from the government, agreed to convert their machinery to roll plate for the navy.[31] The mill continued to operate until Atlanta was burned in 1864 and provided the armor for at least ten ironclads.

The Shelby Iron Company in Columbiana was the only other establishment to roll armor plate for the navy. On September 1, 1862, Flag Officer William F. Lynch wrote the president of Shelby, Albert T. Jones, about producing plate for the navy. Jones agreed to do so, although he was under contract with the War Department at that time "to deliver the entire proceeds of the work [sic] up to 12,000 tons per year."[32] Rolled plate, however, was more profitable than pig or bar iron, and Jones apparently believed that he could fulfill a naval order without breaking his contract with the ordnance bureau of the War Department.[33] After discussions held in Richmond, Jones received a contract on September 27 for four hundred tons of two-inch plate. The agreement stipulated that delivery was to be on or before December 1, 1862.[34]

The company then found itself in difficulty. In the first place Shelby was not prepared to roll two-inch plate, and additional machinery had to be installed. Jones promised to obtain the necessary equipment and to begin operations by the middle of December—a two-week delay. He also became involved in a dispute with the Confederate iron agent for Alabama, Colin J. McRae. McRae was not informed of the order Lynch had negotiated, and found out about it only when his requisitions for iron were not filled. On November 5 he wrote to Jones angrily protesting the Lynch contract: "You are not the proper party to decide the necessity of the government. If such authority rests anywhere outside of the Department charged with making the contracts, it

[31] Elizabeth Bowlby, "The Role of Atlanta During the War Between the States," *Atlanta Historical Quarterly*, V (July 1940), 187; Stephen S. Mitchell, "Atlanta, the Industrial Heart of the Confederacy," *Atlanta Historical Bulletin*, I (May 1930), 22.

[32] Jones to Lynch, September 4, 1862, in Shelby Iron Company Papers.

[33] A new furnace was under construction and Jones believed that the amount of iron could be considerably increased when this furnace was completed. Shelby, like Tredegar, owned its own mines and supplied the Confederate government not only with manufactured iron but with pig iron.

[34] George Minor to Jones, September 27, 1862, in Shelby Iron Company Papers. On September 30 the amount was extended to sixteen hundred tons.

would in this instance be with me as I have been appointed by the proper authority."[35] Two days later he wrote to Colonel Josiah Gorgas, commanding the army's ordnance bureau: "I have been made to play a most ridiculous part in the business and I again ask to be relieved from acting as the agent of the Department."[36] Evidently the trouble was cleared, for the Lynch order retained its priority, and McRae continued as iron agent until early in 1863 when he left on a mission to Europe.

The sixteen hundred tons of plate ordered from Shelby for ironclads under construction at Yazoo City were never delivered. Shelby was still unable to roll the iron when the contract date (December 14) passed. Two days after Christmas, Commander Ebenezer Farrand, superintending the building of two ironclads at Selma, informed an agent of the company that the Selma vessels were nearly ready for armor and that they would have priority over the Yazoo ships. The first plates were rolled early in March, and on the thirteenth Major William R. Hunt of the Nitre and Mining Bureau ordered all orders suspended until the plate for the Selma boats was completed. The suspension order included the Lynch contract.[37] Shelby continued to supply the navy with iron, although little armor plate was rolled because most of the iron was used for casting guns.

The first iron contracts made by the Confederate government called for delivery to be at a given date, but the agreement was rarely kept. Scarcity of iron made it impossible to carry out. By 1863 a new iron policy was adopted in the Confederacy whereby contractors and shipbuilders were to provide their own iron ore, scrap iron, or railroad iron to be manufactured into plate. When this, too, proved unworkable, the department began sending agents throughout the Confederacy searching for iron, and adopted a policy requiring the mills to roll into plate all of the iron the navy was able to obtain.[38] Nevertheless the production of plate iron was never adequate. On March 31, 1863, Tredegar

[35] McRae to Jones, November 5, 1862, *ibid*.
[36] McRae to Gorgas, in Colin J. McRae Papers (Alabama Department of Archives and History).
[37] Hunt to J. A. Wall, March 14, 1863, in Shelby Iron Company Papers; Frank E. Vandiver, "The Shelby Iron Works in the Civil War: A Study of Confederate Industry," *Alabama Review*, I (April 1948), 121-22. See also Joyce Jackson, "History of the Shelby Iron Company, 1862-1868" (unpublished M.A. thesis, University of Alabama, 1948). The Yazoo vessels were destroyed a few weeks later.
[38] Voucher, in Naval Activities at Atlanta file, Subject File, Record Group 45 (National Archives); Chief Engineer Jackson to Catesby ap R. Jones, February 11, 1864, in Naval Activities at Selma file, *ibid*.

temporarily closed down its mill because of an iron shortage. In November 1864, Mallory reported to the President that five ironclads were still unfinished because of insufficient iron for armor. The chief naval constructor in his report of that month stated that there were twelve ironclads awaiting plating. He also made it clear that the rolling mills were available, "but the material is not on hand."[39]

A related "iron industry" vital to the naval shipbuilding program was marine machinery. In August 1862, Mallory asked chief engineer Williamson why progress was not being made in developing facilities for manufacturing marine machinery. Williamson's reply listed everything from insufficient labor to the scarcity of tools, but ended on the encouraging note that the situation would improve in the near future.[40] His report reflected the unsatisfactory condition of the industry at that time. Some effort was made to stimulate the manufacture of ship machinery, but little was accomplished. The building of steam engines and other machinery required experienced and skilled mechanics, and very few of them were available; in addition, the tools and equipment needed to produce the engine parts had to be made or brought in through the blockade. Preparations were made to build marine machinery at the Norfolk yard, but the preparations were still incomplete when the evacuation took place. Mallory contracted also with various firms in New Orleans, Charleston, and other places, but the results were unsatisfactory.[41]

Some progress was made in 1862. The Shockoe Foundry in Richmond was leased by the navy in February. This firm, which had equipment including a fitting shop, turning shop, foundry, and boiler shop, had made steam machinery before the war. It was now leased to build machinery for small wooden gunboats. Later, after the construction of these vessels was dropped, the foundry produced power plants for ironclads.[42]

[39] *Official Records of the Union and Confederate Navies*, ser. 2, II, 753.
[40] *Ibid.*, 240.
[41] Machine shops in Savannah and Charleston built the machinery for the first vessels constructed in those localities. See miscellaneous vouchers, in Construction at Savannah and Charleston files, Subject File, Record Group 45 (National Archives).
[42] Contract signed by Maury with Talbott & Sons, February 8, 1862, in Talbott & Sons Papers; Report of Commander William Radford, April 25, 1865, in Construction at Richmond file, Subject File, Record Group 45 (National Archives). Naval officers abroad also contracted for an undetermined number of engines for vessels constructed in the Confederacy. See *Official Records of the Union and Confederate Navies*, ser. 1, XIII, 644; Bulloch, *Secret Service*, I. 383. and II. 229-30;

All these measures took time. Meanwhile, ships coming off the ways needed machinery, and the department had no alternative but to transfer the engines and boilers from river boats and other vessels to warships. This was considered a temporary measure; the plans were to replace the old power plants with new ones when they became available. Unfortunately, the movement inland of shipbuilding facilities also included machine shops, and this delayed the industry even further.

Charlotte, North Carolina, became the location of an important marine engineering works. In fact it was the only navy yard (after the fall of Norfolk) with the equipment to do heavy forging. A Nasmyth steam hammer, used for making propeller shafts and other heavy work, was obtained by the navy from Tredegar at the war's outbreak and was installed at Charlotte.[43] Complete engines were not manufactured by the Charlotte works, but many of the parts and nearly all of the shafts, propellers, and anchors were made there.

Late in March 1862, an officer attached to the navy's Office of Ordnance and Hydrography was ordered to Columbus, Georgia, to investigate that city's "general fitness for the location of extensive government works."[44] Columbus had been an important industrial center before the war. One of the largest activities there was the Columbus Iron Works, a firm engaged in the manufacture of steam machinery. On the opening of hostilities this company contracted with the War Department to cast small cannon, but in 1862 the Navy Department acquired control of the works. At first the navy planned to continue ordnance work, but in the summer of 1862 it was decided to use the facility to manufacture marine machinery. The establishment was then transferred to the engineering bureau and placed under the command of chief engineer James H. Warner, an experienced and capable officer. Under his direction extensive work was done in expanding the plant, additional land was leased from the city, a small rolling mill and boiler plant were erected, tools and

---

Samuel B. Thompson, *Confederate Purchasing Operations Abroad* (Chapel Hill, 1935), 42.

[43] Anderson to Ingraham, August 30, 1861, in Tredegar Rolling Mill and Foundry Papers; Bruce, *Virginia Iron Manufacture*, 349; Brooke to Mallory, memorandum in Letterbook of Office of Ordnance and Hydrography, Record Group 109 (National Archives); Mallory to the President, January 5, 1863, in Jefferson Davis Papers (Duke University Library). It was apparently the only tool of the kind in the Confederacy.

[44] A. B. Fairfax to Julian Fairfax, in Naval Facilities at Charlotte file, Subject File, Record Group 45 (National Archives).

machinery designed by Warner were made at the works, and additional skilled labor including a large number of Negroes was hired.⁴⁵ By October the engines for several vessels were in various stages of completion in spite of the fact that the facility was still not operating on a full-scale basis. The works continued to expand until they became the most important plant in the Confederacy for the manufacture of marine machinery.

The navy's only ropewalk was located at Petersburg, Virginia. In February 1865 a bill was passed by Congress appropriating $75,000 to remove it to another locality. The final collapse of the Confederacy in the spring evidently prevented this move.

The last major part of the shipbuilding program was the manufacture and procurement of naval ordnance. Of all the bureaus under the direction of Secretary of the Navy Mallory, that of Ordnance and Hydrography probably had the widest range of responsibilities. In addition to the procurement of ordnance and ordnance stores, this office was the one responsible for obtaining iron and other metals needed by the navy. Three officers were successively in charge of the bureau. The first, Duncan M. Ingraham, held office only for a short period. He relinquished it to George Minor shortly after Virginia entered the Confederacy, and Minor was replaced by John M. Brooke in March 1863.

Minor deserves the credit for organizing the bureau and creating the various naval ordnance establishments throughout the Confederacy. He was a quiet, unassuming man—a commander in the old navy—whose obvious abilities have generally been overlooked, possibly because of the fame and flamboyant aggressiveness of his successor. Minor assumed control near the end of April or the beginning of May 1861.⁴⁶ Little or nothing had been done by his predecessor toward organizing the office or establishing ordnance facilities, partly because the navy at that time had more than enough ordnance for its few ships. The evacuation of Norfolk by the Federals on April 20, 1861, resulted in the Confederate navy's obtaining over a thousand pieces of heavy ordnance, including some three hundred Dahlgrens of the latest

---

⁴⁵ Minutes of the City Council of Columbus, Georgia, 1861-1865 (Office of the Probate Judge, City Hall, Columbus); Diffee W. Standard, *Columbus, Georgia, in the Confederacy* . . . (New York, 1954), 43. The rolling mill had to be rebuilt after fire destroyed it on October 24, 1862. Columbus (Ga.) *Inquirer*, October 24, 27, 1862.

⁴⁶ The Minor Family Papers in the Virginia Historical Society include a number of George Minor's letters for the war years. The collection also includes a large number of letters to and from Robert Minor, a brother of George.

type. By the middle of July over 500 of these guns had been shipped to all parts of the Confederacy, and until the Brooke guns (the type developed and most used by the Confederate navy) began to appear, this ordnance was the principal source for arming Confederate warships.

Two other sources, used particularly after the first year of the war, were imports from abroad and manufacture at home. A small number of guns were brought in by blockade runners for the navy, but few were actually mounted on Confederate war vessels. From the beginning it was apparent that the Confederacy would have to develop ordnance manufacture. Despite the fact that gun casting was a relatively unknown industry in the South prior to the war, Southerners proved to be quite successful in gun production.

The policy of contracting, whenever possible, for materials of war, was applied to ordnance and ordnance stores as well as to other war industries. Mallory reported to Congress that, "appreciating the importance of fostering private efforts to manufacture heavy guns for the Navy . . . a contract has been made with two establishments at New Orleans for casting . . . guns."[47] Although the New Orleans venture was generally a failure, agreements with two private establishments in Richmond, Tredegar Iron Works and Bellona Iron Works, were successful. In the fall of 1861, Tredegar signed a contract to cast guns for the navy. Between January 1861 and April 1865, over 1,000 guns were manufactured.[48] The Bellona Iron Works cast both smoothbores and rifles for the navy, but there is no record of their number or type.

In February 1862 a contract was signed between representatives of the war and navy departments and Colin J. McRae, a member of the naval affairs committee and the agent later involved in the dispute with the Shelby Iron Company. McRae, who with several other individuals had recently purchased a foundry at Selma, Alabama, agreed to manufacture heavy ordnance and iron plates for gunboats. The contract stipulated that the first cannon should be cast by September 1 and the plates by the following December.[49] McRae assumed the responsibility of

[47] *Official Records of the Union and Confederate Navies*, ser. 2, II, 53. See also the Robert Minor Diary, 1861-1862, in Robert Minor Papers (Virginia Historical Society).

[48] Bruce, *Virginia Iron Manufacture*, 349. An undetermined number of the guns were for the army.

[49] A copy of the contract dated February 24, 1862, is in the McRae Papers. See also Charles S. Davis, *Colin J. McRae: Confederate Financial Agent* (Tuscaloosa, Ala., 1961), 25. McRae later bought out his associates.

placing the foundry in operation; he purchased machinery and tools, hired workmen, examined the rolling mills in Atlanta and Richmond, and contracted for iron ore and other materials. But the foundry was still inoperative when, early in 1863, he was assigned on a financial mission to Europe. Before agreeing to accept this assignment, McRae made it a "condition of his acceptance" to be relieved of the Selma foundry "without pecuniary loss to himself." By the middle of February the navy and war departments had agreed to purchase and operate the foundry jointly.[50] The joint venture lasted only about three months, for on June 1, 1863, the Navy Department assumed complete control of the works.[51]

When the navy took over, construction on the works had been under way for nearly sixteen months, but the first piece of ordnance was still to be produced. Under the capable supervision of Lieutenant Catesby ap R. Jones, the works were completed and by the end of July the foundry was ready to cast its first gun. This gun, however, and those that immediately followed were experimental, and it was not until January 1864 that the first one was shipped for combat use. From then until the end of the war more than a hundred large naval guns were cast.[52]

Facilities for the manufacture of ordnance stores (shell, shot, fuses, and caps) were also organized by the bureau. One was established at Norfolk and a second one at New Orleans. The one at Norfolk was established by a congressional appropriation of $30,000 in April 1861 but lasted less than a year.[53] On March 26, 1862, Mallory sent a confidential dispatch to Captain Sidney Smith Lee, in command of the Norfolk navy yard, ordering him to prepare for possible evacuation. Tools and machinery not then in use were to be packed, ready for immediate shipment. On May 1 another dispatch to Lee directed that the transportation of the tools, machinery, and stores begin at once. No destination was named, but apparently a decision was made shortly after-

[50] Governor William Gill Shorter to McRae, February 12, 1863, in McRae Papers. See also Josiah Gorgas, "Notes on the Ordnance Department of the Confederate Government," *Southern Historical Society Papers*, XII (January-March 1884), 83, and Frank E. Vandiver, *Ploughshares into Swords: Josiah Gorgas and Confederate Ordnance* (Austin, Tex., 1952), 169.

[51] Catesby ap R. Jones to John K. Mitchell, June 1, 1863, in Naval Facilities at Selma file, Subject File, Record Group 45 (National Archives).

[52] Walter W. Stephen, "The Brooke Guns from Selma," *Alabama Historical Quarterly*, XX (Fall 1958), 464.

[53] Matthew F. Maury to George Minor, May 16, 1861, in Matthew F. Maury Papers (Library of Congress).

ward, for on May 2, Lieutenant Robert D. Minor was ordered to Charlotte to locate and requisition suitable facilities for the Norfolk ordnance stores.[54] This site was chosen not only because of its interior location but also because the city had excellent railroad connections with the coast. The department leased the Mecklenburg Iron Works for the new ordnance facility and there most of the machinery was installed by the end of June. Within six months the establishment was in full operation, manufacturing gun carriages, shot, and other ordnance stores.[55] Some of the Norfolk machinery and tools were also shipped by way of the James River to Richmond where an ordnance store and laboratory were placed in operation.[56]

An ordnance depot at Atlanta was established after New Orleans was threatened with capture and the ordnance facilities there had to be moved. In August 1861, Lieutenant Beverly C. Kennon had been assigned by the naval commandant in New Orleans to organize an ordnance store in the city. Within a month he had contracted for guns, gun carriages, projectiles, and powder, his expenditures amounting to nearly a million dollars. Unfortunately he exceeded his authority. As a result, many of his contracts were canceled, a large number of purchases returned, and Kennon himself transferred. Even so, it was some time before the naval office at New Orleans was out of debt. Kennon's replacement as ordnance officer, Lieutenant John R. Eggleston, did not move so fast, although he did establish a laboratory where fuses, primers, and other ordnance parts were manufactured.[57] In the spring of 1862 the equipment in this laboratory and many of the stores gathered by Kennon were sent to Atlanta. In Atlanta, Lieutenant David P. McCorkle supervised the location of a new ordnance depot. Property and buildings were leased near the Georgia Railroad, and the machinery was installed there. By the summer of 1862 these facilities were in operation.[58]

The movement of ordnance depots inland, to Charlotte, Rich-

[54] Ralph W. Donnelly, "The Charlotte, North Carolina, Navy Yard, C.S.N.," *Civil War History*, V (March 1959), 74. See also George Minor to Robert Minor, May 2, 1862, in Robert Minor Papers.
[55] Few shells were manufactured at Charlotte. On August 25, 1864, Brooke wrote to Catesby ap R. Jones that all attempts to make rifle shells at Charlotte had failed. Naval Facilities at Selma file, Subject File, Record Group 45 (National Archives). Apparently no guns were manufactured at Charlotte.
[56] Robert Minor Diary, in Robert Minor Papers.
[57] *Official Records of the Union and Confederate Navies*, ser. 2, I, 797.
[58] *Ibid.*, II, 250-51.

mond, and Atlanta, was the last major organizational change in ordnance facilities during the war. On June 25, 1863, a circular from the ordnance bureau directed that "Hereafter, as far as practicable Ordnance Stores for the different naval stations will be supplied as follows: Richmond will supply Richmond and Wilmington; Charlotte will supply Charlotte, Savannah, Charleston; Atlanta will supply Mobile and the Gulf Stations."[59] This directive remained in effect until the end of the war, although Sherman's march to the sea resulted in the removal of the Atlanta facilities to North Carolina.

One material of war in which the Confederate navy was nearly destitute was powder. The shortage of powder and powder mills was severely felt at the beginning of the conflict. In fact, with the exception of 60,000 pounds captured at Norfolk, little powder was available. The 1860 census shows only two small powder plants in the states which seceded, one in Tennessee employing ten men, the other in South Carolina employing three. New mills were established by both the army and navy, although there was only one important naval powder factory, located at Petersburg.[60] In 1862 the Petersburg plant was moved to Columbia, South Carolina, where it remained until the end of the war. This mill apparently was adequate for the navy's needs, for no others were built and there are no records to indicate that others were anticipated.[61]

During the last year of the war, naval facilities of all kinds were demolished one by one. Attempts were made to salvage them by moving to other localities, but generally this proved futile. As early as June 1864, the Atlanta Ordnance Works was evacuated, although the city was not threatened until August. The plant was then removed, first to Augusta, Georgia, and later to Fayetteville, North Carolina, where it was destroyed in February 1865.[62] The Atlanta Rolling Mill was taken to Columbia,

---

[59] Naval Facilities at Selma file, Subject File, Record Group 45 (National Archives).

[60] In New Orleans two powder mills were developed jointly by the army and navy. They supplied most of the powder used there.

[61] One authority mentions two small mills belonging to the navy, one at Richmond and the other at Raleigh, North Carolina, but none of the reports of the Office of Ordnance and Hydrography mention them. See Arthur P. Van Gelder and Hugo Schlatter, *History of the Explosives Industry in America* (New York, 1927), 115.

[62] *The War of the Rebellion: A Compilation of the Official Records of the Union and Confederate Armies* (70 vols. in 128, Washington, 1890-1901), ser. 1, XLII, part 2, p. 1264. For the naval evacuation of Atlanta see also Brooke to Mallory,

South Carolina, where it was gutted when the city burned.

During the winter of 1864-1865, General James H. Wilson led a Union cavalry raid through Alabama and into Georgia. His objective was the destruction of Confederate stores, factories, mines, and ironworks in that area, including the naval establishments at Selma, Alabama, and Columbus, Georgia. On April 2, the day Richmond fell, Selma was occupied, and two weeks later the advance units of the Union cavalry reached Columbus. Catesby ap R. Jones at Selma tried to transfer his machinery, but time was not available. At Columbus, chief engineer Warner sent a small vessel down the Chattahoochee to find a suitable site at which to relocate, but by the time the vessel returned it was too late.[63]

Naval facilities located in other cities—Savannah, Charleston, Charlotte, Wilmington, and Richmond—were taken over or destroyed as Federal forces occupied the Confederacy.

The creation of facilities for building and outfitting war vessels, despite the shortcomings of the program, was one of the most impressive accomplishments of the Confederate navy. At least twenty shipyards, five ordnance establishments, two marine machinery shops, a powder works, and a ropewalk were operated by or for the navy. Private foundries and machine works were also successfully operated. Nevertheless, the success of these facilities was seriously curtailed by shortages of labor and deficiencies in certain materials, by inadequate transportation, and by the intense competition for the limited industrial facilities, workmen, and materials.[64] One other factor cannot be ignored—time. Materials needed to complete vessels were delayed because facilities were destroyed or had to be moved in the face of advancing enemy forces. Time and time again uncompleted ironclads and wooden gunboats had to be destroyed to prevent their capture.

January 4, 1865, in Letterbook of the Office of Ordnance and Hydrography, Record Group 109 (National Archives); and Mallory to the President, January 5, 1865, in Jefferson Davis Papers.

[63] For Alabama see Walter L. Fleming, *Civil War and Reconstruction in Alabama* (New York, 1905), 71; John Hardy, *Selma: Her Institutions and Her Men* (Selma, 1879), 204. For Columbus see J. H. Warner to A. O. Blackmar, April 13, 1865, in James H. Warner Papers (Columbus Museum of Arts and Crafts); *War of the Rebellion*, ser. 1, XLIX, part 1, p. 486; James H. Wilson, *Under the Old Flag* ... (2 vols., New York, 1912), II, 267.

[64] See Still, "The Construction and Fitting Out of Ironclad Vessels-of-War Within the Confederacy," especially Chapters V and VI.

PART TWO: GROWING STRONG

From the Civil War through World War I

Once the sad conflict of Blue and Gray had passed, the nation turned to healing its wounds and the restoration of damaged properties. Social hurt and animosity remained, to be sure, but the farms and factories were soon buzzing and gross output quickly surged upward. Before long, things really were better. Most of the country felt it and reacted with a growing optimism as the rewards of material success flowed out to large sectors of a highly receptive populace.

In large part, it was still the land that made it happen. Vast tracts and open spaces beckoned and thousands sought their fortunes to the west. The coming of the railroads then helped bind the country together as food, materials, and ordinary citizens criscrossed the continent on a fast network of profitable and expanding lines. Sometimes, however, things got out of hand as unscrupulous operators watered stock, doctored rates, and in a variety of imaginative ways did their best to get around the customary market process. Other operators brought to bear their experience, came up with a new idea, and rapidly hurled new products into the seas of competition. Much of all this worked, making possible the rise of the great American Corporation, large profits for the entrepreneurs, and, in many cases, genuine benefit for the consuming public as well. Some problems remained, of course. There were still many poor in crowded unhealthy tenements; there was discrimination and little true regard for the plight of the worker; the farmers had troubles and monopolists were commonplace. It all added up to a pressing need for reform, especially in banking, utilities and market competition; and it all came about, too, as the tree that is America leaned against the wind, growing even stronger and pushing herself upward toward a higher level of maturity.

WATER, LAND, AND PEOPLE IN THE GREAT VALLEY

Is It True That What We Learn from History
Is That We Learn Nothing from History?

By Paul S. Taylor

In 1878, scientist and explorer John Wesley Powell stated the problem with typically succinct accuracy in his Report on the Lands of the Arid Regions of the United States: *"All the present and future agriculture of more than four-tenths of the area of the United States is dependent upon irrigation, and practically all values for agriculture inhere, not in the lands but in the water. ... The question for legislators to solve is to devise some practical means by which water rights may be distributed among individual farmers and water monopolies prevented."*

It was perhaps the most important problem the arid West would ever have to face, and one it has had nearly a century to resolve. In a study that takes California's great Central Valley as a case history, economist-historian Paul S. Taylor outlines the mechanics of our failure.

AMERICAN WEST, 1968, vol. 5, pp. 24-29, 68-72.

Too little water for too much land. Water at the wrong time and in the wrong places. And the people? — the People, yes.

Men faced these elements in the American West very early. They could not be avoided if stable communities were to be established and sustained on arid lands. Millennia ago, southwestern Indians found a precarious solution to land and water by diverting flood waters across the sands in shallow ditches. Lacking reservoirs, they remained exposed to seasonal and annual cycles of wet and dry. How, or whether, they divided the waters among their people, we do not know.

The summer of 1847 opened a new era. "Mormon pioneers turned the waters of City Creek upon the alkaline soil of Salt Lake Valley," proudly wrote William E. Smythe in his **Conquest of Arid America**, and Anglo-Saxons embarked upon the "miracle of irrigation." The Mormons distributed land and water with care. Their principle was equity among people.

A generation later the national government, grasping the potentialities of irrigation, began to inquire into ways and means of promoting it. In 1873 Congress authorized a commission, to be composed of two engineers from the Army and one officer from the Coast Survey, for the purpose of examining and reporting on a system of irrigation in the San Joaquin, Tulare, and Sacramento valleys. The Army engineers, joined by Professor George Davidson of the Coast Survey, perceived at once the unity of the watershed assigned for their study, and designated it the "Great Valley of California." Their report became a milestone.

The commission concerned itself with three problems: (1) physical problems of uniting land and water; (2) sources of financing the costly structures essential to its achievement; and (3) ways to forestall bitter conflict between people by assuring equitable distribution of irrigation benefits from the beginning. We have learned since that these problems are perennial.

To illustrate the feasibility of physical union of land and water in the Great Valley, the commission sketched a system of diversion canals from the Sacramento and San Joaquin rivers. Actual construction, of course, would require further engineering investigations and be the work of time.

Then there was the problem of financing the canals. The commission saw three possible sources of funds. First, it noted, the value of land in the driest districts would increase "manyfold" and the lands should be expected, as far as possible, to pay for the construction of the irrigation works. But this source was insufficient because generally the farmers were unable to build. A second source of funds was private enterprise. The commission doubted, however, "that for a long time capital will look upon this kind of investment with favor." A third source was government — both local and state. Each of these units, the report pointed out, "will be directly benefited by ... the increase of wealth in their revenues from taxation."

The commission foresaw a possible financial solution through combining these three sources — the benefiting landowners, private enterprise, and state and local governments. As to the latter, it suggested, "It may be a question for the State to consider whether it is good policy to offer any special inducements in aid of such enterprises."

With these words the commission lifted the lid on an issue that has echoed down through time: should the public subsidize water development for private benefit? If so, how much, and on what terms? But farsighted as it was in perceiving enduring issues, the 1874 report failed to see that no source short of the federal government itself could produce the funds and provide the subsidies necessary to pay the cost of watering the lands of the Great Valley.

Standing at the threshold of development, the commission was mindful of people as well as of water and land. With an awareness gained from its studies of historic irrigation projects in Egypt, India, and the world over, the commission warned that the rights to water had often been handled for the benefit of private interests or particular districts. In the United States, said the commission, the "rights of water which have given so much trouble in other countries . . . can be established beforehand, if not for all time, at least on the principle of 'the greatest good for the greatest number'."

The commission offered suggestions for the protection of people against injury from private interests. The best policy, it said, was for farmers themselves to build and own the canals. Also, "As a matter of public policy... the land and water should be joined together, never to be cut asunder." The purpose behind both ideas was to prevent a division of interest between sellers and buyers of water.

If "private companies" rather than the farmers should undertake construction of irrigation works, these automatic protections to people would not be realized. In that event, the commission had ready a second set of precautions. Arbiters representing "each party in interest" should determine the price of water. In addition, "an association of irrigators should be chartered to administer the works, the company merely selling the water, and having nothing to do with it after it leaves their channels." As final protection for the people, a right to purchase the irrigation works on previously defined conditions should be reserved to the public.

Realism was the foundation of the commission's recommendations. Recognizing that public supervision of private development "will probably be distasteful to the parties concerned" (i.e., to the "private parties"), it nevertheless declared public supervision essential and warned that "its neglect now will bring a fruitful crop of contentions in the future, will delay the development of the country, and . . . by making irrigation unhealthful it may make it odious."

The commission did not stand alone in its insight. Major John Wesley Powell, one of the greatest figures in the history of the West, was equally precise in diagnosis and in prognosis. Barely five years after the commission he wrote in his famous 1879 **Report on the Lands of the Arid Regions** that "the question for legislators to solve is to devise some practical means by which water rights may be distributed among individual farmers and water monopolies prevented." In a moment of political innocence, Powell wrote that "monopoly of land need not be feared."

The commission's 1874 report stirred the Great Valley and gave fresh impetus to the continuing process of land monopolization there. Farsighted, aggressive men responded quickly to passages in the commission's report most relevant to their own private interests. One was explicit: "In the southern end of the valley, between Visalia and Bakersfield, and south of this town, . . . the United States own many thousand acres of land which are capable of irrigation . . . most of this land cannot be cultivated under existing circumstances." This was to be read in light of the forecast that "there is every reason to believe that the value of land in the driest districts will be appreciated manyfold."

Action came fast. "Private parties" moved into the southern end of the Valley so fast that after barely three years the **Visalia Delta** reported and warned: "No one would believe that shrewd, calculating businessmen would invest their money on the strength of land rising in value while unimproved, for even the farmer himself has to abandon it who endeavors to add to its value without water. At the same time, purchasers are not lacking who would add it to their already extensive dry domain and the people . . . will find themselves confronted by an array of force and talent to secure to capital the ownership of the water as well as of the land, and the people will at last have it to pay for."

The era of land-grabbing in California was now in full swing, impelled by the tantalizing prospect of large-scale ir-

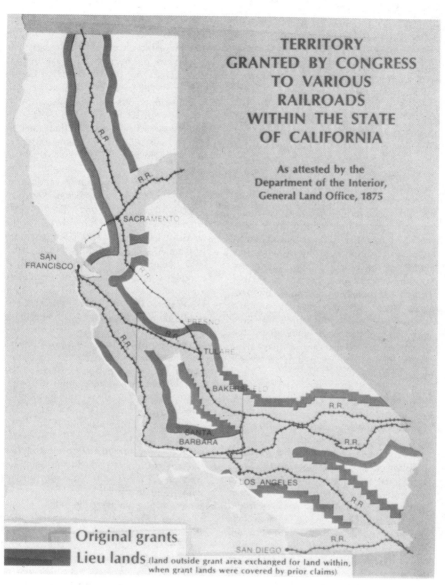

This map, based on a bit of campaign literature distributed by the People's Independent Party ("John Bidwell for Governor") in 1875, reveals one historic base for California's patterns of land monopoly. Some of the land grants shown here were strictly theoretical, since track was never laid on several proposed rail lines; nevertheless, by 1875 the great bulk of arable land lying between the Sierra Madre and the coast range was in the hands of a very few owners. The map on our facing page, showing current landownership in the San Joaquin Valley (an area enclosed within the square above), indicates that historic patterns have been carried into our own times. In fact, the curious checkerboard design apparent in some portions of the San Joaquin Valley map is the direct result of nineteenth-century railroad grants distributed in the form of alternate sections.

rigation. In the same year of 1877, when the **Visalia Delta** spoke its jeremiad, Carl Schurz, the new Secretary of the Interior, suspended all land entries at Visalia and sent an investigator into the field. The **San Francisco Chronicle** published a summary of his 300-page findings. The investigator, it said, "discovered that the [1876] Desert Land Act of Congress was simply a Ring job, and was made the medium for an organized colossal steal by the Ring, to the prejudice of thousands of honest, **bona fide** settlers, against whom it was so used as to prevent them enjoying the benefits of the letter and spirit of the Act. By arrangement and collusion, the thing was so managed as to furnish from Washington to the Ring here the instant information of the Executive approval of the Act, and in less time, by weeks, than it requires to officially communicate the necessary order to give proper operation to an Act of Congress on this coast, the Ring land-grabbers had been allowed by the officers of the Visalia Land Office to list and locate an immense area of the desert tracts."

The following year Commissioner J. A. Williamson of the General Land Office reviewed the matter and applied whitewash. Although the law restricted individual entries to a single section of land, Williamson allowed enterprising parties to obtain large tracts of land by making loans to numerous individual applicants, loans covered by liens on their filings. He found that this circumvention of the law — a method sometimes called "dummy entries"— was a "reasonable and even necessary" means of achieving irrigation on desert land. Williamson certainly reflected one aspect of the spirit of the times. In those days stories went the rounds about men who drove across occasionally overflowed lands, seated in rowboats mounted on horse-drawn trucks, so that they might swear the lands were swamps. A saying — perhaps apocryphal — was attributed to one of the most successful of land monopolists: "God will make more babies but he won't make any more land."

The response to Williamson's review was varied. The **Tulare Times** was entirely uncritical. "All that is desired," it said, "is that these barren plains should be made to blossom as the rose." But neither the **Visalia Delta**, nor the **San Francisco Chronicle**, nor Williamson's successor as Land Commissioner was willing to accept Williamson's official behavior. The **Delta** commented sarcastically that "all that is necessary to make (the barren plains) bloom is to give them away in chunks, the size of whole states, to Carr & Haggin" (forerunners of the Kern County Land Company). The **Chronicle** spoke of "an outrageous decision" by one whose "whole official course [shows] he is always for the wealthy grabber." Land Commissioner William A. J. Sparks said his predecessor's land department had been "very largely conducted to the advantage of speculation and monopoly, private and corporate, rather than in the public interest." The public domain, he said, "was being made the prey of unscrupulous speculation and the worst forms of land monopoly through systematic frauds carried on and consummated under the public land laws." He charged Williamson with either extraordinary ineptness or direct involvement in the frauds.

The verdict of the historian is clear. Paul Wallace Gates, of Cornell University, writes: "The administration of the law, both in Washington and in the field, was frequently in the hands of persons unsympathetic to its principle, and Western interests, though lauding the act, were ever ready to pervert it." Under the cloud of monopoly, says Gates, "land sales and warrant and scrip entries in California were on an enormous scale... in some years comprising well over half of the sales for the entire country." By buying in advance of settlement, he concluded, the enormous monopolization was "virtually thwarting the Homestead Law in California where... homesteaders later were able to find little good land."

Monopoly of water showed its face quickly on the Kern River; it followed naturally on the heels of

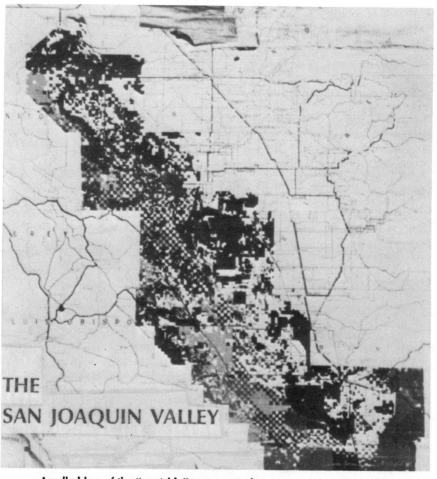

THE SAN JOAQUIN VALLEY

Landholders of the "westside"—an area to be irrigated by the multi-billion-dollar California Water Project.

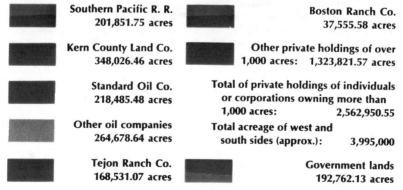

Southern Pacific R. R. 201,851.75 acres

Boston Ranch Co. 37,555.58 acres

Kern County Land Co. 348,026.46 acres

Other private holdings of over 1,000 acres: 1,323,821.57 acres

Standard Oil Co. 218,485.48 acres

Total of private holdings of individuals or corporations owning more than 1,000 acres: 2,562,950.55

Other oil companies 264,678.64 acres

Total acreage of west and south sides (approx.): 3,995,000

Tejon Ranch Co. 168,531.07 acres

Government lands 192,762.13 acres

(From a map compiled by the California Center for Community Development.)

monopoly of land. There, two giant landholders locked horns in the courts — Henry Miller, the cattle king (with Charles Lux his partner), and J. B. Haggin et al. Division of the waters of the river was the issue.

The case hung long in the balance. If the California Supreme Court should decide that the English common law doctrine of riparian rights governed distribution of the waters of the Kern River, these would belong to Miller and Lux.* If the Court should decide in favor of those who actually appropriate the waters for beneficial use, these waters would belong to Haggin and his partners. The Court, by a majority of one, decided for Miller and Lux.

The decision stirred a political hornet's nest. Within a month, on May 20, a convention opened in San Francisco with "anti-riparianism" as its cardinal slogan. The press, from one end of the state to the other, burst with editorials against the alleged injustice of the riparian doctrine. It was attacked as "an obstruction to the prosperity of the commonwealth." The **Alta California** observed: "If the whole bench were vacated by resignation, the operation of putting the court at one with public sentiment would be greatly simplified."

One Judge North, addressing the convention, said, "From the late decision of the Supreme Court, if that is the law of this State, there are thousands of good citizens who have made a great mistake in settling in California. We had better emigrate now to the British Provinces, where riparian law is not in force" [applause].

So the party defeated in the Court carried the contest into the public arena and began to move it toward the next session of the legislature. The victor in the Court, the biggest landowner of them all, brought this move to a halt. In a state where land monopoly had been a battle cry in legislative halls, on the hustings, and in the debates of the Constitutional Convention of 1878–79, Henry Miller "knew that the cry of the multitude was against him; he knew that his opponents were fanning their hatred; he knew that they were buying newspapers, and that petty politicians were being brought under their control." Treadwell, his biographer, records Miller's solution: "Did Henry Miller rest satisfied with the decision which gave him and his associates all the water of the Kern River? Not at all. He immediately said, 'There is more water than we can use, and it does not come at the right time of the year. It comes in a great flood early in the spring, and in the hot months of summer the river is dry.' So he said to his late antagonists, 'You builds me a reservoir, and I gives you two-thirds of the water,' and the difficulty was solved.... The two great interests, now brought close together by the great fight, joined forces."

Public agitation — and public education — on the subject of land and water monopoly promptly subsided. Yet on the pages of the opinion of the State Supreme Court a passage remained, curiously at variance with a decision establishing water monopoly. "It does not require a prophetic vision," said the Court, "to anticipate ... a monopoly of all the waters of the state by comparatively few individuals ... controlling aggregated capital, who could either apply the water to purposes useful to themselves, or sell it to those from whom they had taken it away, as well as to others."

The 1880's closed the first phase of the long history of land, water, and people in California. A commission of the national government has verified the physical potentials of irrigation and pointed to the problems of finding capital and protecting people. Men who foresaw the increase in land values to come with water had quickly grabbed the land. These titans had divided between themselves those waters ready for use with minimum capital investment, as on the Kern. The people were shunned.

The repercussions of the battles among such giants ranged through the courts, the

---

* Owners of land bordering on streams are entitled to the natural and usual flow and are protected against diversion of the waters for use by appropriators.

Congress, the Legislature, and the Land Office and reached down to the ground. People who would lard their own land with their own sweat found access obstructed. Others already on the land found it difficult to remain, their removal encouraged. The United States Senate Committee on Irrigation and Reclamation of Arid Lands heard how poor Mexican farmers were forced out when they refused to be bought out. The spring waters that had served them for irrigation were appropriated and ditched to serve the lands of the appropriators. "For the purpose of freezing them out cement was used to prevent even a drop of water escaping onto their lands, and the poor farmers succumbed." A former State Senator, John M. Day, testified to this after travels in Kern County in 1886.

As a measure of the depopulation caused by water monopoly, Senator Day cited the fact that the average school attendance had dropped from 649 in 1880 to 246 in 1886. "This," he said, "is the effect of private property in water."

The problem drew the attention and disquieting analysis of one of the most acute and penetrating foreign observers ever to study and analyze the United States. In 1889, James Bryce wrote in his classic, **The American Commonwealth:**

> When California was ceded to the United States, land speculators bought up large tracts under Spanish titles, and others, foreseeing the coming prosperity, subsequently acquired great domains by purchase, either from the railways which had received land grants, or directly from the government. Some of these speculators, by holding their lands for a rise, made it difficult for immigrants to acquire small freeholds, and in some cases checked the growth of farms. Others let their land on short leases to farmers, who thus came into a comparatively precarious and often necessitous condition; others established enormous farms, in which the soil is cultivated by hired labourers, many of whom are discharged after the harvest — a phenomenon rare in the United States, which is elsewhere a country of moderately sized farms, owned by persons who do most of their labour by their own and their children's hands. Thus the land system of California presents features both peculiar and dangerous, a contrast between great properties, often appearing to conflict with the general weal, and the sometimes hard pressed small farmer, together with a mass of unsettled labour, thrown without work into the towns at certain times of the year.

It all seemed so natural to some — and so unnatural to others.

Among western states no contrast in the manner of settling people on land was greater than that between Utah and California.

In Utah people came to the land first. There, as farmers on the land with their families, they built their own canals to bring water to their own farms. They left no place for speculators to come between them, their land, and their water.

California reversed this process. Speculators seized the dry land ahead of the coming of people. Then they sought ways to fund the cost of bringing water to their land. The people came last.

Despite this fundamental difference in history and in purpose, Mormon farmers and California speculators shared at least one desire: both wanted more funds to bring more water to more thirsty land. Could this common financial aspiration bring and hold them together? How far could they merge their deep differences in a common desire for public subsidies? An International Irrigation Congress, convening in Salt Lake City in 1891, was the first step. It would be a test.

The cast of characters at the Irrigation Congress quickly revealed a broad spectrum of concern over private interest in water development and public policy. One George Q. Cannon, from Salt Lake City, told the Congress: "We have refrained — I was going to say religiously — from forming large corporations to take possession of water. . . . Settlements have been combined together, and . . . have contributed by their labor in forming dams and digging ditches, to obtain the necessary supply for their acreage.... The poor could take land and could obtain water by their own labor."

Experience with private monopolization of land had aroused public fears in California, too. Delegate Morris M. Estee, from that state, joined with Cannon in warning of the dangers from private distribution of water. "Desert lands will be useless," he said, "if you own the lands and private interests own the water." He followed with a full-scale attack on land monopoly and proposed a legal limitation on landownership. Specifically he recommended "that no man could obtain a title to more than 320 acres of desert land, either directly or indirectly," because "if he has got water on it, [it] is more than any one man can begin to cultivate.... The tendency of the great West, and especially in some of our part of the West is the accumulation of vast estates in land. The object of good government is to stop this.... The large owner of lands only owns it for speculative purposes and not for practical uses, while the small farmer owns it for the purpose of cultivation and for sustaining his family and building up the civilization of the country" [applause].

An Oregon delegate told the Congress that his state had had "a shameful experience. My own experience in connection with the land department there is this, that millions of acres of the very best land we have today... [are] withheld from settlement, by the acquired rights of corporations. Day after day, the honest settler seeking a home for himself and his family comes to my office... and I tell him I could not accept a filing.... The result is, gentlemen, and I am ashamed to say it, that those lands are withheld from settlement, and the prosperity, the increase in population, and the prosperity of the general country... has been called to a halt." He warned "against the perpetuation of any such gigantic frauds in the future."

Senator Frances E. Warren, of Wyoming, pleaded with the Congress: "I do not want to see the law so constructed or passed that it can inure to the benefit of corporations.... I would like to see a restriction that would provide that the acreage should be limited to 160, to 320, or such other amount as may seem best."

No one rose to defend corporate owners of large tracts of land against the repeated charges of monopolization.

But the Irrigation Congress had been convened to promote water development. Discovery of sources of funds was a concern of all, and the eyes of the delegates turned toward government. The nourishing of political strength had become a political necessity.

Among those who stressed the need for public capital to develop water resources — a need foreseen so clearly by the commissioners in 1874 — was W. H. Mills, of California. Delegate Mills spoke forthrightly: "Heretofore agriculture has been classed as one of the non-capitalized occupations among men. The construction of a railroad was the question of large capital. The construction of a banking business, a mercantile business, the building of ships, all these are capitalized occupations requiring the association of capital. Now, for the first time, our race encounters a problem of agriculture involving within its terms the problem of the uniting of capital with the ownership of land. And right at this point comes the great question. There is a conflict between the ownership of the soil and the ownership of the water, which, up to this time has obstructed the development of every irrigation system on the Pacific Coast. It has been one of the great difficulties, to bring the water and the land together, in its financial, not its physical aspect."

Mills clarified his role at the Congress: "I am here by appointment of the Governor of the State of California, as one of the representatives of that State on this floor. I am not here in my capacity as land agent of the Central Pacific Railroad Company, nor of the Southern Pacific Railroad Company, nor of the Oregon & California, nor of the Houston & Texas Central Railroad Companies, of all of which corporations I am the land commissioner. If therefore I should appear on behalf of these, I would not appear by their authority."

Delegate Mills explained that one reason

for the reluctance of private capital to invest in water development was fear of widening political conflict. He sought, doubtless with one eye on the California experience, to sketch lines along which conflict might develop. On the one side were landowners who would buy water from private capital's ditch. "Now they say to themselves," explained Mills, "eventually, the capital invested in fertilizing this land will become oppressive, and we will be the serfs of the capitalists." On the other side were the capitalists, who "have time and time again refused to put their money there for fear of the public. It has been said, and by a gentleman on the floor of this Convention, a gentleman of large capital and large wealth, who is interested himself in the development of these systems . . . that he would not put his money into a system of irrigation, because he was afraid of the numerical strength of the people, who would use the water, and that eventually they would regulate all the wealth-producing capacity out of the capital stock of the ditch."

Faced with this dilemma of conflicting interests, Mills pointed to a way out — the use of public capital to finance water development. Then the public could be an aid to development, not left to stand as an obstacle. To gain access to public capital, the kind of broad public support that the Irrigation Congress could generate was essential. Mills continued optimistically: "Now, you perceive the beginning of the request of the people of the West that the Congress of the United States shall place in their hands a value, which may be devoted to the development of water, so that the public will own the water, and private individuals will own the land, and thus allay forever the conflict, which has obstructed the development of irrigation systems."

Mills was correct in his assertion that public capital was needed for development. Time would prove, however, that the use of public capital would not allay the conflict between public and private interest.

The Irrigation Congress, no matter how deep the division over monopoly and speculation, was not diverted from this political problem of persuading the national government to supply vast amounts of capital to develop western waters. "It will be met with opposition from the representatives from the East and from the Middle West," warned Francis G. Newlands, of Nevada, later to become a co-author of the National Reclamation Act. They "will claim that the United States has no right to expend large sums of money in the development of a particular section."

Governor George C. Pardee, of California, vividly spelled out political hurdles to be overcome. "The objection would come from the east that this was a local affair, and that the people inhabiting this part of the country should act for themselves in this matter under the present land law. They would say there is enough land already in cultivation, the annual agricultural output of which they would say is sufficient. 'We object to any further cultivation of the soil in the West, when the farms of New England are lying idle?'... They feel jealous over this matter, that there are incentives in the great West that induce their young to leave the East and leave their interests there."

The irrigation congresses that began in 1891 ran their course for a decade before they won authorization from Congress to open the national treasury to subsidize western irrigation. During the decade delegates from a score of states met almost annually to exchange views, to reinforce convictions, and to generate enthusiasm. The proceedings of the 1893 Congress in Los Angeles reveal something of the atmosphere in which the congresses functioned. The Los Angeles Chamber of Commerce . . ."took hold of the Irrigation Congress . . .'with as much interest and enthusiasm as though it had been a purely local enterprise.' At the conclusion of the Congress a number of excursions took place. The Southern Pacific placed a special train at the disposal of the delegates for a trip to Chatsworth Park over a line just completed and to Port Los Angeles where there stretches into the ocean the longest wharf in existence, and to the

National Soldiers' Home at Santa Monica. Several days later the same road sent the delegates over its line to Indio. . . . It was not until a week after the adjournment of the Congress that the excursions came to an end and the delegates returned to their homes. . . . It is a source of unlimited gratification to the members of the committee, who gave several weeks of time to the undertaking, that the cause for which the meeting was called and upon which rests in so large a measure the future prosperity of the Middle West, has been aided in so signal a manner by the intelligent deliberation and utterances of the Congress."

The thoughts and feelings exposed so clearly at Salt Lake City in 1891 ran like threads through the pages of the proceedings year after year. There were texts, biblically phrased and idyllic, such as from Genesis 2:10: "And a river went out of Eden to water the. Garden." There was Goldsmith's classic: "Ill fares the land, to hastening ills a prey, where wealth accumulates, and men decay." There were stern warnings of overcrowding in the cities, which it was hoped that irrigation of the countryside might relieve. Alexander H. Revell, of Chicago, stated flatly: "The next great problems we must face are the relief of the crowded conditions of our centers and the amelioration of the distress of the poor in our great cities."

In 1900 the site of the Congress was moved from the suppliant West to Chicago, to reach closer to the heart of national political power, to win over indifference, and to dilute potential opposition.

Exponents of western irrigation by this time had come to agree that state governments could not achieve their ends — that the national government alone possessed sufficient resources and power. The call for the Chicago Congress expressed hope that the deliberations would subordinate controversy and elevate unity of political purpose. "In this session of the National Irrigation Congress," the call read, "all disputation and controversy should be eliminated. Its deliberations should be guided by a high patriotic purpose and a united determination to arouse the whole people of the nation to a realization of the national importance of transforming the western deserts from uninhabitable wastes into fertile and populous territory.... The national government is the only agency through which this can be accomplished."

The Congress did show unity of purpose in appealing to the national government for funds. This unity, however, did not exclude continuing insistence by delegates on clear-cut public assurance that the evils of monopoly would be curbed.

A delegate from Montana, to be sure, gave as his lone opinion that "there is a good deal of bugaboo made of the matter of trusts securing this land, and big corporations." Another from Colorado thought that because of taxes and assessments "no man, no company, with any business judgment or business sense will ever hold onto very much land."

The weight of discussion, however, was heavily on the other side. Delegate Brady, of West Virginia, for example, spoke his apprehensions that "the result will be as it has been in Southern France and Italy — that the land will be owned and controlled, as it is now largely in the West, by large capital..." He cited a report of the U.S. Minister to Italy, stating that "the tendency of irrigation in those countries was to put the land into the hands of wealthy land owners and corporations." Brady added, "That surely is not what we want to do with the arid West."*

In similar vein, Delegate Newberry, of Montana, warned of efforts "to defeat and defraud the very object

---

* Undoubtedly the most vivid attack on land monopoly at the Congress was made by Delegate Ganz, of Illinois, formerly Deputy U.S. Surveyor in California. Referring to the Desert Land Act of 1876, he said: "And I remember, Mr. President and gentlemen of this Congress, that before the ink which flowed from the pen of the President, making that act a law, was dry, in the office of the Register and Receiver of the Land Office in San Francisco every employee of the Wells Fargo Express Co., owned and controlled by Haggin & Tevis, all the employees of the United States Custom House, under the boss-ship of Billy Carr, who formed with Haggin of Haggin & Tevis the firm of Haggin and Carr, and a lot of the employees of the Central Pacific Railroad — each and all appeared

... of this convention." The danger, he said, "is that the government may be at all the expense of conserving this water and that the water itself may ... inure to the profit of the few who can get control of it, and not to the general public in the least. As well might the national government go to work and build a railroad at the expense of construction and turn it over to the management of private enterprise to operate the cars and collect the revenues from the operation."

George H. Maxwell, who during his later years came to be called the "Father of Reclamation," was on his feet immediately to set Delegate Newberry's fears at rest. They were based, he said, upon a complete misunderstanding of existing conditions and even of the resolutions that had been adopted. Under the resolutions "no man can own the right [to use water] for speculative purposes ... water monopoly will be an impossibility." Harrison Gray Otis, of the *Los Angeles Times,* spoke of "this timely movement, looking to the reclamation of Arid America, thus giving 'land to the landless,' ranks in real importance with the foremost public measures of the

---

and deposited his declaration that he knew a certain 160 acres of land in Tulare County and Kern County in California to be desert, from his own knowledge — although they had never been within a thousand miles of the land, that didn't make any difference, they swore to it; and they agreed that they would reclaim these little private 'deserts' contiguous to each other. And as they stepped out with the certificate of the Register in their hands that these 160 acres of land would in the future be their property they transferred it 'for Five Dollars in hand paid me this day', and gave a quit-claim deed to Mr. William B. Carr, the great Republican boss of California at that time. [Laughter] And in that way Carr & Haggin got the possession of hundreds of thousands of acres of land in Tulare and Kern Counties in California. And years afterwards the dear government found that a fraud had been committed and they began suit, and they appointed Mr. Theodore Wagner, who afterwards became United States Surveyor General of California, as their attorney. But somehow Carr & Haggin always knew beforehand what Wagner was going to say before court. [Laughter] And this hundred thousand acres of land which made Mr. William B. Carr the King of Kern County and the Lord of Tulare County, notwithstanding the suit brought by the United States for the recovery of that land, are today still the property of the King of Kern County, and not an acre has ever got back into the hands of the people, whose trustee only the administration is."

times."

Assurances that legislation authorizing the national government to subsidize water development would in fact protect against land and water monopoly began with the opening address, and were given repeatedly thereafter. Substituting for Mayor Carter Harrison, who was ill, Chicago Prosecuting Attorney Howard S. Taylor welcomed the Congress and said:

It is impossible to forecast all of the good things, the fruitage which must follow the reclamation of our arid lands.... If the reclamation of the arid lands of the West shall be accomplished, as I understand is your purpose, your anticipation and desire, by the government of the United States, this great end will be achieved: that the exploitation of these lands by private parties will be prevented, and the further dislocation of society and the erection of servitudes resting upon the necks of unborn generations will be prevented.

Congressman Newlands, of Nevada, surely an authoritative voice, told the Irrigation Congress that "the evils of land monopoly... have afflicted every country and every age, and almost all of the wars of the world have been a protest against... land monopoly." He spoke specifically of Spain, South and Central America, and the Philippines. Then he concluded, "It does seem to me we have got to determine upon some plan... to avoid the evils of land monopoly" [applause]. Two years later he co-authored the famous bill limiting water service for any one landowner to no more than 160 acres.

THE FIFTY-SEVENTH CONGRESS of the United States passed the National Reclamation bill, and President Theodore Roosevelt signed it into law on June 17, 1902. He explained that "every dollar is spent to build up the small man of the West and prevent the big man, East or West, coming in and monopolizing water and land."

The act opened the door of the treasury to reclamation, but only on the condition stated by the President. This condition, in words of the statute, was: "No right to the use of water for land in private ownership shall be sold for a tract exceeding 160 acres to any one landowner."

Debate in the Fifty-seventh Congress was thorough. Congressman Frank Mondell, of Wyoming, who was in charge of the bill, assured the House that the bill was "a step in advance of any legislation we have ever had in guarding against the possibility of speculative land holdings... while it will also compel the division into small holdings of any large areas... in private ownership which may be irrigated under its provisions."

The leading eastern opponent of western reclamation, notwithstanding, was wholly unconvinced that the 160-acre provisions of the law would be so administered as to fulfill the professions of its sponsors. "And so we find behind this scheme," said Congressman George W. Ray, of New York, "egging it on, encouraging it, the great railroad interests of the West, who own millions of acres of these arid lands, now useless, and the very moment that we, at the public expense, establish or construct these irrigation works and resorvoirs, you will find multiplied by ten, and in some instances by twenty, the value of now worthless land owned by those railroad companies, the title to which they obtained through grants from the Government for building great transcontinental railroad lines."

The National Reclamation Act was law for only three years when the 160-acre law faced its first test. The prospect appeared bright. John W. Ferris, C.E., wrote in 1905 in the promotional organ *For California:*

In California much of the best land for Government irrigation is in huge private holdings. It is believed that every great landowner in California will be willing to sign a contract to subdivide in order that the Government may proceed as rapidly as possible to construct irrigation works under the National Reclamation Act. Already owners of more than seventy huge tracts of land have signified their willingness to subdivide their lands for the benefit of intending settlers. This shows which way the wind blows and may be taken as an indication that when the Government is ready to go ahead our patriotic landed proprietors will be willing and ready to cooperate.

As if to verify the optimism of *For California*, large

landowners in the Sacramento Valley assured early construction of a project at Stony Gorge by voluntarily accepting a limitation of forty acres.

Reclamation began to fulfill its promises.

For a quarter century after Stony Gorge, reclamation of the Great Valley slumbered as a practical issue. It awoke during the Great Depression. By then the winds had changed. Once funds began to flow for actual construction of the long-awaited Central Valley Project, giant landholders of the Great Valley, largely corporate, revealed unremitting hostility to the 160-acre law. They overlooked no possible corridor of escape. In 1944, 1947, 1959, and 1960 they demanded that Congress exempt the Central Valley from the law. Each time Congress refused, point-blank. In 1958, they asked the Supreme Court to invalidate the law. The Court refused unanimously, echoing literally the 1874 commissioners' appeal to the "principle of 'the greatest good for the greatest number'."

Blocked in two branches of government, the giant landowners turned to the third, the administrative. Through the landowners' efforts two leading officials of the Bureau of Reclamation who supported enforcement of the 160-acre law — the Commissioner and the Regional Director for Central Valley — were driven from the federal pay-roll on the pretext that they were not "engineers." The Eightieth Congress removed them; the Eighty-first, following re-election of President Harry Truman, restored them.

Undaunted, large landholding interests kept up the pressure on those charged with administering the law until, in time, the administrators succumbed. In 1964 Secretary of the Interior Stewart L. Udall gently confessed for himself and some of his predecessors that the executive branch had "on occasion exhibited a degree of concern for the excess-land owner which may be difficult to reconcile with the policies embraced by the excess land laws."

The difficulty in reconciling administration with law, to which the Secretary referred, can be illustrated by two examples from the Central Valley. As early as 1944 *Business Week* reported that "big landowners" had prepared alternative tactics to sidestep the 160-acre limitation in the event — soon realized — that Congress should deny exemption from the 160-acre law. Among these tactics was a "proposal, said to have originated among the big landowners of Fresno County... for the State of California to take over the Central Valley Project, paying the entire bill." Later, the state water project did assume this task, the state bonded itself for $1.75 billion, and is now confronting a huge deficit, its size uncertain. The state's official spokesmen advised Congress in the late fifties that the state water project will ultimately entail a state expenditure of $11 billion.

A second example is at Westlands, in the San Joaquin Valley, where four hundred thousand acres are ineligible under federal reclamation law to receive the benefits from public expenditures approaching a half billion dollars, because their owners have not complied with the 160-acre law. Giving point to his own earlier observation that upon occasion it has been difficult to reconcile execution of the law with the policies the law embraces, the Secretary of the Interior has made no move to exact compliance. A single Westlands landowner, the Southern Pacific, owns 120,000 acres — 187 square miles. The size of the ineligible area equals about two-thirds the area of the state of Rhode Island.

The prophecy of Congressman Ray, of New York, in opposing national aid to western reclamation in 1902, is fully validated two generations later at Westlands.

Clearly, large landowners have felt morally justified in pressing for an administration of the 160-acre law favorable to their interests, and have said so. One of their spokesmen testified to Congress in 1944 that Bureau of Reclamation officials, seeking a matching fund of $25,000 from owners on the west side of the valley to study ground waters, had indicated at the time that "the 160-acre provision was not to be taken seriously." The landowners, after contributing their share, naturally were displeased that Bureau officials later (under orders from

Secretary of the Interior Harold L. Ickes) gave open support to the 160-acre law. Thereupon the spokesman, now head of the Westlands Water District, testified, "It seems to me that the Bureau was completely in bad faith in taking the $25,000, knowing that our district could not accept that."

No two historical figures, poles apart, have epitomized better than Theodore Roosevelt and Clair Engle the fundamental questions of law, policy, and politics that inhere in the 160-acre law. Roosevelt defended the law in 1911 before the Commonwealth Club of California with these words: "Now I have struck the crux of my appeal. I wish to save the very wealthy men of this country and their advocates and upholders from the ruin that they would bring upon themselves if they were permitted to have their way. It is because I am against revolution; it is because I am against the doctrines of the Extremists, of the Socialists; it is because I wish to see this country of ours continued as a genuine democracy; it is because I distrust violence and disbelieve in it; it is because I wish to secure this country against ever seeing a time when the 'have-nots' shall rise against the 'haves'; it is because I wish to secure for our children and our grandchildren and for their children's children the same freedom of opportunity, the same peace and order and justice that we have had in the past."

Congressman (later Senator) Clair Engle spoke these contrasting words to Congress in 1955: "I grant you, you start kicking the 160-acre limitation and it is like inspecting the rear end of a mule: You want to do it from a safe distance because you might get kicked through the side of the barn. But it can be done with circumspection, and I hope we can exercise circumspection."

In the long history of the uniting of water to land, politics has given law and policy a hard run. The commissioners of 1874, reviewing the history of irrigation in Italy, France, Egypt, and India had warned of trouble over the rights of water and spelled out ways to forestall it. Their foresight of the "clashing of private interests" in the Great Valley of California was not sufficient to prevent the pushing of the people into the shadow.

The service of the irrigation congresses of the nineties in opening the doors of the national treasury for western water development has not been forgotten. Large landowners have held great political power. In 1947 the Bureau of Reclamation pointed out that thirty-six landholders in the southern and western Central Valley owned three-quarters of a million irrigable acres. The corporate character of most of the ownerships gives them eternal life and hence a sustaining power that extends beyond the human generation. Large landowners have been as assiduous in exploiting the initial financial success of the Irrigation Congress as they have been in seeking to circumvent the controls over speculation and monopoly that the national Congress attached to its appropriations. To the original grant of interest-free money for irrigation, they have progressively been able to add further subsidies. They have won contributions from the national taxpayer in the name of flood control and contributions from users of hydroelectric power and from municipal and industrial users of water. The original Central Valley Project cost allocations assigned 63 percent to irrigation and asked irrigators to repay only 17 percent. They assigned 33 percent to California power users and asked them to repay 72 percent. Stanley Davison has called reclamation "irrigation-at-any-price," and Laurence Moss has described it as a "subsidy machine."

History has validated the 1877 prophecy of the *Visalia Delta* that "the people . . . will find themselves confronted by an array of force and talent to secure to capital the ownership of the water as well as of the land, and the people will at last have it to pay for."

The sources available to historians of development of the Great Central Valley of California are infinitely varied and rich. In addition to official documents and formal proceedings, a wealth of intensive studies have been produced treating one phase or another. Each of these is generous in citation of their sources. What remains unaccomplished is a comprehensive story of the Valley's development. The closest to this achievement is Robert de Roos,

The Thirsty Land, published by Stanford University Press in 1947. The citations that follow are classified by main types. They are fragmentary, but sufficient to indicate the broad scope of materials drawn upon.

Nineteenth century primary sources: *Irrigation of the San Joaquin, Tulare, and Sacramento Valleys, California*, H. Exec. Doc. No. 290, 43rd Cong., 1st sess. (1874); *Proceedings of the State Irrigation Convention . . . San Francisco . . . with Extracts from the Press on Irrigation* (1886); the proceedings of the irrigation congresses, which were held nearly annually, beginning in 1891 at Salt Lake City.

Studies of western land acquisition prior to irrigation: Margaret Cooper, *Land, Water and Settlement in Kern County, California, 1850-1890* (unpublished thesis, University of California, Berkeley, 1953); Paul Wallace Gates, *Homestead Law in an Incongruous Land System*. *American Historical Review*, 41: 655 (1936); also "Pre-Henry George Land Warfare in California," California Historical Society *Quarterly*, XLVI (1967), 121.

Histories of irrigation and works on reclamation law: William E. Smythe, *Conquest of Arid America* (1905); Alfred R. Golze, *Reclamation in the United States* (1961); Sheridan Downey, *They Would Rule the Valley* (1947); United States Department of the Interior, Bureau of Reclamation, *Landownership Survey on Federal Reclamation Projects* (1946).

Congressional Committee Documents: *Hearings before Subcommittee of Senate Military Affairs Committee on Central Valley Water Project...San Francisco, April 7, 1944* (mimeo.); Acreage limitation (reclamation law) review: *Hearings before Subcommittee on Irrigation and Reclamation of Senate Interior and Insular Affairs Committee*, 85th Cong., 2nd sess., on S. 1425, S. 2541, and S. 3448 (1958); Westlands water district contract: *Hearing before Senate Interior and Insular Affairs Committee*, 88th Cong., 2nd sess. July 8, 1964; *Acreage Limitation Policy: A Study Prepared by Department of the Interior Pursuant to a Resolution of the Senate Interior and Insular Affairs Committee*, 88th Cong., 2nd sess. (1964); *Hearings on Westlands Water District Contract, before Senate Interior and Insular Affairs Committee*, July 29, 1966 (not printed, in committee files).

Articles by present author containing extensive documentation *(inter alia)*: "Excess Land Law: Execution of a Public Policy," *Yale Law Journal* 64: 477 (1955); "Excess Land Law on the Kern? a study of law and administration of public principle vs. private interest," *California Law Review* 46: 153 (1958); "Excess Land Law: Pressure vs. Principle," *California Law Review* 47: 499 (1959); "Excess Land Law: Secretary's Decision? A Study in Administration of Federal-State Relations," *UCLA Law Review* 9: 1 (1962); "Excess Land Law: Calculated Circumvention," *California Law Review* 52: 978 (1964).

The Troy Case:
A Fight Against Discriminatory Freight Rates

By KENNETH R. JOHNSON

IN THE 1880s a number of new cities sprang up in Alabama, and many of the existing communities took on new life. It was an optimistic era in many respects. Businessmen dreamed of expanding activity and increased profits. Whole communities dreamed of becoming industrial and business centers. These dreams were supported by efforts to attract more tourists and to secure an ever-increasing population. This dream was often expressed collectively by the actions of the boards of trade, chambers of commerce, and local governments. Progress was to a great degree measured by the increase in business activity. Numerous individuals and towns in Alabama, as in other parts of the nation, eagerly looked forward to the growth and expansion of business and agricultural activity.[1]

One influential factor in determining which community and its businessmen would prosper was the freight rates charged by the railroads. The lack of uniform freight rates often resulted in new and expanded business opportunity for merchants in favored trade centers while opportunities were restricted for the businessmen in other, less favored communities. Freight rate discrimination between persons and places, and the resulting inequality of opportunity, was one of the factors contributing to the passage by Congress of the Inter-

---

[1] Albert Burton Moore, *History of Alabama* (University: University Supply Store, 1934), 521-27.

state Commerce Act in 1887.[2] Section three of the Commerce Act made it unlawful for the carriers to give unreasonable advantage "to any particular person, company, firm, corporation or locality. . . ." Section four of the act was intended to eliminate the long-short haul rate discrimination practice, in which railroads charged more for a short haul than a long haul on the same line. This was a common method by which carriers discriminated against one community in favor of another. Section four of the Commerce Act states:

> That it shall be unlawful for any common carrier subject to the provisions of this act to charge or receive any greater compensation in the aggregate for the transportation of the passengers or of like kind of property, under substantially similar circumstances and conditions, for a shorter than for a longer distance over the same line, in the same direction, the shorter being included within the longer distances.[3]

In the first years after the Commerce Act was passed, the railroads made little effort to conform to section four, and the Interstate Commerce Commission did not enforce a strict compliance. The long-short haul discriminatory principle had been followed for many years by the railroads in rate making, but the Commission expected the carriers to gradually abolish it.[4] A few carriers complied with the Commerce Act, but the railroads generally continued to follow the established practices of rate making without regard to the prohibitions in section four. The carriers generally justified the continuation of the higher rates per mile for a short haul by claiming that the conditions and circumstances were dissimilar between trade centers and intermediate points. They maintained that condi-

---

[2] Edward A. Purcell, Jr., "Ideas and Interests: Businessmen and the Interstate Commerce Act," *Journal of American History*, LIV (December, 1967), 562-64. For an excellent discussion of the fixing of railroad rates in Alabama and the application of those rates to local conditions see James F. Doster, "Trade Centers and Railroad Rates in Alabama, 1873-1885: The Cases of Greenville, Montgomery, and Opelika," *Journal of Southern History*, XVIII (May, 1952), 169-92. Also see testimony given by W. L. Bragg, former president of Alabama Railroad Commission, in the "Report of the Committee on Interstate Commerce," *Senate Reports*, 49th Congress, 1st Session (1885-86), no. 46, part 1, Appendix, 50.

[3] Copy of the Interstate Commerce Act can be found in Henry Steele Commager, ed., *Documents of American History* (New York: Appleton-Century-Crofts, Inc., 1949), II, 129-32.

[4] *Seventh Annual Report of the Interstate Commerce Commission* (1893), 33.

tions were dissimilar between two given points when railroad competition was different at the two points, when water transportation existed at one point and not the other, and when the volume of business to be competed for by the carriers was different.[5] In cases where there was dissimilarity, they argued, the railroads were free to establish rates as they saw fit, even if the rates did favor one community over another.

The Interstate Commerce Commission indicated its disagreement with the carriers' interpretation of section four and, after careful studies of the rate structure in the South, expressed strong disapproval of the continuation of the long-short haul practice. It readily admitted that water competition might influence the railroads' rate structure, but generally the commission held that common carriers were justified in fixing rates higher for a short haul than a long haul only in "rare and peculiar" cases.[6] This conflicting interpretation made clear the need for clarification of the meaning of various provisions of the Commerce Act before much success could be expected in eliminating rate discrimination.

The continuation of the old system of rate making after passage of the Commerce Act brought an increasing number of complaints from the numerous shippers and small communities. They suffered a commercial disadvantage due to rate discrimination.[7] Troy, the county seat of Pike County, Alabama, was one of the communities which felt the disadvantages of discriminatory freight rates. The businessmen of Troy had dreams of their town becoming an industrial, tourist, and commercial center in east Alabama.[8] Advantageous freight rates would greatly facilitate this ambition.

---

[5] William H. Joubert, *Southern Freight Rates in Transition* (Gainesville: University of Florida Press, 1949), 83-87.

[6] *Seventh Annual Report of the Interstate Commerce Commission* (1893), 33.

[7] Joubert, *Southern Freight Rates in Transition*, 91.

[8] Dudley S. Johnson, "Early History of Alabama Midland Railroad Company," *The Alabama Review*, XXI (October, 1968), 276-77. The population of Pike County increased from 24,423 in 1890 to 29,172 in 1900, and many counties neighboring to Pike grew even faster. The growth of Alabama from 1880 to 1900 indicates that the ambitions of the businessmen of Troy were grounded in the overall progress of the state: manufacturing establishments increased from 2,070 to 6,602. Capital in manu-

In 1892, Troy was served by the Central Railroad and Banking Company of Georgia and the Alabama Midland Railroad. The Alabama Midland had been completed in 1890 and was operated briefly by a group of Troy and Montgomery businessmen. Because of local control over the Alabama Midland and its competition with the Central of Georgia, there had been hope that Troy would receive more favorable freight and passenger rates. However, the Alabama Midland shortly passed from the local owners into the hands of Henry B. Plant and became a part of the "Plant System." And Troy businessmen soon found that they were paying rates higher in most cases than the shippers in Montgomery and Columbus, Georgia paid for the same service. Their obvious recourse was an appeal to the Interstate Commerce Commission.

On June 27, 1892, the Troy Board of Trade filed a complaint before the Interstate Commerce Commission against the Alabama Midland Railroad Company and the Central Railroad Banking Company of Georgia and their connections. The general ground for complaint was that the railroads were violating the Interstate Commerce Act and discriminating in their rates against Troy in favor of Montgomery. More specifically, the charges were as follows:

1. That the railroads collected a higher rate on goods shipped from New York, Baltimore, and other eastern points to Troy than on similar shipments through Troy to Montgomery.

2. That the railroads collected $3.22 per ton on phosphate rocks shipped to Troy from South Carolina and Florida and only $3.00 per ton on such shipments from the same fields over the same roads through Troy to Montgomery.

3. That the rate on cotton shipped to the Atlantic seaports was 47 cents per hundred pounds from Troy while the rate from Montgomery, the greater distance, was only 40 cents over the same railroad.

4. That special rates on goods intended for export shipped to Atlantic seaports favored Montgomery over Troy.

5. That Troy was unjustly discriminated against in being charged on shipments of cotton via Montgomery to New Orleans the full local

facturing establishments increased from $9,668,008 to $70,370,081. Value of manufactured products increased from $13,565,504 to $80,741,449. See United States Census Office, *Abstract of the Twelfth Census of the United States 1900* (Washington: Government Printing Office, 1902), 136, 149, 331.

rate to Montgomery plus the through rate from Montgomery to New Orleans.

6. That the rates on "class" goods shipped from northwestern cities to Troy were unjust and discriminatory against Troy when compared with the rates to Montgomery and Columbus.[9]

This complaint by the Troy Board of Trade, while tremendously important to local interests, also developed into a case of national significance. It became the basis for a test of the Interstate Commerce Commission's power to prevent the long-short haul rate discrimination practice.

The commission conducted an investigation and held hearings in Montgomery on the Troy complaint. Facts seemed to confirm the charges made by the Trojans. The rates on certain classes of merchandise from Louisville, Kentucky to Troy were $1.40, yet the same goods could be shipped to Columbus, Georgia, a greater distance, for only $1.07. The rates from Louisville to Montgomery were, in terms of distance, more favorable to Montgomery than to Troy. The same rate structure existed with merchandise shipped from other northwestern cities such as Cincinnati and St. Louis.[10] The railroad rate on cotton from Troy to Atlantic seaports was the sum of the local charge from Troy to Montgomery and the through rate from Montgomery to the seaports. Rates on commodities destined for Troy from the seaports were made on the same basis. But cotton sent from Montgomery through Troy to the seaports paid only the through rates.[11] All charges made by the Troy Board of Trade against the railroads were shown to be essentially correct. There could be little doubt that the rates collected by the railroads were more friendly to the Montgomery shippers than to those in Troy.

On August 15, 1893, the commission rendered its decision. Every important aspect of the Troy case was decided in favor of the Troy interests. The railroads were directed to conform to the following:

[9] *Interstate Commerce Commission* vs. *Alabama Midland Railroad Company*, 168 U.S. 144 (1897).
[10] Joubert, *Southern Freight Rates in Transition*, 92-94.
[11] *Ninth Annual Report of the Interstate Commerce Commission* (1895), 87-88.

1. No higher rates should be charged on goods shipped from Louisville, St. Louis, or Cincinnati to Troy than were collected on such shipments to Columbus, Georgia, and Eufaula, Alabama.

2. The rates on cotton from Troy to New Orleans through Montgomery should not exceed 50 cents per hundred pounds.

3. No higher special rates should be charged on cotton shipped from Troy to the seaports for export than were charged on such shipments from Montgomery.

4. No higher rates should be charged on cotton shipped from Troy to the ports of Brunswick, Savannah, or Charleston than were collected on similar shipments from Montgomery.

5. No higher rates should be charged on goods shipped from New York and eastern points to Troy than were collected on similar shipments through Troy to Montgomery.

6. No higher charge should be made for shipping phosphate rock from South Carolina and Florida to Troy than was made for similar shipments to Montgomery.[12]

The Troy business interests felt that they had won a decisive victory. The Trojans were not interested in mounting a broad-scale attack on the railroads, the long-short haul principle, or the system of rate making commonly followed in the South. They simply wanted rates at least as favorable as those received by the Montgomery and Columbus shippers and preferably a rate structure more advantageous than that of neighboring communities around Troy.

The commission's order to the railroads, however, went further than just protecting a bit of local interest. It constituted an attack upon the whole system of rate making in the South. The order would have removed from Montgomery its special rate advantage or given Troy the same advantage. Such success on the part of the Troy interests would surely have encouraged numerous other small towns to make similar complaints. The result would have been a broad-scale lowering of rates at small, noncompetitive points or an increase in rates at the large trade centers where more competition existed. From the carriers' point of view, either course would be disastrous in that revenues would be reduced at a time when

[12] *Interstate Commerce Commission* vs. *Alabama Midland Railroad Company*, 168 U.S. 144 (1897).

many carriers were already operating at a loss. Thus the question of the authority of the Interstate Commerce Commission was brought into sharp focus. Under what conditions did the Commission have the authority to order an end to the long-short haul rate discrimination? Or under what conditions would the carriers be left free to establish rates as they saw fit? While these questions in no way motivated the complaint by the Troy Board of Trade, its case and the future effectiveness of the commission over similar cases depended upon the federal courts' interpretation of the Interstate Commerce Act.

The Alabama Midland and other carriers chose to ignore the commission's order. The commission, having no authority to enforce its own decisions, filed a Bill of Complaint in the United States Court for the Middle District of Alabama, seeking a court order directing the carriers to comply.[13]

In the case of the *Interstate Commerce Commission* vs. *Alabama Midland Railroad and others*, Judge John Bruce presided.[14] The attorneys for the commission presented their evidence showing rate discrimination against the Troy interests and in favor of Montgomery. They wanted a court decision based on that evidence. If the court had concerned itself only with the question "Did the railroad freight rates favor the Montgomery interests over the Troy interests?" then the case could have been closed quickly. Such action by the court would have presupposed the authority of the commission to order the carriers to alter their rates. The court was unwilling to make this presupposition.

Actually, the carriers did not deny the facts as presented by the commission, although they did introduce some "new evi-

---

[13] *Ibid.*; *The Railroad Age*, a magazine friendly to the railroad interests, suggested that in the Troy Case a convincing explanation of the rate structure would be so difficult that "a railroad is likely to find it best to yield without arguing the matter." See *The Railroad Age* (Sept. 8, 1893), 673.

[14] Judge John Bruce had been a Republican carpetbagger in Alabama during the Reconstruction Era. Originally from Iowa, he was appointed to the bench by President Grant and reflected the more conservative outlook of the Republican Party in the late 19th century. See Thomas A. Owen, *History of Alabama and Dictionary of Alabama Biography* (Chicago: The S. J. Clarke Publishing Co., 1921), III, 239-40.

dence" which the commission had not considered before.[15] The attorneys for the railroads simply denied that the commission had the authority to issue the order. They pointed out that section four of the Commerce Act states that no more shall be charged for a short haul than a long haul when the transportation is under "substantially similar circumstances and conditions."[16] They claimed that conditions at Montgomery were dissimilar from the conditions at Troy. If the court upheld this claim, then the commission would be denied authority over the rates at Troy, Montgomery, and some other surrounding points, and the carriers would be left free to set rates as they saw fit.

In claiming that conditions were dissimilar at Troy and Montgomery, the carriers put forth three basic arguments. First, it was pointed out that Montgomery was an old established commercial center. The volume of business to be competed for by the carriers was much greater than at Troy. The commission could hardly deny the obvious fact that Montgomery was a larger commercial center than Troy, but it denied that the volume of business to be competed for would justify discriminatory rates. The commission claimed that:

> The fact that one city is much larger and has more important and extensive business interests than another . . . is no justification for discrimination in favor of such city. The object of the Act to regulate commerce was to eradicate the existing system of rebates and unjust rate discriminations in favor of particular localities, special enterprises, and favored individuals.[17]

If the more extensive business interests at Montgomery could justify lower rates there than at Troy, then the Commerce

---

[15] The Commission was very critical of the Court for permitting introduction of "new evidence" and of the carriers for withholding information from the Commission during its investigation and then bringing that information before the court as "new evidence" at a later time. The Commission claimed that such "new evidence" was often inconsequential, but as long as the carriers could indulge in this practice, the powers of the Commission would be curbed. See *Ninth Annual Report of Interstate Commerce Commission* (1895), 89.

[16] The leading attorney for the Alabama Midland Railroad was A. A. Wiley, a prominent businessman and attorney in Montgomery, who was for a number of years associated with Henry B. Plant.

[17] *Ninth Annual Report of the Interstate Commerce Commission* (1895), 88.

Act, which was partially to prevent discrimination between localities, could actually be used to justify the discrimination that it was intended to prevent.

Secondly, the carriers claimed that rail competition for the business in Montgomery was much more strenuous than in Troy, thus creating dissimilar conditions. Again the commission readily admitted that more railroad lines served Montgomery than Troy but pointed out that rail competition could exist at Troy with two lines just as well as at Montgomery with many. But more basically the commission argued that competition between carriers could not justify lower rates for one community than for another when the effect of such lower rates was to destroy natural competition between communities or to discriminate between persons. It stated that:

> In so far as competition of carriers promotes the welfare of persons and places without undue injury to other persons and places it shall be encouraged; but when such competition plainly operates to destroy or prevent the growth of one town and build up another, it should be justly regulated. Due observance of the fourth section will largely accomplish both purposes; it will encourage legitimate and restrain illegitimate competition in the carrying trade.[18]

One of the purposes of the Interstate Commerce Act was to reinstitute competition among the railroads in order that fair rates and equal opportunity might result. If the existence of competition at Montgomery could justify lower rates there than at Troy, then the act could be used to justify the discriminatory rates that it was enacted to prevent. The commission expressed the opinion that "one of the principal causes for insufficient revenue on many lines is the charging of unreasonably low rates upon competitive traffic," and that fair and proper adjustment of rates might prevent the further financial losses the roads had been reporting.[19]

Thirdly, the carriers claimed that there was a threat of water competition at Montgomery that did not exist at Troy.

---

[18] *Eighth Annual Report of the Interstate Commerce Commission* (1894), 20.
[19] *Seventh Annual Report of the Interstate Commerce Commission* (1893), 38. For the financial losses of the Alabama Midland see *Poor's Manual of Railroads* (1893), 8.

The Alabama River was open all the year. Goods could be carried by water from Montgomery not only to southern seaports but to their ultimate destination, if the railroad rates were not kept at or below the level of profitable carriage by water. The carriers claimed that at an earlier time when railroads had higher rates, active water competition had developed on the Alabama River.[20] The railroad rates at Montgomery were then reduced to a point competitive with the water rates, and the volume of carriage on the river became comparatively small. The railroads concluded that the controlling power of water competition remained in full force and must ever remain in force as long as the river continued to be navigable.[21] Thus rail rates at Montgomery were based on dissimilar conditions from those at Troy.

The Interstate Commerce Commission did not produce evidence to refute this statement but denied that the mere threat of water competition could justify discrimination in freight rates. The commission merely pointed out that the railroads did not prove that water competition actually existed between Mobile and Montgomery or that it justified lower competitive railroad rates at Montgomery than at Troy.[22]

The carriers thus concluded that due to the dissimilarity of "circumstances and conditions" at Montgomery and Troy, they were not bound by the prohibitions in section four of the Commerce Act. If this was true, then the commission's order was invalid because it simply had no authority to issue such order.

Judge Bruce agreed with the carriers' logic. The bill of complaint was dismissed and the carriers were not ordered by the court to comply with any part of the commission's order. The decision did not confirm or deny that the freight rates

[20] The carriers were probably referring to a line of steamers formed in 1885 to operate between Montgomery and New Orleans. It operated for a short time and did succeed in forcing a reduction of railroad rates to and from Montgomery. See Joubert, *Southern Freight Rates in Transition*, 75.
[21] *Eleventh Annual Report of the Interstate Commerce Commission* (1897), 38; *Interstate Commerce Commission vs. Alabama Midland Railroad Company*, 168 U.S. 144 (1897).
[22] Joubert, *Southern Freight Rates in Transition*, 93-94.

discriminated against the Troy interests. In fact the existing rates, however discriminatory, apparently had no influence on the court decision. It merely concluded that conditions at Montgomery were sufficiently dissimilar to justify a departure from the prohibitions in section four of the Commerce Act, thereby leaving the carriers free to set rates as they saw fit. The commission protested that "This construction of the law by the courts amounts to an endorsement of the Southern carriers' practically assumed right to divert trade in the South through such channels and to benefit such localities as they may desire."[23]

The commission, listing twenty-five distinct grounds of error, appealed its case to the Fifth Judicial Circuit Court of Appeals in New Orleans. In June, 1896 the appeals court upheld the lower court's decision. The appeals court stated that:

> The volume of trade to be competed for, the number of carriers actually actively competing for it, a constantly open river present to take a large part of it whenever the railroad rates rise up to the mark of profitable water carriage, seem to us, as they did to the circuit court, to constitute circumstances and conditions at Montgomery substantially dissimilar from those existing at Troy, and relieve the carriers from the charges preferred against them by its Board of Trade.[24]

The Interstate Commerce Commission further appealed the case to the Supreme Court, which rendered its decision on November 8, 1897. The lower court's decision was upheld by a vote of eight to one. This decision of course confirmed the fact that there would be no relief for the rate discrimination against the Troy business interests. But more importantly, the decision emasculated section four of the Commerce Act. If a higher rate could be justified for a short haul than for a long haul every time the volume of railroad business differed between two points or when rail competition differed, then the long-short haul discriminatory principle could be justified in almost every conceivable situation. Despite the Supreme

---

[23] *Ninth Annual Report of the Interstate Commerce Commission* (1895), 87-88.
[24] New Orleans *Times-Democrat*, June 3, 1896; *Interstate Commerce Commission* vs. *Alabama Midland Railroad Company*, 168 U.S. 144 (1897).

Court's protest to the contrary, section four was a dead letter and the commission's power to prevent the long-short haul rate discrimination practice was almost completely abolished.[25]

Associate Justice John Marshall Harlan, the lone dissenter in this case, wrote that:

> Taken in connection with other decisions defining the powers of the Interstate Commerce Commission, the present decision, it seems to me, goes far to make that Commission a useless body for all practical purposes, and to defeat many of the important objects designed to be accomplished by the various enactments of Congress relating to interstate commerce.[26]

The commission agreed with Justice Harlan and went so far as to state that unless the country was willing to undergo a recurrence of all those practices which existed before the passage of section four, it must be reenacted by Congress in a different form.[27]

In the next few years there continued to be numerous complaints to the Interstate Commerce Commission against the long-short haul practice. But little could be done in the way of relief due to the commission's limited authority.[28] The Congress was requested on many occasions to strengthen section four, but nothing was done in this direction until 1910, when the Mann-Elkins Act was passed. This act amended section four, making it mandatory that railroads receive permission from the commission before a higher rate could be made for a shorter than a longer haul. In the first eight months after the Mann-Elkins Act was passed, the commission received 5,723 applications for permission to continue the long-short haul practice.[29] This large number of applications seems to demonstrate that conditions about which the Troy Board of Trade complained were typical throughout the nation.

---

[25] *Interstate Commerce Commission* vs. *Alabama Midland Railroad Company*, 168 U.S. 144 (1897); D. Philip Locklin, *Economics of Transportation* (Chicago: Richard D. Irwin, Inc., 1947), 546-47.
[26] *Interstate Commerce Commission* vs. *Alabama Midland Railroad Company*, 168 U.S. 144 (1897).
[27] *Eleventh Annual Report of the Interstate Commerce Commission* (1897), 45-46.
[28] Joubert, *Southern Freight Rates in Transition*, 94.
[29] *Twenty-fourth Annual Report of the Interstate Commerce Commission* (1910), 45-46; Locklin, *Economics of Transportation*, 546-49.

By Patrick G. Porter

Origins of the American Tobacco Company

During the 1880's and 1890's, the innovations of James Buchanan Duke first disrupted and then rationalized the American tobacco industry. Duke's career and the early history of his American Tobacco Co. serve as case studies in both the history of business administration and in the coming of "big business" to the United States.

Like many other American industries processing agricultural products for urban, mass consumption, the cigarette business underwent great changes during the 1880's. Innovations in production processes caused supply to outrun demand and drove manufacturers into severe competition. Packaging and advertising became the major competitive weapons as producers vied to market relatively undifferentiated products that were saleable only within a narrow price range. Despite increased advertising and organizational integration, however, the industry's growth rate declined. Then, the man who had initiated most of the revolutionary innovations — James Buchanan Duke — succeeded in leading his industry into combination by founding one of the first great holding companies in American history.[1]

When Duke's American Tobacco Co. was listed on the New York Stock Exchange in the summer of 1890, it signaled the combination of the major producers in its industry. Its five constituent companies produced approximately 90 per cent of the cigarettes made that year in the United States, and from this base, the company's growth was phenomenal.[2] By 1909, its equity capitalization had increased from $25,000,000 to over $316,000,000. In the two decades after its founding, American Tobacco absorbed approximately 250 separate companies and came to produce about 80 per cent of the cigarettes, plug tobacco, smoking tobacco, and snuff made in the United States.

[1] A part of the research for this study was made possible through the financial support of the Institute of Southern History at Johns Hopkins University and the Ford Foundation.
[2] *Commercial and Financial Chronicle*, September 13, 1890; U.S. Bureau of Corporations, *Report of the Commissioner of Corporations on the Tobacco Industry* (Washington, 1909–1915), I, 64.

Only the cigar industry eluded James Duke.[3] By the time of its court-ordered dissolution of 1911, the Duke empire included a vertically integrated network of companies engaged in wholesale and retail distribution of tobacco products, in the production of leaf tobacco, packaging, and various "smokers' supplies," as well as a national monopoly in licorice, an important material in the manufacture of cigarettes.[4] American Tobacco did business in every part of the tobacco industry.

Because Duke combined forces of innovation and amalgamation, he revolutionized the tobacco industry in a decade and influenced its course for many more years. His methods of purchasing, sales, and distribution were forced upon his competitors in the 1880's. He was the leading factor in the introduction of machine production, an innovation which, more than any other, made for change. In a time of productive overcapacity, high sales costs, declining growth rates as the market leveled off, and brutal competition, he overcame the obstacles to cooperation and succeeded in rationalizing the industry. He is, therefore, of considerable importance in the history of American business administration.

But Duke's career and the origins of the American Tobacco Co. have significance beyond the history of administration or the isolated story of the tobacco industry. The company's early history is also an important chapter in the rise of the large corporation in the American economy. Thus, this article presents Duke and his company as a case study in the coming of one such corporation and advances several hypotheses about the firm and the industry. First, that the experience of the cigarette industry exhibited several important differences from the general pattern of combination first developed in the petroleum industry: significant vertical integration occurred in cigarettes prior to combination, unlike the case of Standard Oil; integration proved effective in assuring a reliable flow of raw materials and in rationalizing distribution and marketing, so producers never tried the loose associations which proved so unworkable in the oil industry; overproduction came later in the lifespan of the cigarette industry than in petroleum and not until the experience of other companies — such as the railroads and the oil manufacturers — had shown pools and associations to be too weak an organizational form. Second, that cigarette manufacturers did not utilize the trust because a new legal device, the holding company, had appeared by the time producers were ready for combination. They took advan-

---

[3] Bureau of Corporations, *Report on Tobacco*, I, xxxi.
[4] Bureau of Corporations, *Report on Tobacco*, I, 13.

tage of the holding company device and of the New Jersey general incorporation law of 1889. Like Standard Oil and other companies, however, the cigarette combination soon turned to increased vertical integration and organizational consolidation. And, again in line with the Standard Oil pattern, the creation of an overproductive capacity in the face of a limited market was the key force in causing consolidation.

I

The dual symbols of American tobacco consumption in 1850 were the cigar and the spittoon. In succeeding decades, however, chewing tobacco and cigars were slowly being supplanted as the dominant forms of tobacco use by the cigarette. For a variety of reasons — its low price, its popularity with women, its advertising, and its role as what one perhaps overwrought observer has termed a "natural accompaniment of the creeping neurasthenia of urban existence" — the cigarette grew enormously in popularity after the Civil War.[5] American businessmen soon moved to take advantage of this growing market.

In most cases, the firms which later came to dominate the industry began cigarette production as an extension of an already-existing tobacco business. Francis S. Kinney, founder of the Kinney Tobacco Co. of New York, moved into cigarettes in 1869 as an addition to his smoking tobacco business.[6] William S. Kimball, president of W. S. Kimball & Co. of Rochester, had begun tobacco manufacture during the Civil War and became a cigarette producer in 1876.[7] The firm which later became the largest manufacturer in the industry, W. Duke, Sons & Co. of Durham, North Carolina, had produced smoking tobacco since 1866, but did not make the move into cigarettes until 1881.[8]

[5] The quotation is from Jerome E. Brooks, *The Mighty Leaf* (Boston, 1952), 252. The number of cigarettes on which internal revenue taxes were paid jumped from slightly under 20,000,000 in 1865 to over 3,300,000,000 in 1895. See U.S. Bureau of Internal Revenue, *Report of the Commissioner of Internal Revenue* (Washington, 1895), 378–79. The product was apparently introduced into the United States in the 1850's, though this is not entirely clear. Robert K. Heimann, *Tobacco and Americans* (New York, 1960), 204, presents a persuasive case that the cigarette first appeared in the 1850's. Charles D. Barney & Co., *The Tobacco Industry* (New York, 1924), 19, places the date of appearance at about 1860; William W. Young, *The Story of the Cigarette* (New York, 1916), 8, says "about 1866;" and Charles E. Landon, "Tobacco Manufacturing in the South," in William J. Carson (ed.), *The Coming of Industry to the South* (Annals of the American Academy of Political and Social Science, Philadelphia, 1931), 44, asserts that the cigarette was introduced in 1867 from England.

[6] Testimony of Francis S. Kinney, U.S. Department of Justice MSS relating to the case of *U.S. v. American Tobacco Co.*, U.S. Circuit Court, Southern District of New York, Equity Case Files E 1–126, 1908, Records Group 60, National Archives. Hereafter cited as Circuit Court MSS.

[7] Blake McKelvey, *Rochester: The Flower City 1855–1890* (Cambridge, Mass., 1949), 236–38; *Tobacco*, February 25, 1887.

[8] Testimony of James B. Duke, Circuit Court MSS.

All the companies which turned to cigarettes faced a common production problem, and all met it in the same way. The problem was that cigarettes were hand-made by skilled rollers, who were scarce in the United States. The solution which the cigarette companies found was either to import skilled immigrants or to hire away the immigrant laborers already working for New York firms making the expensive Turkish and Egyptian brands.[9] The Kinney Tobacco Co. induced East European cigarette rollers to immigrate, and the Dukes brought 125 immigrant rollers from New York to Durham.[10] Often these skilled laborers worked at several successive American firms.[11] They usually supervised cigarette production and trained large numbers of young women and girls in the art of rolling cigarettes by hand. These factory girls provided a less expensive source of labor than male workers and were the industry's main line of defense against labor unions after they had replaced their instructors in the factories.[12]

Once domestic firms solved the labor problem, they began to expand their markets, driving sales of imports and of the expensive Turkish and Egyptian brands downward by means of lower-priced cigarettes. By the latter part of the 1870's, it was clear that the domestic, bright-leaf tobacco cigarettes were becoming the dominant type. Domestic producers were acquiring an ever-increasing superiority in American markets.[13]

But the cigarette remained a kind of orphan in the family of American tobacco manufacturing. As a trade journal later pointed out:[14]

> For about fifteen years . . . the cigarette business did not attract much attention in the trade, nor develop as rapidly [as it did during the decade of the 1880's]. This was due not so much to . . . small profits, as to the fact that established tobacco manufacturers took up the cigarette as a side issue, but, having organized departments for making and selling them,

---

[9] The importation of immigrant labor was a common practice for other domestic industries in need of skills. See Edward C. Kirkland, *Industry Comes of Age* (New York, 1961), chapter XVI.

[10] Heimann, *Tobacco and Americans*, 206, 212.

[11] For example, J. M. and David Siegel, immigrant brothers from Kovno, Russia, worked for Goodwin & Co. in New York, were hired by the Dukes to supervise the Duke cigarette department in Durham for a time, and later set up their own company. See Hiram V. Paul, *History of the Town of Durham, N.C.* (Raleigh, North Carolina, 1884), 111-12.

[12] In 1878, a trade journal observed that "in the last great strike of segar makers, the cigarette makers did not participate. Three years ago men were generally employed, and a strike took place — women were substituted, and no trouble has since occurred." *U.S. Tobacco Journal*, November 2, 1878. The cigarette industry was an early example of the kind of semi-skilled, industrial labor force which the American Federation of Labor later found so difficult to assimilate. After the advent of machine production in the 1880's the labor skills required in the cigarette industry were further lessened, thereby further diminishing the possibility of unionization.

[13] See *U.S. Tobacco Journal*, June 19, 1877.

[14] *Tobacco*, January 31, 1890.

found little difficulty in putting them on the market along with a line of smoking tobaccos, and under these conditions the growth and development of the cigarette in its early stages was slow.

The industry in these early years was, thus, relatively small-scale and unimportant. But with the opening of the 1880's, a new competitor appeared on the scene who was to revolutionize the entire industry in a decade — James Buchanan Duke.

II

"Buck" Duke was a shrewd and tough businessman, ambitious and fiercely competitive. He drove W. Duke, Sons & Co. to the top of the cigarette trade in less than a decade. More than any other individual he was responsible for the formation of the American Tobacco Co., and he ran that vast combination for a score of years after its founding. When the courts dissolved the company, the only man who understood the complex interrelationships of the combination well enough to dismantle it rationally was James Duke. His Horatio Alger story is genuine, and his biographers have done him more than justice.[15]

James Duke's father, Washington Duke, had founded the business after the close of the Civil War. The Dukes ran their business from the family farm outside Durham until 1873, when they built a factory in the town. In 1878 a five-man partnership was created. Washington Duke, James Duke, James' brother Benjamin N. Duke, Richard H. Wright (a local tobacco manufacturer), and George W. Watts (a Baltimore businessman) each contributed $14,000. The partnership ended in 1885 and the firm then incorporated under its previous name, W. Duke, Sons & Co.[16]

The Duke firm did make some profits, mostly in granulated smoking tobacco, but not enough to satisfy the ambitious James Duke. He felt that as long as the company stayed in the production of smoking tobacco it had no real future. The predominance of the "Bull Durham" brand, manufactured by W. T. Blackwell & Co., also of Durham, was apparently unshakeable. Duke allegedly remarked, "my company is up against a stone wall. It can't compete with Bull Durham. Something has to be done and that quick. I am going into the cigarette business." The other partners were less certain of the

[15] See John Wilber Jenkins, *James B. Duke: Master Builder* (New York, 1927); John K. Winkler, *Tobacco Tycoon: The Story of James Buchanan Duke* (New York, 1942); and Watson S. Rankin's pamphlet, *James Buchanan Duke (1865-1925): A Great Pattern of Hard Work, Wisdom, and Benevolence* (New York, 1952).
[16] Testimony of James B. Duke, Circuit Court MSS.

wisdom of the decision, but Duke ultimately persuaded them. In 1881, the company began the production of cigarettes.[17]

For the first two years after the shift into cigarettes, the older partners had good reason to regret their decision, for the move had all the earmarks of a rousing disaster. The Duke firm made little headway because the government at that time was considering a reduction of the cigarette tax from $1.75 to 50 cents per thousand.[18] The bill was not passed until March 1883, the reduction to take place the following May. The Duke brands were not established in the market and many dealers refused to buy them and take the chance of losing the difference in tax, should the tax be lowered. Duke thus found himself in a very unenviable position: his factory was forced to close, his warehouse bulged with unsold cigarettes, and his brands made little progress.[19]

When the government finally reduced the tax in March 1883, Duke made the first in a long chain of bold decisions. He immediately reduced the price of his cigarettes from 10 cents to 5 cents per pack of ten cigarettes. He declared that jobbers' orders would be filled at the lower price, provided that at least three-quarters of the goods were delivered after the tax reduction in May.

The Duke products became the lowest priced ones on the market, and in the two months before the tax reduction went into effect, Duke sold his backlog of cigarettes, though at a loss. His factory reopened. He firmly established his brands in the trade through a combination of low prices and advertising.[20] He had caught his competitors napping.

Duke's increased advertising at the time of the tax reduction taught him a lesson he did not forget. He continued to use advertising to stimulate sales so that he could keep production costs down.[21] He established offices and a factory in New York to be nearer his markets and to secure better advertising facilities.[22] That same year he bought 380,000 chairs and had painted on the back of each an advertisement for his "Cameo" brand cigarettes. The chairs were placed in cigar stores throughout the nation. Duke "was an aggressive advertiser, devising new and startling methods which dismayed his competitors; and [he was] always willing to spend a proportion

---

[17] Jenkins, *James B. Duke*, 65.
[18] Excise taxes on tobacco products originated during the Civil War.
[19] Testimony of James B. Duke, Circuit Court MSS.
[20] Testimony of James B. Duke, Circuit Court MSS; Neil H. Borden, *The Economic Effects of Advertising* (Chicago, 1942), 221; and Jenkins, *James B. Duke*, 70–72.
[21] Borden, *Economic Effects of Advertising*, 221.
[22] Alfred D. Chandler, Jr., *Strategy and Structure* (Cambridge, Mass., 1962), 27; American Tobacco Co., "Sold American!" (n.p., 1953), 20.

of his profits which seemed appalling to more conservative manufacturers."[23]

Duke showed a consistent willingness to innovate and to move quickly in order to obtain the maximum benefit from innovations in the industry. An interesting example of this can be seen in his handling of packaging. One of the problems in increasing the market for cigarettes was that they were sold in loose, fragile paper packages which caused the cigarettes to break readily. Duke introduced a stiff, sliding box for cigarettes, and when another inventor produced a better version, Duke immediately ordered 50,000 of them.[24]

It is hard to avoid the conclusion that James Duke was the leading innovator in the American cigarette business during the 1880's. He made entrepreneurial contributions in marketing, in purchasing, and in production which were the driving forces for change. In other industries supplying either new or established consumer goods for the expanding urban markets, innovation often came through the entry of a competitor who had little previous experience in the business, like Gustavus Swift in meat-packing and John D. Rockefeller in petroleum.[25] Duke's fulfillment of the entrepreneurial function fits the same pattern — that of economic change wrought by an innovator not thoroughly grounded in the previous competitive methods within an industry.[26]

Duke's vigorous, imaginative merchandising put his company among the five dominant cigarette producers during the 1880's. The others were Allen & Ginter of Richmond, W. S. Kimball & Co. of Rochester, and the Kinney Tobacco Co. and Goodwin & Co., both of New York. These five companies followed the same basic pattern in production, in distribution, and in the means of acquiring their leaf tobacco. In each area, however, the Duke company appeared to achieve greater efficiency and to display more interest in innovation.

This was apparent in Duke's handling of his firm's purchasing problems. The major producers obtained most of their leaf tobacco through tobacco brokerage houses in the bright-leaf belts of the South. These brokers purchased the leaf at warehouse auctions, stored and dried it in their own warehouses, and then resold it to the manufacturers.[27] This situation made it possible, especially in years

[23] "The Beginnings of a Trust," *Collier's* XXXIX (August 10, 1907), 15–16.
[24] Jenkins, *James B. Duke*, 68; "The Beginnings of a Trust," 15.
[25] See Alfred D. Chandler, Jr., "The Beginnings of 'Big Business' in American Industry," *Business History Review*, XXXIII (Spring, 1959), 6–9.
[26] For an excellent analysis of the entrepreneurial function, see Joseph A. Schumpeter, *The Theory of Economic Development* (Cambridge, Mass. 1934), chapter II.
[27] On the warehouse system, see Nannie M. Tilley, *The Bright-Tobacco Industry 1860–1929* (Chapel Hill, North Carolina, 1948), 191–308.

when there was a shortage of crops, for speculators and rehandlers to make handsome profits.[28] Some of the manufacturers began in the 1880's to create their own purchasing departments in a half-hearted attempt to reduce these raw material costs. Only the Duke company, however, made a really successful effort to eliminate the middleman. Duke, as he later testified, appreciated the value of reducing the role of the "speculator who had been . . . buying and selling to the manufacturers, with the exception of Duke's Sons and Co. We had been buying a good part of our tobacco in the loose warehouses direct from the farmer."[29] The Dukes had their own warehousing facilities almost from the beginning of their business. The warehousing and the use of company-employed leaf buyers marked important steps toward vertical integration in purchasing, well in advance of the combination of the leading producers. The middleman was a costly liability for the manufacturers, and Duke led the way in his elimination even before the creation of the American Tobacco Co.

Here an important difference from the experience of Standard Oil is clear: vertical integration backward into an extensive purchasing network occurred before the combination in cigarettes and not in petroleum. Overproduction of petroleum came very early, and manufacturers tried associations in an attempt to rationalize the flow of crude oil to the refineries. Overproduction of leaf tobacco did not occur so rapidly, and the leaf markets did not join in unwieldy associations, as did producers of crude. Consequently, Duke could assure a steady and less costly flow of raw materials for cigarettes by integration through company buyers and warehouses.

Significant vertical integration also appeared in the means of distribution, as well as purchasing. From its earliest years, the cigarette industry sold its products by means of traveling salesmen. The drummer was the means through which the companies made potential consumers and retailers aware of their product. These drummers traveled all over the United States and abroad, attempting to stimulate demand for cigarettes. They took orders from wholesalers and retailers, wholesalers probably taking the larger share. Even company officials sometimes acted as drummers.

James Duke had been an effective salesman from the beginning of the Duke company, and continued to serve in that capacity from time to time until about 1885.[30] During the sales push at the time of

[28] U.S. Bureau of Corporations MSS relating to its investigation of the tobacco industry, File 4766, sections 1 and 2, Record Group 122, National Archives. Hereafter cited as Bureau of Corporations MSS. See also testimony of tobacco broker John B. Cobb, Circuit Court MSS.
[29] Testimony of James B. Duke, Circuit Court MSS.
[30] See Jenkins, *James B. Duke*, chapters III and IV.

the cigarette tax reduction, Duke worked hard to drum up sales: "Packing a bag with samples, he made one of his dashing trips through the country, taking orders everywhere . . . practically every keeper of a tobacco shop ordered in large quantities."[31] This basic method of distribution to jobbers and retailers through drummers continued until the formation of the American Tobacco Co. in 1890.

During the 1880's, however, significant organizational advances occurred in the distribution system. Each of the five leading producers continued to sell its products by having its traveling salesmen take orders for goods, title passing to purchaser on delivery.[32] But, with the Dukes again in the forefront, manufacturers organized and maintained a system of independent distributing centers in the principal cities in order to expand the market. Connected with these sales agencies were generally a manager, a city salesman, and one or two traveling agents. This organizational innovation was another indication of vertical integration, this time forward into sales and distribution. The cigarette found no ready market; it was a relatively new product of no intrinsic value to the consumer. Producers therefore had to devise a system of sales and distribution to make consumers aware of their products and to see that wholesalers and retailers stocked them. There is a clear contrast with the petroleum industry which, in its early years, utilized established marketing channels for coal oil to meet a ready market.[33]

In production, as in purchasing and distribution, the major firms followed similar patterns, patterns usually set by the Duke company. Until the 1880's, they produced all cigarettes by hand labor. The factory girls were virtually human machines, but the manufacturers sought a reliable mechanical means of mass producing the cigarettes. Like many another American invention, the cigarette machine was an example of induced innovation, called forth by the needs and rewards for machine production. During the 1870's, a wave of more or less useless contraptions designed to make cigarettes appeared.[34] A young Virginian, James Bonsack, invented the earliest practical machine in America. He patented his device in 1881 and improved it during the next two years. The newly-formed Bonsack Machine

---

[31] "The Beginnings of a Trust," 15.
[32] Defense stipulation marked Government Exhibit A, Circuit Court MSS.
[33] *Western Tobacco Journal*, July 15, 1889, cited in Bureau of Corporations MSS, File 4766, sections 1 and 2. See Chandler, "Rise of 'Big Business'," 8. On petroleum, see Harold F. Williamson, Arnold Daum, and others, *American Petroleum Industry: Age of Illumination 1859–1899* (Evanston, Illinois, 1959), chapter 13.
[34] *Tobacco News and Prices Current*, May 24, 1879; *U.S. Tobacco Journal*, June 7, 1879. See also statement by J. E. Bonsack on the invention of a cigarette machine by his uncle, James Bonsack, J. E. Bonsack MSS, Duke University Library.

Co. had its machines tried commercially in 1883.³⁵ Allen & Ginter received the first of the Bonsack machines for their factory at Richmond. They tried the device for a short time but soon decided against using it. There was, they apparently thought, a strong public prejudice against machine-produced cigarettes, and besides the machine did not function perfectly.

James Duke leased some of the Bonsack machines later that same year. Duke ignored the reasons for which Allen & Ginter had rejected the machines. He put his mechanics to work to improve the operation of the machines and brushed aside the question of consumer prejudice.³⁶ As he later stated: ³⁷

> We commenced to use the machines . . . largely earlier than any other manufacturer. The others could not make them go and they were also afraid that the cigarettes made on the machines the public would be prejudiced against them because they were machine-made. . . .
> I think Allen and Ginter started — Allen and Ginter had had the machine . . . as early as 1883 . . . but they did not do much with them; were afraid of them, but after they saw we had made a success of them in 1887 they took them up. The Kinney Tobacco Company took them up in 1888.

By the time the other manufacturers saw the error of their ways, Duke had stolen a long and quiet march on them. His use of the machines had put the Bonsack Co. in his debt, since he was willing to take the risk of trying full-scale production when others would not. In 1885, Duke secured a very favorable contract with the Bonsack company. The contract noted that: ³⁸

> the manufacturers of cigarettes who use the Bonsack machines . . . have so far declined to put the machines on their fine brands, for the reason that they fear that there may be a prejudice against machine-made work which might injure the sales of their goods, and . . . W. Duke, Sons & Co. are willing to put the machines on their best brands, and to do all their plain work on the Bonsack machines.

In return for this, the contract provided that Duke should get the use of the machines at a rate of 24 cents per thousand cigarettes rather than the usual rate of 30 or 33 cents per thousand. Duke was to get his discount through rebate checks. Further, the arrangement was to be permanent unless the Dukes divulged the provisions of the contract or unless they failed to use the machines on their better brands.³⁹ Just over six months later, an addition to the con-

---

³⁵ Paul, *History of the Town of Durham*, 207-208.
³⁶ Heimann, *Tobacco and Americans*, 212.
³⁷ Testimony of James B. Duke, Circuit Court MSS.
³⁸ *Bonsack Machine Co. v. S. F. Hess & Co.*, 68 F. 125 (1895).
³⁹ Ibid.

tract provided that W. Duke, Sons & Co. should always get the machines at a rate 25 per cent below that given any other manufacturer.[40]

The Duke company began to produce most of its cigarettes by machine in 1885, encountering little of the consumer resistance its rivals had anticipated. Duke's application of the Bonsack machines revolutionized the business of making cigarettes, and the profits of the Duke company rose during subsequent years.[41]

The machines brought about a tremendous reduction in the cost of manufacturing. By 1884, the Bonsack machine was producing from 100,000 to 120,000 cigarettes per day, the equivalent of the production of forty to fifty hand workers. The exact amount by which production costs fell is unclear, but one scholar estimates that the cost of manufacture was reduced from 80 cents per thousand to 30 cents.[42] Government estimates do not include the 1880's, but they do show the labor cost differentials from the handmade products of the 1870's to the machine-made ones of the 1890's. In 1876, labor costs were about 96 cents per thousand; by 1895, labor costs for the same cigarette were slightly over 8 cents per thousand.[43] Duke, like other producers, initially overcame any popular prejudice against the machines in a very simple way: he used them in the greatest secrecy and the public remained unaware of their widespread application for years.[44]

The machines played a key role in bringing about a high degree of concentration and eventual combination in the cigarette industry. Duke's introduction of machine production was clearly the most significant innovation he made in the industry. A few men controlled the best machines through patents, which led to concentration of production in a few large companies.[45] The machines meant that these few companies could produce at relatively equal costs tre-

---

[40] *Wright v. Duke*, 36 N.Y. Supp. 855 (1895).
[41] See deposition by William H. Butler of the Kinney Co., *Bonsack Machine Co. v. S. F. Hess & Co.*, 68 F. 126 (1895) and *Wright v. Duke*, 36 N.Y. Supp. 856 (1895).
[42] Borden, *Economic Effects of Advertising*, 493.
[43] Bureau of Corporations, *Report on Tobacco*, I, 63.
[44] The feared prejudice may have been overestimated by the manufacturers. This prejudice is a bit difficult to understand. Perhaps the consumer would have regarded a machine-made cigarette as too artificial, as somehow not genuine. There was also a constant clamor by anti-tobacco leagues that the product was poisonous, particularly the paste used to hold the cigarette together, though paste was used in the hand-rolled products as well. The manufacturers may well have contributed to public fears through the nature of some of their advertising: Goodwin & Co., in pushing one of its brands, claimed a great advance in the kind of rice paper used to enclose the tobacco, and stated that smokers "have heretofore . . . been inhaling one of the deadliest poisons known." *Tobacco News and Prices Current*, February 15, 1879; *U.S. Tobacco Journal*, September 7, 1878. Such maneuvers were hardly calculated to inspire public confidence. On the secrecy of the use of machines, see *Tobacco*, February 25 and June 10, 1887, June 8, 1888, January 17 and January 31, 1890.
[45] Bureau of Corporations, *Report on Tobacco*, I, 63.

mendously increased numbers of what were basically similar products. Because the products usually came to the consumer in a pack of ten cigarettes for 5 cents, even an advantage like Duke's rebates from the Bonsack Co. could make no significant inroad in price to the consumer. Reduced costs even of about 10 cents per thousand in the cost of leasing machines, and slightly reduced leaf costs could not cause price competition since the price for consumers was already at such a low level. Overcapacity also became a real problem as soon as machine production was introduced, and the manufacturers fought fiercely to preserve or enlarge their share of the limited market.

As was the case in some other industries, such as patent medicines, the competition expressed itself chiefly in the form of expensive and sometimes elaborate advertising. Advertising and sales costs became almost identical. The advertising grew more bizarre as the decade progressed.[46] As competition came to center around packaging and gimmickry, advertising costs rose. Profits were squeezed as the major companies spent huge sums to outdo each other. In 1889, the last year before the companies combined, the Duke company spent $800,000 on advertising. This amounted to 20 per cent of gross sales and provided one impetus toward combination.[47]

The advertising flood represented, of course, a desperate struggle among the leading firms for a share in a market in which supply had outrun demand. Machine production came at a time when the industry's growth rate was declining. Although the demand for cigarettes grew almost every year, the rate of increase fell off sharply in the 1880's and 1890's. In the five years from 1879 through 1884, the number of cigarettes on which internal revenue taxes were paid increased by 281 per cent. In the following five years, 1884–1889, the figure rose by only 137 percent, indicating a trend which was confirmed by the increase of only 48 per cent in the period 1889–1894.[48] The market clearly was leveling off (see Figure 1). As a result, the dominant firms were running out of maneuvering room and competed all the more fiercely.

III

As the industry moved toward the end of the 1880's, producers faced overcapacity in production, increasingly costly and wasteful

---

[46] For an indication of the strange and wondrous advertising, see *Tobacco*, May 13 and December 23, 1887, January 20, April 6, and August 10, 1888.
[47] Testimony of James B. Duke, Circuit Court MSS.
[48] These data are drawn from the *Report of the Commissioner of Internal Revenue* for 1895, pp. 378–379. Figure 1 is derived from information in the same source.

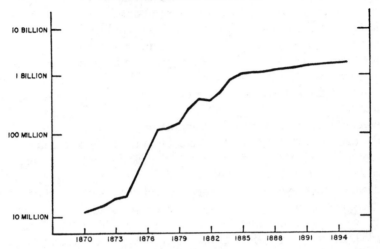

Figure 1

NUMBER OF CIGARETTES ON WHICH INTERNAL REVENUE TAXES WERE PAID, 1870-1895

(Semi-Logarithmic Scale 1 x 60)

competition through packaging and gimmickry, and, perhaps most important of all, a leveling off of demand for their product. Cooperation became an ever more appealing and logical solution, and, once more, James Duke led the way.

Duke tried to effect a loose combination between his company, Allen & Ginter, and the Bonsack Co. early in 1888. He sought to keep the Bonsack Co. from leasing its machines to the Kinney Co. and wanted a contract giving his firm and Allen & Ginter exclusive use of the machines. A large stockholder in Bonsack wrote Francis S. Kinney: "Whether Duke is paying enough to prevent . . . further efforts to put the machines in your factory and secure his determined effort to form this monopoly I do not undertake to say and firmly believe that unless you take some action soon you will be too late." [49] The attempt to keep the machines out of Kinney's hands failed, though some kind of agreement was signed that year between Duke, Allen & Ginter, and the Bonsack Co.[50]

By the summer of 1888, however, the forces of combination were gathering in earnest.[51] In that year, the Bonsack Co. bought the

[49] Richard H. Wright to Francis S. Kinney, March 30, 1888, Richard H. Wright MSS, Duke University Library. See also Wright to James A. Bonsack, May 11, 1888 and May 28, 1888, Wright MSS.

[50] James B. Duke to D. B. Strouse (president of the Bonsack Co.), December 12, 1889, James B. Duke MSS, Duke University Library.

[51] Combination was discussed as early as 1885 at a meeting in Florida, and Duke and

patents of one of the two other machines in use — the Emery machine used by Goodwin & Company.[52] The Bonsack Co. tried to effect a compromise with the owners of the other machine, the Allison.[53] The Allison patents were secured late in 1889, when final arrangements were under way for the organization of the American Tobacco Co.

A pooling agreement on leaf purchases reached in 1889 by four of the big five manufacturers paralleled the combination built around the machines. The old informal system of each company purchasing most of its leaf in different markets was formalized and more rigidly pursued. Cooperation between the manufacturers grew, and the efforts to effect a combination increased.[54]

The producers held a series of meetings at the end of the 1880's in an attempt to work out a combination, but the clash of personalities and the wounds remaining from previous fierce competition made negotiations difficult. As a participant later commented, "there were some pretty stormy times . . . they would meet and break up and wouldn't speak, and then they would get together again . . . there was a great deal of friction among these men and they were very difficult to keep talking any length of time together." [55]

In addition to the personal relations involved, the negotiators faced other difficult problems. These problems lay in three main areas: the reluctance of the individual companies to give up their identities and join a combination; the recurrent doubts about the legality of forming a trust; and disagreements about how to divide the combination's stock among the companies forming it.

Each of these businessmen took pride in his own company and was hesitant to see it subsumed into a combination.[56] As a result, it was necessary to work out a compromise which allowed the constituent companies to maintain the facade of individuality for a time. There was an interregnum of two or three years after the founding of American Tobacco, during which the companies used their old names, styling themselves the branches of the American Tobacco Co. Until firm lines of authority and control were established in American Tobacco's New York headquarters, the different firms operated independently as separate branches.[57]

---

[52] Kinney had considered an amalgamation in 1887. Testimony of Charles G. Emery, digest of evidence in the case of *John P. Stockton v. American Tobacco Co.*, 55 N.J. Eq. 352 (1895), File 3017, Bureau of Corporation MSS.
[53] Richard H. Wright to Charles Watkins, June 30, 1888, Wright MSS.
[53] Richard H. Wright to James A. Bonsack, May 11, 1888, Wright MSS.
[54] *Western Tobacco Journal*, October 28, 1889, December 23, 1889, cited in File 4766, Bureau of Corporations MSS.
[55] Testimony of Francis S. Kinney, Circuit Court MSS.
[56] Statement by James B. Duke, File 3077, Bureau of Corporations MSS.
[57] File 4766, Bureau of Corporations MSS; *Tobacco*, March 21, 1890. Company

The second problem was the manufacturers' reluctance to form a trust because they feared this organizational form might soon be declared illegal. All parties to the negotiations except James Duke expressed such doubts.[58] Lewis Ginter especially was opposed to the trust. The manufacturers discussed the alternative of a pool, but rejected it because it was too weak a form of combination and left each producer free to violate any agreements under a pool.[59] The recently developed device, the holding company, was finally selected as the most sensible form, despite its possible legal complications. American Tobacco was one of the first giant holding companies in American industry.

The last remaining problem was the determination of how the stock of American was to be divided among the five constituent companies. This, understandably enough, was the source of much contention among the negotiators. They finally broke the logjam and made the decision. A participant later reported that:[60]

> This was arranged by each gentleman writing on a slip of paper what proportion he thought each party should have. To this he signed his initials and they were all put in a hat. They were then spread on the table and all read and the average was taken from the ballots. An agreement was reached, the major difference being between Mr. Duke and Major Ginter. Major Ginter claimed a little more than he finally got and Mr. Duke claimed that he should get as much as the Ginter interests. Another ballot was taken, with slight variation . . . Finally [Charles G. Emery of Goodwin and Company], said: "Major, I believe this thing. We all seem able to agree upon it, all but you here. Now let Duke have his way about it."
> In that way a satisfactory division was reached.

As a result of these negotiations, the American Tobacco Co. was incorporated early in 1890. After an unsuccessful attempt to organize the company in Virginia, a charter was secured in New Jersey.[61] American Tobacco was capitalized at $25,000,000 — $10,000,000 in preferred and $15,000,000 in common stock. W. Duke, Sons & Co. and the Allan & Ginter firm each received $3,000,000 in preferred and $4,500,000 in common. The Kinney Co. got $2,000,000 in preferred and $3,000,000 in common; W. S. Kimball & Co. and Goodwin & Co. each received $1,000,000 in preferred and $1,500,000 in common.

reluctance to sink identity into a combination was a common problem, and an interregnum often occurred between combination and consolidation. For example, see the description of the formation of International Harvester in John A. Garraty, *Right-Hand Man* (New York, 1960), chapter VII.
[58] Testimony of James B. Duke, File 3017, Bureau of Corporations MSS.
[59] Testimony of Charles G. Emery, File 3017, Bureau of Corporations MSS.
[60] *Ibid.*
[61] The recently passed New Jersey general incorporation law made the acquisition of a charter in that state very easy.

James Duke was elected president, and central offices were located in New York.[62]

Savings accrued immediately from the combination's exclusive contract with the Bonsack Co., secured when final arrangements about the amalgamation were made at the end of 1889. That agreement effected a reduction in the costs of production to the various firms of 15 to 25 cents per thousand cigarettes.[63] In 1893, the secretary of the American Tobacco Co. estimated that the costs of making cigarettes with the machines had fallen to around 5 cents a thousand.[64]

The cigarette companies soon moved to secure the advantages for which they had effected the combination. The need for a more rational organizational structure, which had been felt especially by Duke, was met in the following few years. Duke installed an elaborate system of cost accounting in 1890; less profitable brands were abandoned and small, less efficient factories closed. After the interregnum between incorporation and centralization had passed, advertising and sales were coordinated from the New York office. Through consolidation the company achieved reduced costs per unit sold during the early 1890's.[65]

IV

In the years after 1890, American Tobacco began the policy of expansion which ultimately led to the court-ordered dissolution of the firm. Producers of plug and smoking tobacco were acquired (usually through exchanges of stock and occasionally through cash purchases) and the company started the so-called "Plug Wars" of the 1890's. American Tobacco used its economic size to full advantage by selling various "fighting brands" of plug below cost, sacrificing "several millions" of their cigarette profits in order to force a combination of plug manufacturers with the company.[66] In December 1898, an agreement was reached with the major independent plug producers, and their firms were joined with the plug businesses of American Tobacco to form the Continental Tobacco Co. The largest competitor brought into the Continental was the Union Tobacco Co., owned by a group of financiers including P. A. B. Widener,

[62] Bureau of Corporations, *Report on Tobacco*, I, 66.
[63] File 4766, sections 1 and 2, Bureau of Corporations MSS. See also the agreement of November 11, 1889 between the Bonsack Machine Co. and the constituent firms of American Tobacco, United Cigarette Machine Co. MSS, Duke University Library.
[64] File 4766, sections 1 and 2, and File 4711, Bureau of Corporations MSS.
[65] Testimony of James B. Duke, Circuit Court MSS; Files 4711 and 4766, sections 1 and 2, Bureau of Corporations MSS.
[66] Bureau of Corporations, *Report on Tobacco*, I, 2.

Thomas F. Ryan, William C. Whitney, A. N. Brady, and Thomas Dolan.[67]

The formation of the Continental Co. led to a shift in the membership of the board of directors of American Tobacco; the financiers acquired positions on the board, and all the original founders of the company except James B. Duke left the board. Duke had backed the policy of expansion wholeheartedly, but the others were less eager to venture beyond cigarettes; they disposed of most of their holdings and had all retired from the board by the end of 1898. This shift of control from the manufacturers to financiers occurred in other industries as well at the turn of the century — the role of investment bankers in the history of such firms as United States Steel, International Harvester, and U.S. Shipbuilding is well known. In the case of American Tobacco, the financiers apparently shared the goals of James Duke, for he remained president of the firm. The strategy of increasing control over all branches of the tobacco industry continued.

The purchase of independent concerns (often in secrecy) and the competitive practice of selling below cost brought more and more companies into the tobacco combination. The same competitive methods which succeeded in centralizing a majority of the plug and smoking tobacco output worked equally well in the case of snuff, culminating in the formation of the American Snuff Co. in 1900. In addition, American Tobacco acquired control of the majority of licorice output in the United States and achieved a near monopoly in tin foil through its subsidiary, MacAndrews Forbes. The cigar industry proved more difficult to dominate, but the effort was made through the creation of the American Cigar Co. in 1901.[68] That same year brought a further cementing of the union between American Tobacco and Continental through the formation of a holding company, the Consolidated Tobacco Co., which acquired nearly the entire amount of the common stock of both.[69]

The strategy of expansion was pursued in foreign markets as well as in the United States. In 1901, American Tobacco bought Ogden's, Ltd., a leading British manufacturer, and initiated competitive warfare in the English market. Thirteen large British producers responded to the American challenge by uniting to form the Imperial

[67] This group of financiers had formed the Union Tobacco Co. through purchase of major producers of plug (such as Liggett & Meyers) and smoking tobacco (such as Blackwell's Durham Tobacco Co.).
[68] Cigars were produced in very simple machines in small lots and could not be mass-produced by machine as could cigarettes. The cigar industry was characterized by many relatively small competitors, which made the task of acquiring a majority of output very difficult.
[69] Bureau of Corporations, *Report on Tobacco*, I, part 1.

Tobacco Co. After two years of costly competition, James Duke went to London to negotiate a settlement. The deal was completed in the fall of 1902 and provided that Imperial Tobacco would have exclusive manufacturing and sales rights in Great Britain and that American Tobacco would enjoy similar sway in American and Cuban markets. In addition, the companies agreed to form the British-American Tobacco Co. (with two-thirds of the stock controlled by the Americans, one-third by the British), and the British-American would conduct operations in the rest of the world's markets.[70] Expansion proceeded apace.

The growth and the competitive tactics of American Tobacco brought it increasingly under the scrutiny of the Bureau of Corporations and the Department of Justice. The company was, naturally, acutely aware of this problem and paid close attention to the workings of the antitrust division. After the Northern Securities Co. case of 1904, the holding company (Consolidated Tobacco) was scrapped and Consolidated, American, and Continental were all merged under the name American Tobacco Co. This failed, however, to bring the desired immunity from prosecution, and the Department of Justice brought suit against the company on July 19, 1907, asserting that American Tobacco was in violation of the Sherman Act. After lengthy trials and appeals, the Supreme Court, shortly after its ruling against Standard Oil, ordered the dissolution of American Tobacco in 1911 in an effort to restore competition in the tobacco industry.[71] After several years the old combination was divided into several separate companies. These firms, created as a result of the court actions, include the (new) American Tobacco Co., R. J. Reynolds, Liggett & Meyers, and P. Lorillard. The antitrust action destroyed the near monopoly built from the base of the original American Tobacco Co. and replaced it with the oligopolistic competition which continues to characterize the modern tobacco industry.

[70] Duke wrote Oliver H. Payne (a financier and a member of the board of directors) from London after the completion of the agreement, saying he had made "a great deal with British manufacturers covering the world." Duke to Payne, September 26, 1902. James Duke MSS, Duke University Library.
[71] See *U.S. v. American Tobacco Co.*, 164 F. 700 (1908), 164 F. 1024 (1908), 221 U.S. 106 (1911), 191 F. 371 (1911), and *Federal Anti-Trust Decisions 1890–1917* (Washington, 1917), IV, 168–251.

Poverty in the Urban Ghetto

By John F. Bauman

IN THE MID-NINETEENTH CENTURY, the founder of the Children's Aid Society of New York, Charles Loring Brace, described the denizens of that city's alleys and rookeries as the "dangerous classes." To people like Brace and Robert Hartley, early pioneers of American social work, the poor were social pariahs abetted by evil tobacco, "demon rum," and ramshackle tenements. Out of the low dives and filthy hovels inhabited by these classes crept not only the thief, the cutthroat and other brigands, but also—so the thinking of the times went—cholera, smallpox, and the effluvia borne on the hot summer breezes which wafted from the city's blighted core. Long after the genteel Lady Bountiful became the professional social worker, cities still measured the impact of poverty in terms of sanitary conditions, crime and the level of morality.[1]

Despite the emergence over the years of sophisticated sociological treaties dealing with slum pathology, each wave of immigration has fallen heir to the stigma of the ghetto, and with each ethnic succession society has found a new group to label vicious and thus to fear. What the affluent American has yet fully to comprehend is that he fears not the "turbulent Irish" or the "militant black," but the poverty of these less fortunate groups.

America's old urban poor, the ethnic poor of European extraction, challenged social workers from Robert Hartley and Mary Richmond in the nineteenth and early twentieth centuries to Harry Hopkins in the 1930's. The closing of the floodgates against foreign immigration in the 1920's, followed by the coming of age of American labor unionism in the 1930's and the New Deal's social legislation, all served to lift these old poor out of their inner city habitat into the so-called mainstream of middle class America. Those who remained ghetto-side were the "new poor," people whom Michael Harrington has called "rejects." Among them were the old and the unskilled, but a particularly large number of them were the newest arrivals to the urban ghetto, the black Americans.[2]

Gilbert Osofsky and Allan Spear have both described the making of the black ghetto. The steady drift of black migration northward from the South during the 1890's reached rush proportions during World War I and then slowed somewhat in the late 1920's and the 1930's. According to Osofsky,

---

[1] For the history of the growth of American thought about poverty and its treatment see Robert Bremner, *From the Depths: The Discovery of Poverty in the United States* (New York: New York University, 1969); and Roy Lubove, *The Progressives and the Slums* (Pittsburgh: University of Pittsburgh, 1962).

[2] See Michael Harrington, *The Other America: Poverty in the United States* (New York: Penguin, 1963).

by 1930 Harlem had already become a black ghetto. Like World War I, World War II attracted thousands of southern blacks because of the prospect of decent jobs in the defense industries. The postwar years witnessed an acceleration of the northward trend largely attributable to the effect of federal agricultural policies which rewarded affluent farmers for taking marginal land out of cultivation and using modern farm equipment. This policy led throngs of displaced blacks (about 1.5 million between 1950 and 1959) to seek greener pastures in promised lands like New York, Chicago and Philadelphia.

The census of 1960 delineated the urbanization of black America. By that year, almost three-fourths of the nation's black population lived in the cities. The trend continued into the 1960's. In that decade, the black population of America's large cities increased 20 per cent from 9.7 million to 12.1 million people.

Equal in importance to the migration of blacks into urban America was the exodus of whites during the same years. While 1.8 million blacks moved into the nation's 12 largest cities during the 1960's more than 2 million whites packed up and fled to their patch of greenery in the all-white suburbs. This dual process ultimately created a large island of poverty and social decay in the heart of almost every large American city.[3]

POVERTY AND RACE

Today, urban poverty remains, significantly, black poverty. The report of the President's Commission on Civil Disorders called particular attention to the destitution affecting black America. In black America, in 1966 the medium income was only 58 per cent of white income; in the midst of national prosperity one-fifth of the black population was making no significant economic gain. Today, half of these black poor, over 2 million people, live in center city neighborhoods. In the Hough ghetto of Cleveland, the prosperous 1960's brought a rise rather than a fall in the percentage of black poor, from 31 per cent to 39 per cent.

Recent census figures indicate a slight drop in the number of families living in urban poverty areas (places characterized by average incomes under $3,335). According to the 1970 census, the years 1960–1968 witnessed a decline in the number of center city families living in poverty from 4.8 million to 4 million. The greater part of this drop came after 1967, spurred, no doubt, by rioting, rising crime rates and troubled urban school systems. Nevertheless, while these statistics suggest that some blacks are moving to the suburbs, they also indicate that twice the number of whites left the city. Furthermore, although only 14 per cent of the whites who remained are considered poor, still a sizable 31 per cent of the non-white residents are designated poor. Urban poverty remains part of the dilemma of being black in white America.[4]

The principal impact of the new urban poverty is observable in the climate of hate and despair it breeds. Within the heart of almost every large city in the United States, the ravages of racism and poverty have produced a "no-man's land" which, although characterized by many of the same problems indigenous to slum environments of the past, has the additional dimension of hopelessness. This atmosphere of hopelessness corrodes every aspect of ghetto life and makes that which was tolerable for the old ethnic poor intolerable for the denizens of today's slums.

Today, America's ghetto population exists in a world foreign to the experience and imagination of the middle class. Ghetto dwellers from Harlem, Hough and North Philadelphia—children as well as adults—must contend with the darkened hallways

---
[3] For discussion of the black migration north and its impact see Gilbert Osofsky, *Harlem: the Making of a Ghetto* (New York: Harper and Row, 1966); and also Louise V. Kennedy, *The Negro Peasant Turns Cityward* (New York: Columbia University Press, 1930); and Allan Spear, *Black Chicago: The Making of a Negro Ghetto, 1890–1920* (Chicago: University of Chicago Press, 1967).

[4] *Report of the National Advisory Commission on Civil Disorders* (New York: Bantam Books, 1968), pp. 251–259; Anthony Downs, "Alternative Futures for the American Ghetto," *Daedalus*, 97, fall, 1968, 1331; Jeanne R. Lowe, *Cities on a Race With Time: Progress and Poverty in America's Renewing Cities* (New York: Random House, 1967), p. 279.

and shadowy alleys which often conceal unconscious drunks or junkies "shooting up." Roaches and rats in the ghetto are nuisances to be coped with, not shrieked at. Despite the frequency of garbage and trash collection, slum dwellers find these services unequal to the high population density of the area. Then, too, the frequently stolen trash and garbage containers are rarely replaced promptly by the landlords. The results are rotting garbage and dirty streets.

But filth only exacerbates already miserable health conditions. Poor diets (a 1964 Chicago study found that 76 per cent of infants in poverty stricken families were anemic before age two) compounded by a paucity of effective medical services has produced a health crisis especially among the non-white poor. Infant mortality is 58 per cent higher among non-whites; tuberculosis, pneumonia and cancer of the cervix are twice as common in black America as in white. Mental illness likewise abounds among blacks. Dr. Robert Coles of Harvard, who made a study of the state of mental health in the Boston ghetto, found psychiatric problems there as commonplace as rotting teeth; he regarded both conditions as badges of slum occupancy. According to the National Advisory Commission's Report on Civil Disorders, 30 per cent of the American poor earning less than $2,000 a year suffer from a chronic health condition affecting employment. Yet despite these deplorable conditions in the ghetto, affluent America expects a middle class performance from its "other half." The American welfare structure, whether private or public, has traditionally been the agency to exact this performance.[5]

The welfare structure which the United States has shaped since the 1930's is today a two-edged sword. Not only does its administration often contribute to the despair of ghetto life (a point to be discussed later), but the skyrocketing costs of rising welfare caseloads constitute a significant cause of the fiscal drain which helps make city government the least solvent member of the federal structure.

THE WELFARE CASELOAD

Welfare caseloads in cities across the United States rose astronomically in the 1960's. New York City had over a million people on its rolls in 1969, a rise of about 300,000 from 1968, although not half so high as the rise of over 800,000 during the preceding year. Currently, with the tight job market induced by anti-inflation measures, New York City is adding almost 5,000 new welfare recipients each month.

The principal increase has been in the Aid to Families with Dependent Children Program (A.F.D.C.) commonly called A.D.C. (Aid to Dependent Children). The rise in A.D.C. dramatically reflects the character of the cityward migration of the past decades. New York City welfare director Mitchell Ginsberg has observed that it has been the young, disadvantaged Puerto Ricans, southern blacks and poor whites who have sought the American dream in the big city, and their frustrations have been graphically portrayed in the growing A.D.C. rolls. In New York City, the nation's traditional Mecca for the poor who are seeking opportunity, the number of A.D.C. clients rose 300 per cent during the 1960's; but the rolls also rose elsewhere: 293 per cent in Los Angeles; 281 per cent in Baltimore; 250 per cent in Newark, New Jersey. Although 90 per cent of the country's welfare programs—general assistance excluded—are at least partially funded out of federal funds, welfare still represents one of the cities' largest costs. New York City spent slightly under $3 billion for welfare in 1969.[6]

This increased cost of welfare and other

---

[5] One of the best comprehensive discussions of the pathology of ghetto life is Kenneth Clark's, *Dark Ghetto: Dilemmas of Social Power* (New York: Harper and Row, 1965); see also Orletta Ryan, "If I Get One More U My Mother's Goin To Beat Me Till It Rains," *The New York Times Magazine*, May 13, 1966, p. 26; Robert Coles, "What Poverty Does to the Mind," *The Nation*, June 20, 1966, p. 746; Report by the Citizen's Board of Inquiry Into Hunger and Malnutrition in the U. S., *Hunger, U.S.A.* (Boston: Beacon Press, 1968). On the state of sanitary services in the ghetto see George J. Kupchik, "Environmental Health in the Ghetto," *American Journal of Public Health*, 59, February, 1969, 220–225.

[6] See Mitchell Ginsberg and Jack Goldberg's explanation for New York City's rising caseloads in *The New York Times*, June 30, 1970, p. 23.

services to the poor, such as day care centers, clinics and police protection, must come out of the already overused city property tax. For both industry and the encumbered middle class, the escape from taxation and the problems of welfare has been to relocate in the suburbs, leaving the city with an increasing demand for better services and a dwindling tax base to provide them.

The final impact of the vicious fiscal squeeze upon the city is the loss of urban autonomy. Edward M. Kaitz and Herbert H. Hyman, in a recent study, *Urban Planning for Social Welfare,* point out that as the city relies more and more upon federal sources of assistance, its power to decide its own future erodes. What is more, conclude Kaitz and Hyman, the welfare decisions of federal planners notoriously by-pass the needs of the city's growing population. Other experts, such as Herbert Gans, Charles Abrams, and Jan Jacobs, have catalogued the past failure of urban renewal to aid the poor.[7]

Much as in the past, the ordinary citizen experiences the impact of urban poverty through reading of, hearing about, or actually being victimized by city-bred crime or violence. Studies in some cities indicate that as many as 43 per cent of the people are afraid to walk the streets at night. Poverty therefore gnaws at the very fabric of urban society, for the citizenry's growing fear of the poor, whether it be of their crimes, their riots, or their seemingly alien life-style, is primarily responsible for the feeling of some observers that a death sentence has been pronounced upon urban life.

The fact that poverty and slum life breed crime, which was vividly portrayed earlier in the literature of writers like Charles Dickens and Theodore Dreiser, has more recently been documented by statistics. As Roy Lubove pointed out in his excellent study of tenement house reform, urban reformers like Alfred T. White, Jacob Riis and Lawrence Veiller were all motivated in part by the prospect of making the city safe from slum-fostered crime. Yet it is the modern dimension of slum pathology that makes the present day violence of the ghetto so formidable. In addition to low income, dependency, racial and ethnic concentration, unemployment and poor education, the modern ghetto breeds a dangerous level of frustration especially among the young.[8]

Two authorities on delinquency, Richard Cloward and Lloyd Ohlin, both of the Columbia School of Social Work, ascribe juvenile crime essentially to a society that encourages certain aspirations among all its members, and then withholds from some the possibility that they can achieve these goals legitimately. Spurned by opportunity, constantly subjected to degradations of the ego by one agency of society after another, the slum youth sees little alternative to "making it" in the ghetto. For him, the flashy "hustler" is the most meaningful model of success. Claude Brown and Malcolm X have both eloquently linked narcotics pushing, prostitution, burglary and other "hustles" to the dark ghetto. Yet while many of the victims of these hustles may be outsiders, the ghetto itself sustains much of its impact. Crime is another symptom of the disease of poverty in the inner city.[9]

POVERTY AND RIOTING

Since 1964, the frustrations of the ghetto have manifested themselves in another more ominous manner which has served to alert

---

[7] A discussion of the relationships of poverty to the city tax structure is found in Edward M. Kaitz and Herbert H. Hyman, *Urban Planning for Social Welfare: A Model Cities Approach* (New York: Praeger, 1970), p. 5. On the shortcomings of urban renewal see especially Charles Abrams, *The City is the Frontier* (New York: Harper, 1965), pp. 19–101.

[8] According to the President's Commission on Law Enforcement and the Administration of Justice, *The Challenge of Crime in a Free Society* (Washington, D.C.: Government Printing Office, 1967), p. 38, "it is one of the most documented facts about crime that serious crimes that worry people most . . . happen most in slums of cities": see also *Report on Civil Disorders,* pp. 263–268; on housing reform and social control see Lubove, *op. cit.,* pp. 245–251.

[9] See for example Richard A. Cloward and Lloyd E. Ohlin, *Delinquency and Opportunity: A Theory of Delinquent Gangs* (New York: Free Press, 1960). Two excellent accounts of the impact of slum life on youth are Claude Brown, *Manchild in the Promised Land* (New York: Macmillan, 1965), and Malcolm X, *Autobiography* (New York: Grove Press, 1966).

America to the problems of race and poverty. In 1964, the Cleveland ghetto revolted against its sordid, isolated condition and since then, among most middle class Americans, areas like Watts, Newark and Harlem are most clearly identifiable as the riot scenes of the long hot summers.

According to a recent sociological study of urban rioting, violence-prone people are people who feel threatened. Moreover, the tendency to riot is intensified among persons who have suffered some damage to their self-esteem, especially at the hands of such impersonal authorities as loan agencies, "the welfare," and the police. Even some improvement in the economic condition of ghetto dwellers does not necessarily restrain their inclination to riot, not at least while their opportunity to participate freely in American society is confined by the walls of racism.[10] A quantitative study of the relationship between deprivation and protest by Don Bowen, Elinor Bowen, Shelden Gawises, and Louis Masotti found rioting considered "acceptable behavior" among those disadvantaged individuals with rising expectations as well as among those who expected little change in their conditions.[11] E. L. Quarantelli sees the riot in another way. In the midst of the riot, he argues, property is redefined in collective terms. Rioting can be seen as a "new process of collective bargaining, and in this sense the looter is temporarily a full participant in the mass consumption society."[12]

Poverty relief in urban America, like American social welfare in general, has been premised upon America's traditional beliefs about individual worth and redemption. Many of the concepts which dominate our thinking about welfare are traceable to the nation's Calvinist heritage. The traditional Puritan work ethic was reinforced by the country's inherited Elizabethan poor law. In essence, this Calvinist-Elizabethan view of welfare preaches that an individual's godliness is measured by his ability to labor and produce. Accordingly, all non-productive members of society are seen as unwanted burdens which society should care for only in the most miserly manner. It follows that the whole purpose of welfare should be to force the thriftless, intemperate, sinful individual to work.

However, with the increasing sophistication of social work at the end of the nineteenth century, poverty was attributed more to environmental causes and less to moral depravity or original sin. Professionalized social workers described the poor as victims of family maladjustment in need of psychiatric counseling and other services to overcome their inability to fit into the middle class mold. Few social workers, even in the 1930's, saw poverty as a problem of income deprivation.

THE NEW DEAL WELFARE STRUCTURE

The passing of the voluntary welfare agency and the burgeoning of federally supported public welfare in the 1930's meant little for those made incomeless by technological change or discrimination. Despite the protests of a number of social workers in the early 1930's, the welfare structure framed by the New Deal after 1935 only reinforced the Calvinist-Elizabethan view of poverty and the aversion of the private social agency to cash instead of service assistance. The unemployment benefits of social security, for example, were premised upon the existence of a job; general assistance, for several years (1933–1935) governed by federally established standards, was turned over to the states to be doled out as sparingly as possible. The welfare system that finally emerged under public auspices still assumed that joblessness and poverty were attributable to a social defect best treated by casework. Some closely guarded financial aid might be allowed, but only a pittance, for too much

---

[10] Kurt Land and Gladys Lang, "Racial Disturbances As Collective Protest," *American Behavioral Scientist*, Vol. II (March–April, 1968), p. 11.

[11] Don Bowen, Elinor Bowen, Sheldon Gawises and Louis Masotti, "Deprivation, Mobility and Orientation Toward Protest," *American Behavioral Scientist*, Vol. II (March–April, 1968), pp. 2–23.

[12] E. L. Quarantelli and Russell A. Dynes, "Looting and Civil Disorders: An Index of Social Change," *American Behavioral Scientist*, Vol. II (March–April, 1968), p. 7.

would be detrimental and would promote the "habit of dependency."[13]

THE "NEW POOR"

The depression poor, however, were the old ethnic poor, people subject to the whims of a relatively unregulated and unchallenged capitalism. It was in the early 1960's that Michael Harrington publicized the "new poor," the old, the blacks, the domestics, the hospital employees, casual workers—categories unaffected by the union movement and passed over in the social welfare advances of the 1930's.[14]

In the face of this new clientele, public welfare remains highly bureaucratized and committed to providing services to the poor. It is still geared to an older type of poverty, still so conditioned by a middle class work-oriented value structure that it is helpless to combat a type of poverty alien to middle class experience.

Today, the American welfare system is discovering that its casework technique cannot penetrate the new poverty. Individualized services are not relevant to a situation of mass impoverishment, nor can they respond effectively to such mass handicaps as being black or being unskilled.

Yet it is precisely the forces of racism and technology that are largely responsible for today's largest body of welfare recipients, the A.D.C. mothers. The typical A.D.C. family is urban (75 per cent live in population centers with over 2.5 million black residents), with three children headed by a 30-year-old mother. According to recent census figures, the number of black families headed by a woman increased 83 per cent from 1960 to 1969, while black families headed by a male increased only 15 per cent.

This A.D.C. phenomenon tells us as much about the limited employment opportunities for black unskilled men as it does about the welfare structure which perpetuates the alienation of the black family. Statistics clearly show that rising A.D.C. caseloads are symptomatic of rising levels of ghetto unemployment. Unemployment in the ghetto has consistently reached twice the national rate. Even when blacks are employed, they are generally subemployed in low-status jobs. More serious still is the high level of unemployed black teenagers. While the non-white unemployment rate in cities stood at 7.6 per cent in 1968, a study that year of unemployment in the 20 largest metropolitan areas showed the teenage jobless rate among blacks at 33 per cent.

The impact of these figures is alarming. In a nation where holding a job, as Patrick Moynihan observes, is the primary source of group identity, the ghetto harbors a depression-size unemployment rate. Not only does this black joblessness foster family disruption, but it also greatly increases the chance of unemployment in the offspring. Jobless youths then become part of the street corner society, and from there they may add to statistics on delinquency, crime and violence. Hence, despite the persistent myth that A.D.C. principally caters to slothful, dishonest and promiscuous women, there is equally persistent evidence that the welfare problem stems from a condition of joblessness fostered and compounded by the alienation and hopelessness of the dark ghetto.[15]

Presently, the welfare system not only fails to remedy the root causes of ghetto pathology, but in fact contributes to the further aliena-

---

[13] Bremner, *From the Depths*. See also Roy Lubove, *The Professional Altruist: The Emergence of Social Work as a Career, 1880–1930* (Cambridge: Harvard University Press, 1965), for an excellent discussion of the growth of psychiatric casework. Lubove also discusses the contempt of the private charity agency for public welfare, an important aspect of the professionalization of social work. See also Lubove, "Social Work and the Life of the Poor," *The Nation*, 202 (May 23, 1966), p. 611.

[14] Harrington, *The Other America*; also Harrington, *Toward a Democratic Left: A Radical Program for a New Majority* (New York: Macmillan, 1968), Chapter III.

[15] On A.F.D.C. see *Report on Civil Disorders*, p. 457; see also Alvin Schorr, "The Family Cycle and Income Development," *Social Security Bulletin* (February, 1966); *Business Week*, (November 11, 1967), p. 72; Eveline Burns, "The Poor Need Money," *The Nation*, 200 (June 7, 1965), p. 613. On the employment potential of A.F.D.C. mothers see Genevieve W. Carter, "The Employment Potential of A.F.D.C. Mothers," *Welfare in Review*, 6 (July–August, 1968), pp. 1–11. For a discussion of joblessness in the ghetto see Paul O. Flaim, "Jobless Trends in Twenty Large Metropolitan Areas," *Monthly Labor Review*, 91 (January–June, 1968), pp. 16–18.

tion of those families served. In 1968, Mitchell Ginsberg, the head of New York City's department of welfare, declared that the welfare system was bankrupt as a social institution. That same year, the Commission on Civil Disorders found that the present system of public assistance "contributes materially to the tensions and social disorganization that have led to civil disorders."[16]

True to affluent America's fear that decent levels of relief would encourage dependency at the taxpayer's expense, A.D.C. grants have provided only part of the client's needs; in 1965, New York City (where welfare is considered particularly generous) met only 88 per cent of a welfare family's minimum needs.[17] In return for relatively meager grants, most cities still subject the client to humiliating investigations followed by brash impersonal checkups. The investigations, the gaps in coverage and the low level of payments only exacerbate the plight of the poor, reminding the recipient that society considers him untrustworthy, ungrateful, promiscuous and lazy. Urban poverty in America, which forces an individual to subsist on welfare, thus invites a loss of the elemental privilege of privacy and dignity. It imposes upon the poor a badge of inferiority very little less debasing than the white letter "p" sewn on the cheap wool uniforms worn by the inmates of nineteenth century English workhouses.

The earlier charitable agencies tried to overcome poverty by reconstructing sundered families and inculcating middle class values. In contrast, the present system actually contributes to the trauma of the ghetto and in this way further generates the conditions which increase the welfare rolls.

The cure for today's urban poverty eludes the remedies of the social worker, a fact recognized today by many social workers themselves, by sociologists and by economists. More than ever, as Martin Meyerson observes, the enormity of the welfare crisis is rallying the white middle class reformer and the black militant. Their demand, supported by the Welfare Rights Organization, is for some type of income guarantee for all Americans. Conservative opinion on the subject of guaranteed income insists that a work incentive must be incorporated into any income maintenance program. Accordingly, President Richard Nixon's proposal for a family assistance program is essentially a watered-down income-maintenance scheme which its designer, Patrick Moynihan, appropriately labels "workface." On the other extreme stands the Welfare Rights Organization which calls for a minimum guaranteed income of $5,310.[18]

Significantly, both camps agree that decent welfare provisions must be regarded in the United States "as a right." Support for the needy should be as easily granted as old age benefits under Social Security. An income maintenance program structured to provide a decent level of existence to needy families could restore that ingredient of dignity so essential if families are to remain socially and psychologically sound.

There is little doubt that some form of income maintenance is imminent. However, in President Nixon's current Family Assistance Program proposal the minimum income is not only too low to provide meaningful relief to the urban poor, but the work incentive still presumes that the unskilled jobless individual, male or female, can find employment if he looks for it, and that a decent job awaits him when he reforms and makes the effort to find work.[19]

In any event, providing income to the poor

---

[16] Ginsberg quoted from *U.S. News and World Report* (July 17, 1967), p. 45; *Report on Civil Disorders*, pp. 458–459.

[17] *Hunger, U.S.A.*, p. 73.

[18] Martin Meyerson, "Urban Policy: Reforming Reform," *Daedalus*, 97 (fall, 1968), pp. 1418–1419. See Patrick Moynihan's discussion of family assistance in "One Step We Must Take," *Saturday Review* (May 23, 1970), p. 20; Yale Brozen discusses at some length all the various income maintenance proposals in "Toward an Ultimate Solution," *Saturday Review* (May 23, 1970), p. 30. In his article "Guaranteed Minimum Income Proposals and the Unfinished Business of Social Security," *Social Service Review*, 41–42 (June, 1967), pp. 166–178, George Rohrlich disagrees that income maintenance obviates the need for social services.

[19] For a critique of President Nixon's F.A.P. see Richard Elman, "If You Were on Welfare," *Saturday Review* (May 23, 1970), p. 27, and John Hamilton, "Will Work Work?" *Saturday Review* (May 23, 1970), p. 24.

can be considered only a start toward solving the problem of urban poverty. In his 10-point program for uplifting the cities, Moynihan has spoken of the need for restoring the vitality of the inner city neighborhoods. This approach might have some validity. One of the positive results of the Office of Economic Opportunity's experiment with Community Action Programs (C.A.P.) has been to spur interest in the community as a source of neighborhood reconstruction. Today, a wide variety of parent groups, women's groups, black power organizations and youth organizations are arousing concern in the ghetto for (among other things) school lunch programs, improved education and better urban medical and sanitary services.[20]

But the city, with its blighted areas, can never be revitalized until it ceases to be a prison for the victims of poverty and racism. This does not mean that a federal outpouring of money for improved inner city sanitation, schools and hospitals is unnecessary. Nor can we disapprove of Moynihan's call for a federal policy to transform "the urban lower class into a stable community based on dependable and adequate income flows, social equality, and social mobility."[21] But concurrent with a program aimed at stable inner city neighborhoods must come a broadening of opportunity for both jobs and housing in the sacrosanct white suburbs outside the blighted core. The freedom to move out of the ghetto into the better surroundings beyond has been a hallmark of upward mobility in the United States. Its denial to black America has been to a large degree responsible for the cycle of despair in the ghetto. Opening suburbia to blacks, even locating attractive low cost housing units in the suburbs and closer to the new industrial parks, will do as much to end poverty as billions spent for slum clearance and renewal.[22]

However, suburban and city planners and politicians must first recognize the mutuality of interests which join their two worlds. Just as the city planners of today measure the cost of unplanned suburban sprawl upon the ecology, so they must weigh the equally high cost of walling off poverty in the city. Admittedly, a revolution in thought is demanded. The prospect of open housing in the suburbs will be greatly enhanced only when the middle class in both city and suburb becomes aware that anyone's poverty impinges upon everyone. The existence of poverty and its concomitant discrimination gnaws at the fabric of everyman's society, and the resultant crime, riots, youthful alienation, and rebellion are all manifestations of this most glaring of the unfulfilled promises of American life.

---

[20] See Daniel P. Moynihan, "Toward a National Urban Policy," in the introduction to *Report of the National Commission on the Causes and Prevention of Violence, Violent Crime: Homicide, Assault, Rape and Robbery* (New York: Braziller, 1969); for an excellent critique of the Community Action Programs see Kenneth B. Clark and Jeannette Hopkins, *A Relevant War Against Poverty: A Study of Community Action Programs and Observable Change* (New York: Harper and Row, 1969).

[21] Moynihan in *Violent Crime*.

[22] See Nathan Glazer, "Slum Dwellings Do Not Make a Slum," *The New York Times Magazine* (November 21, 1965), p. 54; Herbert Gans, "The Failure of Urban Renewal," *Commentary*, 39 (April, 1965), pp. 29–37.

WHO KILLED THE ALDRICH PLAN?

By Andrew Gray

Worse than disastrous, it was preposterous: America entered the twentieth century without a central bank. For a time, it is true, the National Banking Act of 1863 had promised to blossom into a uniform banking system under federal supervision. But as the growing use of checks made the privilege to issue notes less crucial, this promise vanished; and while national bank notes secured by federal bonds circulated freely, their volume was responsive to fiscal policy rather than to the demands of trade. Since the National Banking Act also imposed rigid reserve requirements on every national bank, deposit currency and note issues were equally inelastic. Bonds bearing the note-issue privilege were retired whenever the Treasury wished to reduce its cash surplus, and repeated bank-note shortages, coupled with the resulting deflationary pressures, inevitably provoked the Populist revolt and the Greenback-silver crusade.

After the panic of 1907, a month-long uproar in which banks warred with each other to obtain and hoard cash, no one, in the White House or Wall Street or even in the offices of William Jennings Bryan's *Commoner*, denied the need for banking reform. But agreement ended there. Although free silver and easy money had been beaten at the polls, the struggle had left conservatives so badly frightened that they tended to cling to those evils they knew. Similarly, any major proposal for reform - or even the whisper of a need for a Central Bank - invited Bryanites to raise anew the banner of Andrew Jackson. And if Bryan saw a Central Bank as a means of consolidating the power of Wall Street, Republican bankers feared that such an institution might furnish a Democratic President and Congress with a dangerous instrument for political tinkering with currency and banking.

The obvious solution, therefore, appeared least likely to be adopted. Moreover, the task of reconstructing a banking system is complex and delicate, calling for a blend of technical and political skills seldom combined under any one brow. As Henry Cabot Lodge put it to his fellow senators early in 1908, "It is easy enough to find gentlemen who can sketch banking systems. It is very rare to find men who can not only put a banking measure into the shape of law, but fit a measure designed to meet an exigency into laws already upon the statute books. That requires not merely a knowledge of banking and banking systems, but the skill of a practical lawmaker and a familiarity with existing law which very few men possess."

How it was done has remained, in part, an untold story. According to Carter Glass and Parker Willis, the Federal Reserve System was conceived and born in the House Banking and Currency Committee in 1913. According to Paul Warburg and Nathaniel Stephenson, the principal biographer of Nelson Aldrich, the system grew organically from the work of the National Monetary Commission. One of the prime movers in the banking reform movement of those years, Piatt Andrew, special assistant to the National Monetary Commission, editor of all its publications, and Assistant Secretary of the Treasury during the campaign to enact its recommendations into law, has remained to this day an unheard witness. This account of these events is not, however, intended to rekindle the long-extinguished debate as to the paternity of the Federal Reserve System, but rather to suggest the deep historical roots of today's debates concerning the future of our major financial institutions and the U.S. banking system itself.

Shortly after the panic of 1907 subsided, Congress began the series of hearings which culminated in the Aldrich-Vreeland Act of May 1908. Inevitably, it was a stop-gap measure, permitting national banks belonging to "clearinghouse associations" to issue emergency currency secured by some of their less liquid assets. But its most important provision created the National Monetary Commission. Organized under the Chairmanship of the Senate majority leader, Nelson Wilmarth Aldrich, the Commission was asked to produce a report to end all reports on banking reform, and make a formal recommendation to the Congress for remedial legislation. Aldrich immediately hand-picked eight fellow Senators and nine Congressmen for the Commission, and embarked for Europe that August to interview European central bankers. Hitherto profoundly opposed to a central bank for America, Aldrich had been converted, as it were, on the road to Damascus. Determined to secure the most competent assistance available for such an inquiry, he asked Charles W. Eliot, President of Harvard, to recommend a special assistant for the Commission. Eliot chose a thirty-five year old Assistant Professor of Economics, Piatt Andrew.

On board ship, the members of the new Commission were subjected to daily seminars in banking theory under the aegis of their chairman. But despite these pedagogical efforts, few of the legislators participated effectively in the ensuing fact-finding interviews with foreign central bankers. Most members of the Commission returned to America later that August, leaving Aldrich and his experts to carry out the more important research at the Banque de France and the Reichsbank in September, and thereafter, to begin work on the formal report.

The National Monetary Commission, nominally bipartisan, was completely under the thumb of its chairman. In many respects Senator Aldrich was not ideally situated to conduct an impartial survey of banking reform in the United States. The fiercely partisan majority leader of the United States Senate, he was embroiled in the bitter tariff disputes of those years which culminated in the Payne-Aldrich Tariff of 1909. A convinced protectionist, Aldrich was soon at loggerheads on the tariff not only with the Democratic minority in the Senate, but also with President Taft and Western Republicans as well. On the other hand, there was no comparable figure in the United States capable of his astonishing tour de force in mastering such an elusive subject as currency reform in such a sovereign manner in so short a time. Moreover, Senator Aldrich did not regard the subject as a partisan issue and was unfettered by any regional loyalties or party obligations in considering its various aspects. But the old leopard could not shed his spots, and his identification with the crusade did not enhance its political prospects.

By the Autumn of 1910, Aldrich and Andrew had completed the 23-volume Commission report that gathers dust today in many a financial library. The next step was to draft the remedial legislation itself. But this could not be done openly. Countless self-proclaimed experts had drawn up banking plans. It had become a national pastime. And so Aldrich, at Harry Davison's prompting, arranged the most famous duck hunt in American history, inviting five men to accompany him to a small island off the Georgia coast and lock themselves away in seclusion at the Jekyll Island Club until they had produced a draft bill. Ostensibly off on a vacation jaunt, five men surreptitiously boarded the Senator's private car at a secluded Jersey City rail siding on November 10, 1910, laden with hunters' accoutrements. As the train rolled southward, Aldrich disclosed the true quarry and swore all the participants to inviolable secrecy, a vow scrupulously observed by all five men until long after the Senator's death in 1915.

The hunters were well equipped for their task. Paul Warburg of Kuhn, Loeb & Co., was the expert in international finance, Frank Vanderlip of National City Bank in commercial banking, Harry Davison in investment banking, and Piatt Andrew, by then Assistant Secretary of the Treasury, in monetary theory and banking history. Shelton was the Senator's secretary. Aldrich himself, seventy years old and without a specialist's knowledge of the subject, was nonetheless fully capable of guiding the debate. In order not to inhibit discussion, nothing was committed to paper until agreement on a given topic had been achieved.

To stand the remotest chance of enactment, any plan would have to overcome the resistance not only of those critics who opposed a central bank with the same traditional but often irrational objections that had doomed the two Banks of the United States, but also of the bankers themselves, whose distrust of sudden change was emblematic of their profession. The Jekyll Islanders therefore sought to remedy the most obvious defects of American banking, the inflexibility of the bond-secured currency and the immobility of bank reserves, with as little dislocation of existing banking practices as possible:

RESERVE ASSOCIATION OF AMERICA

Charter and Location

It is proposed to charter the Reserve Association of America which will be the principal fiscal agent of the Government of the United States. The authorized capital of the Reserve Association shall be approximately $300,000,000. The length of its charter shall be 50 years. The head office of the Association shall be in Washington, D.C.

The Country shall be divided into 15 districts, and a branch of the Reserve Association shall be located in each district.

The Reserve Association and its branches shall be exempt from State and local taxation, except in respect to taxes upon real estate owned by it....

Simple and obvious? Perhaps. But America had been without a central bank for more than three generations, and the political landscape was still strewn with the painfully visible wreckage of the first two.

EXECUTIVE OFFICERS OF THE RESERVE ASSOCIATION

The executive officers of the Reserve Association shall consist of a governor, two deputy governors, a secretary, and such subordinate officers as may be provided by the by-laws. The governor and deputy governors shall be selected by the President of the United States from a list submitted by the board of directors. The governor shall be subject to removal by the President of the United States for cause. The term of office of the deputies shall be seven years, but the two deputies first appointed shall be for terms of four years and seven years, respectively....

As always, control was the crux. How could regional interests be protected, while control remained sufficiently centralized to permit effective management? How could political manipulation be prevented, while still retaining federal fiscal prerogatives? These objectives were not easily reconciled. There was, first of all, the harsh fact of presidential power. The President would clearly have to appoint the Association's principal officers; otherwise it could scarcely function as fiscal agent of the United States Government.

Senator Aldrich was willing to invite the President for cocktails, but not for dinner. The plan proposed that the President should select the governor and his deputies from a list to be submitted to him by the Association's board of directors. But this limitation was inconsistent with the chief executive's constitutional appointive powers. Previous attempts to limit these powers had not proven fruitful, nor did this one. On the other hand, the Jekyll Islanders forsaw that presidential appointees should dominate the Executive Committee of the Board of Directors, which included the governor, the two deputy governors, and the Comptroller of the Currency as ex-officio members. But they also tried to preserve at least a facade of initiative and responsibility in the branches, hoping that regional loyalties would yield to national considerations when setting monetary policy. But the very size of the board made it obvious that the real power would vest in the Executive Committee:

LOCAL ASSOCIATIONS OF NATIONAL BANKS

All subscribing banks shall be formed into associations of national banks, to be designated as local associations. Every local association shall be composed of not less that 10 banks, and the combined capital and surplus of the members of each local association shall aggregate not less than $5,000,000.

All the local associations shall be grouped into 15 divisions, to be called districts. The territory included in the local associations shall be so apportioned that every national bank will be located within the boundaries of some local association. Every subscribing national bank shall become a member of the local association of the territory in which it is situated....

The National Banking System was nearly fifty years old in 1910. In the years immediately preceding its establishment in 1863, banks in the thirty-four states were differently organized, enjoyed disparate privileges, and were subject to varying regulations. Note issues proliferated wildly. Some were secured by the pledge of bonds, some by cash reserves, some utterly unsecured, and virtually all were easy to counterfeit. To survive, businessmen had to subscribe regularly to a "Bank Note Reporter" which gave purported current values of the issues of nearly 1,500 different banks, and contained lengthy lists and descriptions of the multitudinous counterfeits. When a bank found its notes widely counterfeited, it would simply destroy the old plates and engrave new ones, with the consequence that several varieties of genuine notes were often in circulation by the same institution simultaneously.

It took a war to put an end to this banking anarchy. Soon after the first battle of Bull Run, the U.S. Treasury and the banks suspended specie payments, and Lincoln's government had to resort to direct issue of legal tender notes, the famous "greenbacks." To make matters worse, the Treasury encountered great difficulty in placing its loans, and when Treasury Secretary Salmon P. Chase presented his bill to charter national banks with the privilege of issuing notes secured by government bonds, it proved politically irresistible. Chase believed that such national banks would subsidize the Treasury by purchasing government bonds in order to expand their own note issues. The new banks, he hoped, would also provide a national currency to supersede the "greenbacks," and would help to hold in check the issues of other banks.

The "Act to Provide a National Currency" of 1863 stipulated that "national" banks purchase and deliver to the Treasury government bonds in an amount equal to one-third of their capital stock. Such national banks could then issue notes secured by these bonds in amounts up to ninety percent of their par value, plus an amount equal to their capital. The Secretary expected no immediate relief for the Treasury through the sale of these bonds, and indeed the war had ended before the deposit of bonds reached $100,000,000. So gradual was the organization of banks under the Federal law that the act was revised in 1864, and permission given to reorganize state banks under national charter. Nationalization of state banks was further abetted by the levy of a tax of ten percent upon the notes of state banks paid out after July 1, 1866. This harsh stimulus produced a rapid transformation of state banks into national banks and soon thereafter the U.S. had a stable banking system for the first time in two generations.

The national banking law contained one highly significant provision unlike any prevailing in European banking law, its only prototype being in an obscure state law of Louisiana. This provision established a minimum proportional reserve to be maintained against deposits. The percentage requirements varied between "central reserve cities" and country districts, but the reserve had to be maintained inviolate, come what may. The possibility of expanding loans to meet industrial demand was thus subject to inflexible restrictions, similar to those imposed by the bond-secured currency on note issue.

Under the National Banking Act, the government's balances were distributed among national banks at the discretion of the Secretary of the Treasury. During the 1870's, proceeds of bond sales were left with the banks for brief periods while old bonds were being redeemed, but public opinion always opposed depositing ordinary revenues with the banks. From 1880 to 1887 the public deposits with the banks averaged less than $15,000,000; in the latter year, however, money began to accumulate in the Treasury in uncomfortable proportions, and President Cleveland and his Treasury Secretary Fairchild became anxious to return it to circulation. Cleveland asked Congress to reduce the excessive revenue by lowering tariffs, but the Senate blocked every such proposal, and Fairchild proceeded to redeem all bonds subject to Government call privilege, also buying government bonds in the open market as a means of getting rid of the embarrassing surplus. Some of the bonds he purchased did not mature for another twenty years, and he had to pay premiums as high as twenty-nine percent. Yet even by such means the surplus was not sufficiently dissipated, and at last he was obliged to increase his deposits in the banks. Howls of protest ensued, and the deposits of the Government with the banks were again reduced to small working balances. In the late 1890's, when the Spanish War brought higher tax revenues, the Treasury was again burdened with a mounting surplus, once more posing the vexing question of its proper role in the American banking system.

Theodore Roosevelt's Treasury Secretary, Leslie Shaw, went to the other extreme and used government deposits as instruments of monetary policy, rewarding and punishing banks as he wished, intervening to block sales of gold to foreign banks, and generally behaving as if the Treasury itself could function as a central bank. The Jekyll Islanders proposed to remove the government from the banking business entirely:

> *The Government of the United States and those national banks owning stock in the Reserve Association shall be the sole depositors in the Reserve Association. All domestic transactions of the Reserve Association shall be confined to the Government and the subscribing banks, with the exception of the purchase or sale of Government or State securities or securities of foreign governments or of gold coin or bullion.*
>
> *The Government of the United States shall deposit its cash balance with the Reserve Association, and thereafter all receipts of the Government shall be deposited with the Reserve Association or (when necessary) with such national banks as the Government may designate for that purpose in cities where there is no branch of the Reserve Association. All disbursements by the Government shall be made through the Reserve Association.*
>
> *The Reserve Association shall pay no interest on deposits.*
>
> *The Reserve Association may rediscount notes and bills of exchange arising out of commercial transactions, for and with the endorsement of any bank having a deposit with it.*
>
> *The rate of discount of the Reserve Association, which shall be uniform throughout the United States, shall be fixed from time to time by the executive committee and duly published....*

To most people in those days, the note issue question appeared overriding. For half a century, national banks had been issuing their own notes against purchase of equivalent amounts of two percent U.S. Government bonds, and these banks were naturally jealous of this prerogative and apprehensive about tinkering with a system that had enabled them to earn money on both the buying and selling end of the note issue function. Without this "circulation privilege," the outstanding two percent bonds would become a $700 million albatross in the hands of the national banking system, and so the proposed reserve association would inevitably have to take them over at or near par. But who was to absorb the loss? The Jekyll Islanders proposed to split the difference with the Treasury:

NOTES ISSUES

There is hereafter to be no further issue, beyond the amount now outstanding, of bank notes by national banks. National banks may, if they choose, maintain their present note issue, but whenever a bank retires the whole or any part of its existing issue it will permanently surrender its right to reissue the notes so retired.
The Reserve Association must, for a period of one year, offer to purchase the 2 percent bonds now held by national banks and deposited to secure their circulating notes. The Reserve Association shall take over these bonds with the existing currency privilege attached and assume responsibility for the redemption (upon presentation) of outstanding notes secured thereby. The Reserve Association shall issue, on the terms herein provided, its own notes as fast as the outstanding notes secured by such bonds so held shall be presented for redemption, it being the policy of the United States to retire as rapidly as possible, consistent with the public interest, bond-secured circulation and to substitute therefor notes of the Reserve Association....

These provisions were the core of the lengthy draft bill brought north by Senator Aldrich. Its existence, of course, was as secret as the Jekyll Island meeting itself, and the Senator was initially at a loss how to reveal it to the public. He could scarcely disclose that he had taken five men to a club controlled by Pierpont Morgan to devise a new central bank for the United States without triggering an avalanche of conspiracy charges and allegations that Wall Street had dictated its text verbatim. While the Senator pondered his next move, Paul Warburg vented his personal impatience in a handwritten letter to Andrew on December 10, 1910, using the Jekyll Island pseudonym for Aldrich:

I wrote to Mr. Nelson while he was in New York sending him the suggestions, of which you received a copy. I had hoped to see him because I should like to have received some hints as to how to proceed. When I came back I found a lot of invitations to speak which I have declined. The Ohio bankers, however, did not give in; they began to write individual letters and telegraphed and telephoned till I finally said I would try to address them in January. Meanwhile the Chamber of Commerce has made me chairman of the committee of the monetary conference in Washington on January 18th (Mr. LaLanne's fool convention) and they want me to make a speech there. I have managed to keep the Merchants' Association quiet for the time being. The Produce Exchange made up a resolution more or less from my United Reserve Bank article, and when they submitted the report to me, I managed to keep them on the right line as you may see from their resolutions. I hope very much that Mr. Nelson can manage to come out with his plan till we meet. I should like to talk his plan now, but unless I can do that I shall have to talk mine, which I dislike thoroughly since that should be relegated to the background. But something we must take as a basis for our discussions and my plan being published I cannot get away from it till we substitute Mr. Nelson's. There is no harm done since in principle the two plans are in full accord, but I should like to do all I can to give Mr. Nelson's plan all the prominence and not let it lose in originality by the continued publications of plans very similar as that of the Produce Exchange.

Will you please find out from Mr. N. what he wants me to do and what he intends to do. I have not called together my committee up till now but I cannot delay that very much longer.

Things are moving splendidly. Not a day passes without some new evidence that the Central Bank is wanted. Just read the report of the Produce Exchange Monetary Committee and see what they have learnt. That is only one of many!

Meanwhile, Andrew addressed the American Academy of Political Science at Philadelphia, and although he could not reveal details of the cure, his diagnosis was unequivocal:

"The provision of our national banking law requiring a proportional minimum of cash to be maintained inviolate against all deposits fixes an uncompromising limit to the expansion of loans and discounts and prevents the reserves of our banks from really serving as reserves. From the instant the required reserve levels are touched, no matter how critical the need, no further loan accommodation can be granted. There are no such restrictions in other important countries, where the accepted and unimpeded practice in periods of stress is to lend and discount freely to all who have legitimate requirements. One of the German bankers whom we interviewed last year described our system very aptly by comparing it to the regulations with regard to cab stands in the City of Berlin, where the law demands that there shall always be at least one vehicle at each of the appointed stands. Under this law a man returning home in the course of a wet and windy night and finding a cab on a street corner may be denied its use because the law provides that there must always be a reserve of one cab at each of the appointed stands...

"We are confronted with a great opportunity for constructive legislation. Bearing in mind our own history and learning from the centuries of experience of other countries, we are on the way to rid ourselves of the one obstacle which prevents this country from taking its place in the financial forefront of the world. You know all too well how the business of the country has suffered from the lack of organization and coherence in our banking system; you know the deliberate and scientific methods by which the problem of finding a remedy is being approached; you know the masterful quality of the mind responsible for framing this legislation, and those of you who have met the Chairman of the Monetary Commission during the last two years can sense his profound ambition and desire to achieve this banking reform. But if this question is to be rightly resolved everyone who is interested will have to cooperate magnanimously. Those who have preconceived plans as to what should be done will have to put them again into the crucible, and those who have particular interests which may be partially affected for a time will have to sacrifice some of the smaller and more temporary benefits for those which are larger and more enduring."

On Sunday, January 15, 1911, all the Jekyll Island veterans except the Senator held a clandestine meeting in Andrew's Washington apartment to begin the campaign for enactment of the plan. Senator Aldrich, in precarious health, had been summarily ordered by his doctors back to Jekyll Island, but this time his trip was solely for recuperative purposes. The date of their leader's return uncertain, Andrew, Davison, Warburg, and Vanderlip rushed a transmittal letter over his signature to the Monetary Commission by which the draft bill, already the subject of nationwide curiosity, could at least be made public.

From the opening skirmishes of the campaign, Andrew made every effort to enlist bipartisan support and to carry the proposal forward under the banner of nationwide consensus. But the Jekyll Island bill was universally dubbed the "Aldrich Plan" and bore that name from then on. By and large it received an enthusiastic reception. One shrill dissenting voice was Parker Willis, at that time Washington correspondent of the Journal of Commerce, who attacked it as a "scheme" purportedly "concocted" and then "hatched" at a White House conference that was not a "lovefeast" but an "acrimonious controversy." Later a prime mover with Carter Glass in drafting the Federal Reserve Act, Willis was evidently disgruntled at the unwillingness of the National Monetary Commission to utilize his services.

Despite the absence of Senator Aldrich, who did not return to Washington until mid-March, the campaign got off to a flying start with a resounding endorsement by the American Bankers' Association at a meeting in Atlantic

City in February. In countless speeches throughout 1909-1910 on behalf of banking reform, Aldrich, Andrew, Vanderlip and Warburg had prepared the terrain well, and the country's leading commercial bankers, all hitherto hesitant to confer any of their powers to a central bank, pledged their support to the Aldrich Plan.

Aldrich had accomplished one of the great educational feats of the age. Bankers had begun to grasp the implications of the changed components of the money supply. As long as banks had continued to extend their credit in the form of notes, supervision of the note issue had been the most important regulatory task of government. But as credit increasingly took the form of bank-created deposits, the note issue problem evaporated. The intellectual challenge in banking reform demanded a deeper understanding of national reserve banking. Yet a central bank empowered to adjust the quantity of money and its rate of growth to the actual productivity of the economy was beyond the horizon of all but a few. An avowed statement of such purpose would probably have frightened many a potential friend of the Aldrich Plan, and it was necessary for Andrew and his fellow proponents to speak softly and underplay the potential scope and impact of the proposed reserve association. While the power to raise the percentage of required reserves, the nuclear weapon of monetary policy, was implicitly vested in the highly centralized National Reserve Association, the Aldrich Plan proposed no specific reserve percentages and left the matter to future negotiations.

In its provisions governing the rediscount mechanism and the creation of bankers' acceptances, the Aldrich Plan sought to foster a liquid secondary market in acceptances to provide a much-needed means of exchange between member and non-member banks. But in order to make these provisions palatable, country bankers had to be convinced that the National Reserve Association would not be subject to periodic pillage by the big banks that would inevitably create the bulk of these acceptances; and big bankers had to be reassured that open market operations of the association would not be inimical to their interests. As Andrew knew, the Aldrich Plan could not possibly chart an unalterable course for future monetary policy or provide reassurance against the inherent mutability of the banking business. The goal was not to produce a technically perfect blueprint but to begin construction of the edifice.

One obstacle was the ill-starred Commission itself. As originally constituted, the National Monetary Commission consisted of nine senators and nine congressmen, five Republicans and four Democrats from each house. Within three years, two of the senators, Allison of Iowa and Danield of Virginia, had died, and four, Burrows of Michigan, Hale of Maine, Teller of Colorado, and Money of Mississippi, had lost their senatorial togas. The two deceased senators had been replaced early in 1911, without the advice or consent of the absent Aldrich, by two other ex-Senators, Taliaferro of Florida and Flint of California. Apart from the Chairman, the only Senator able to contribute effectively to the Commission's work was Theodore Burton of Ohio, although at Andrew's recommendation to Aldrich, Flint was eventually replaced by the weighty Boies Penrose of Philadelphia. The House delegation had also lost heavily through attrition, Overstreet of Indiana having died and Smith of California having retired in poor health. Bonynge of Colorado and McLachlan of California had been defeated for re-election in their districts. Apart from the Vice-Chairman, Edward Vreeland, John W. Weeks was the only effective member of the Commission in the House. The replacement of Flint was the last change in its membership. Short of a mass purge, there was no way to transform it into an effective body, and indeed Aldrich had never intended it to be more than a legislative phalanx, a role it failed lamentably to perform.

The 1911 Summer White House was at Beverly, Massachusetts, a stone's throw from Andrew's cottage in Gloucester; and so Andrew got his chance, unusual for a mere Assistant Secretary, to belabor the President directly. Taft was sympathetic enough to the Aldrich Plan and made several speeches on its behalf, but he was not willing to accord its enactment high political priority. Andrew tried to prod the 340 lb. chief executive:

"Dear Mr. President:
   In our conversation of the other day you suggested that it might be helpful to know how far the changes in the plan as originally proposed have been tentatively agreed to by those who are most concerned with the preparation of the report. Of course the Commission as a Commission has not formally approved of anything as yet.
   But Senator Aldrich has freely and publicly endorsed several amendments to the tentative plan as originally promulgated.
   As you know, practically everyone is agreed that state banks and trust companies must be allowed to share the privileges of the Reserve Association, but in extending these privileges it is necessary to prescribe certain standards which must be complied with by them. These have not been worked out definitively, but it is safe to say that participating state banks or trust companies will have to submit to some sort of prescribed examination, will have to publish certain prescribed reports, will have to have capital stock of an amount proportional to the size of the city in which located, and will have to conform, as regards its reserves against demand liabilities, to regulations similar to those governing national banks.
   At the same time, it is clear that the powers of the national banks must in several respects be enlarged, so

that our federal institutions will not be driven out of existence because of the larger privileges extended to banks under state charters..."

Andrew's direct access to the President was anathema to his immediate superior, Treasury Secretary Franklin MacVeagh. It was an intractable situation. Older than Aldrich, MacVeagh was a dour fellow, fiercely attentive to detail but without a sense of wider objectives. He and Andrew were an incongrous pair and, despite contiguous offices at the Treasury Department, communication between them was infrequent and strained:

August 23, 1911

"Dear Sir:
  In reply to yours, just received, about your going with Senator Aldrich to Europe, permit me to say it turned out yesterday forenoon that the question had been decided last Sunday by the President. When I learned this from the White House I, as a matter of course, quit considering the question. Mr. Curtis then expressed himself willing to make arrangements between himself and Mr. Bailey which would keep one Assistant Secretary at his post and make my longer presence in Washington unnecessary. The matter, therefore, is decided; and I am telegraphing Senator Aldrich that the President's wishes will, of course, be carried out.
                    Very truly yours,
                    Franklin MacVeagh
                    Secretary of the Treasury"

The Aldrich Plan tide reached its high-water mark in mid-1911 and slowly, imperceptibly, began to ebb that Autumn. Its proponents had not succeeded in making it a non-partisan issue, no doubt a hopeless task from the beginning. Realizing this, Andrew made a determined bid for the support of the Governor of New Jersey, Woodrow Wilson. As a former pupil in Wilson's political economy classes at Princeton in 1892, he tried a back-door approach:

November 23, 1911

"Dear Mr. Wilson:
  I am taking the liberty of sending you a set of the publications of the National Monetary Commission, in the editing of which much of my time has been occupied during the past two years. I do not imagine that you will be able to read many of the volumes, but I hope you may perhaps consider them worthy to occupy a place upon your shelves and may possibly from time to time find occasion to refer to them in your study of the financial problem. If at any time you desire information on this, or for that matter on any other subject connected with the departments in Washington, I should

be glad if you would call upon me. As perhaps you know, I have worked on the plan at every stage of its development, and, though if you recall me at all it is probably as one of your former and very callow students in Princeton, it is possible I might be able to throw some light upon questions which occur to you with regard to the plan. I am immensely anxious that your great influence should rest on what I believe to be the right side of this question, and I want you to feel free to call upon me at any time for any service in connection with it that may occur to you."

But Wilson, already eyeing the Democratic presidential nomination, was much too cagey to engage in a tête-à-tête with a high official of the Taft administration and rebuffed Andrew summarily with a rubber-stamped note:

November 27, 1911

"My dear Mr. Andrew:
   Thank you sincerely for your kindness in sending me a set of the publications of the National Monetary Commission. They will be interesting and useful for reference, even if I cannot go very extensively into them. I am very anxious indeed to study the monetary question as thoroughly as possible and shall value your guidance in the matter very much indeed.
         Cordially and sincerely yours,
         Woodrow Wilson"

One of the principal objectives of the Aldrich Plan was to boot the U.S. Treasury out of the banking business, but Secretary MacVeagh was in no hurry to surrender his banking prerogatives and adopted delaying tactics, requesting frequent opinions from Attorney General Wickersham concerning the constitutionality of the plan. This was too much for Andrew, who finally resorted to a public forum to bring the issue to a head. In a speech to the Louisiana Bankers' Association, one of the most insubordinate ever delivered by a subcabinet Treasury official, he took Secretary MacVeagh to pieces. He ended by saying:

"...And so the Treasury will rid itself at last of two branches of activity for the conduct of which it has always been ill-equipped and which it ought never to have assumed. One of these, a bank-note circulation, was undertaken only as a measure of war, but has lagged on for a half-century thereafter; the other, the independent Treasury, though it has been on the statute books a longer time, collapsed in reality years ago, and for at least a decade has maintained only a limping existence. We shall see them both go, with benefit to the country and profit to the Government."

Late in February 1912, Theodore Roosevelt formally

challenged President Taft for the presidential nomination, and from then on the House Banking and Currency Committee, under Carter Glass, refused to touch the Aldrich Plan. Their prospects of regaining the White House dramatically enhanced, Democratic legislators had little incentive to endorse a bill bearing the Aldrich imprimatur, whatever its merits. Aldrich himself on the verge of retirement from the Senate, had little remaining political leverage; and the President, lukewarm in the first place, was left to the ministrations of his seventy-five year old Treasury Secretary, Franklin MacVeagh, who accorded enactment of monetary legislation very low priority. Theodore Roosevelt would stand at Armageddon and battle for the Lord, but among the cadavers littering the field would be the Aldrich Plan. Meanwhile, uncertain whether Roosevelt might succeed in his unprecedented bid, Andrew and Warburg made a desperate effort to save the Plan from destruction at the hands of the House Banking and Currency Committee.

As the schism in the Republican party widened, however, party discipline collapsed, and western senators began attacking the plan, often merely for publicity and on the flimsiest grounds, claiming that the Plan would do nothing for Agriculture or would further enhance the already legendary powers of Pierpont Morgan. Since Aldrich himself had yielded the Senate whip to younger colleagues, it fell to Andrew to lead a final desperate charge.

The last hope for the plan was a strong endorsement in the Republican Party platform, reversing course by making the plan a partisan issue in view of Democratic unwillingness to consider the Monetary Commission's recommendation as a basis for legislation. Secretary MacVeagh, well aware of this fact, saw fit to forbid Andrew to attend the Chicago Convention in June. Andrew chose to disobey, and attended the Convention to make a final plea on behalf of the Plan.

But the party leaders had other worries, and Andrew's plea fell on ears deafened by the uproar over the contested seating of Taft delegates from the Southern states. Relentlessly gavelled by Elihu Root, the Convention renominated Taft at the cost of a sundered party. The Adrich Plan received no formal endorsement. His expedition a failure, Andrew returned to Washington to face the music:

June 24, 1912

Dear Sir:
It will greatly accommodate me to have you tender your resignation as Assistant Secretary of the Treasury; and I trust it will be convenient to you to have it take effect at once.
Very truly yours, Franklin MacVeagh

June 28, 1912

Dear Doctor Andrew:
On the 24th instant, I wrote you suggesting your

resignation as Assistant Secretary of the Treasury; and in conversation that day I assured you I could not change my judgment in the case. As you have not complied with my suggestion, I am writing to say I must expect you to take action today.
      Very truly yours, Franklin MacVeagh

            June 29, 1912
Dear Mr. MacVeagh:
 It was agreed, as I understood, in our conversation on Monday that there was no necessity for precipitate action. I am giving the matter which we discussed the most careful consideration and will reply to you at the earliest opportunity.
      Sincerely yours, A. Piatt Andrew

 Secretary MacVeagh, by then the oldest man ever to sit in the Cabinet, was a hardy fellow and ultimately lived to be 96, but Andrew's use of the word "precipitate" must have raised his blood pressure to dangerous levels.

            June 29, 1912
Dear Sir:
 I have just received your note of this date in reply to mine of yesterday. Your statement, that you understood from our conversation on Monday that there was no necessity for precipitate action I will not wait to comprehend. I may have made the mistake in my letters of being over-conventional, and I did not fail to give you, in the conversation, an opportunity to consider your resignation with a view to phrasing it to suit yourself, and to preserve, if you choose, the initiative. But, of course, I wanted a prompt resignation. It is not, however, important to discuss this for you will know now that I want your resignation, without delay, to take effect at once.
 Your present attitude of vagueness I cannot accept. It is embarrassing to my plans, both official and personal, to wait longer in so plain a matter; and it is imperative that I should insist upon an immediate reply, saying whether you will tender your resignation to take effect immediately or whether you decline to do so. This is not a case that can require any further consideration.
        Very truly yours,
        Franklin MacVeagh

            The White House
            July 2, 1912
My dear Mr. Andrew:
 I have had a long talk with the Secretary of the

Treasury on the subject matter of our interview this morning. It is perfectly evident to me that your relationship with the Secretary can not continue, and that either you must resign or I must have a new Secretary of the Treasury.

Mr. MacVeagh has conducted the Treasury in a way very satisfactory to me. He has effected many reforms and many economies, and while his method of reaching conclusions has sometimes been different from my own and more deliberate than seemed to me necessary, I am bound to say that the general result in the Treasury Department is a very strong vindication of his course. I can not think that you were well advised in deliberately disobeying the plain direction or intimation from Mr. MacVeagh that you should not absent yourself for the purpose of going to the convention in Chicago. Of all things that can not be countenanced in the department, disobedience is the chief, and I am obliged to say that in spite of the extended explanations which you have made with reference to your attitude and your conduct in the Department, I heard no attempted justification of this act of insubordination.

I am sorry on many accounts to part with you, because I believe you to be an able economist and familiar in a scientific way with financial matters; but the issue is such that I can not avoid it, and I have therefore written to the Secretary of the Treasury authorizing and directing him to ask for your resignation. I write this letter in response to your request that you be advised before formal official action is taken in the matter.

Sincerely yours,
Wm. H. Taft

That was the end of the Aldrich Plan. It never again received serious legislative consideration by friend or foe. The Bull Moose Convention ignored it. The Democratic Convention denounced "the Aldrich Plan for a central bank," a phrase which an unseen hand revised to read "the Aldrich Plan or a central bank" in the platform as published in its final form. During the ensuing campaign, Congress inevitably deferred all action on the monetary question, and after Wilson's election, Carter Glass opened a new series of hearings before the House Banking and Currency Committee with a flat statement precluding consideration of the Monetary Commission's proposal.

Throughout 1913, while Glass and Willis rough-hewed the Federal Reserve Act, Aldrich and Andrew were relegated to the role of onlookers. But in May, Andrew finally obtained the long-postponed meeting with his former history teacher and argued strongly but vainly to limit the number of Federal Reserve Banks to a maximum of eight.

With greater success, he pleaded to lengthen the terms of the Federal Reserve Board's appointive members, and to phase out all ex-officio members as soon as possible. Indeed, the Act as eventually passed differed in many fundamental respects from that originally introduced in Congress, or as it passed the House in its original form. During the last few hours of Senate debate, it was further modified to resemble the Aldrich Plan even more closely, and it is untrue, as President Wilson and others subsequently stated, that the act was passed "over the objections of the bankers."

During the bill's vicissitudes, Andrew published frequent criticisms in the columns of the *Boston Evening Transcript*. These criticisms may have cost him an appointment to the first Federal Reserve Board. After President Wilson signed the act into law on December 23, 1913, both Davison and Vanderlip were approached to go on the board but declined. Unemployed at the time, Andrew would probably have accepted. As it turned out, Paul Warburg became the only veteran of Jekyll Island to participate in the actual construction of the edifice designed there three years before.

In basic structure, the Aldrich Plan was superior to the Federal Reserve Act, because it constituted an avowed central bank, with branches to support its monetary policies. The National Reserve Association was to be incorporated at Washington and to have all powers not specifically delegated to its branches. Conversely, the Federal Reserve Act provided for the incorporation of twelve separate reserve banks and stipulated that all powers not expressly vested in its Board of Governors would remain with the reserve banks themselves. Each Federal Reserve Bank, for example, received the right to conduct open market operations on its own initiative, a provision which made a farce of any concerted monetary policy until subsequent establishment of the open market committee to coordinate these activities. Indeed, the first years of the system were devoted in large part to closing such holes in the dome, for the act itself contained many concessions to the Bryan wing of the Democratic party, whose obsession with decentralization gave the system a sectional imprint it has retained to this day.

Inevitably, the Federal Reserve Act restored to the President the full appointive powers the Aldrich Plan had sought to limit. Neither Senator Aldrich nor the leading bankers had been enthused at the prospect of confiding the destiny of the banking industry to any occupant of the White House. The Federal Reserve Act, however, wisely fixed terms for the appointed members out of phase with presidential elections. Andrew's advocacy of longer terms for the board's appointive members had demonstrable influence on a number of senators, who amended the Glass bill in this direction to guard the system from partisan embroilment. The note issue and reserve provisions of the

Federal Reserve Act also differed sharply from the Aldrich Plan. Federal Reserve Notes became direct obligations of the United States. While this step was an apparent concession to friends of fiat money, it had no tangible inflationary effect, for note issue had ceased to be the principal instrument of Federal monetary policy.

The Federal Reserve Act was superior to the Aldrich Plan in its treatment of reserve requirements, fixing the percentages applicable across the board to member banks, expressly asserting the right to adjust the reserve percentages as and when deemed advisable, and doing away with the split reserve provisions of the National Banking Act. In order not to frighten his supporters in the banking industry, Senator Aldrich had skirted this issue, preserving the right of member banks to keep part of their reserves with each other. On the other hand, the Federal Reserve Act denied banks the right to count federal reserve notes held as vault cash toward fulfillment of the reserve requirement. This provision was another gap in the dome, as it induced banks to retain gold coin and pay out their Federal Reserve Notes, thereby impeding the desired mobilization of gold by the reserve system. The Federal Reserve Act was also superior to the Aldrich Plan in requiring all national banks to become members of the system, but was vitiated by the provisions permitting each reserve bank to fix its own rediscount rate. The Aldrich Plan called for a uniform rate throughout the country, a farsighted provision possibly not palatable politically at the time, but in the long run far more consonant with the dictates of effective monetary policy.

What if the Aldrich Plan had become law? Would the subsequent financial history of the United States have been different? The branch network proposed by the Aldrich Plan might well have proved more flexible in the face of demographic change than today's twelve venerable but ponderous banks comprising the Federal Reserve System. There might also have been economies of scale, avoiding duplication of research and ancillary services by the quasi-independent reserve banks. And the emphasis on self-regulation by the banking industry implicit in the Aldrich Plan might have set a pattern for future regulatory agencies, limiting direct incursion of federal authority over other industries. But these were marginal differences. The spirit of the National Reserve Association is embodied in the Federal Reserve System and shares its history.

As for Piatt Andrew, he did not resume his career as an academic economist. Instead, he went on to become the founder of the American Field Service in France in World War I. Returning to his home in Gloucester, he was elected to Congress and represented Massachusetts' 6th District in the House until his death in 1936. He never claimed credit for his role in the creation of the Federal Reserve System. In part, his silence was rooted in loyalty to Senator Aldrich, whose biographers have attempted to give his version, and in part in reluctance to summon up the memories of such an overwhelming defeat. But the best thing about an accomplishment is the doing of it; and historical laurels, like battlefield medals, do not always go to the most deserving. Shortly before his death he relented somewhat, however, and at the urging of Everett Case, future Chairman of the Federal Reserve Bank of New York, gave him the following dispassionate account of the rise and fall of the Aldrich Plan:

"The Monetary Commission was a one-man show. Not that Senator Aldrich started with any prejudged notion of what was ailing our banking system, or any preconceived remedy that he wanted to see adopted. But he saw an opportunity to solve a great problem through a constructive piece of financial legislation, and he hoped to bequeath to the United States something akin to what Peel had given to England in 1844, or what Hamilton had given our own country in the first decade of American history.

"He expected little help from the members of the commission, most of whom had little to offer in the way of scholarship or experience in financial matters and all of whom he knew he could control. He sought advice outside the Commission's membership. The people he turned to first of all were Paul Warburg, Harry Davison, George Reynolds and to some extent to Frank Vanderlip, Barton Hepburn and James B. Forgan. He later consulted with financial writers and economists in all parts of the country. He never called upon Parker Willis for adivce in spite of the fact that I urged him to do so. He had some deep-rooted antipathy to Willis, perhaps because he had found some of his articles offensive. Whatever the reason, he was obdurate on that subject. So far as the Commission itself was concerned, the Senator's principal idea was to keep its members happy until he had a bill ready and then get their approval.

"After the tariff was out of the way, he would call meetings of the Commission every two months or so to discuss some new phase of the subject. These meetings, usually held in some such place as the Hotel Plaza in New York, were designed to keep the members happy and reassure them they were not being left out of the picture. These meetings were not, however, important from any other point of view. Occasionally some member would have an idea to which the Senator would listen patiently, but following some general discussion one of his friends on the Commission would usually move that 'the matter be left to the Chairman with power to act.' This was the formula which had long been followed in Aldrich's finance committee on tariff questions.

"The first rough draft of a bill evolved at the Jekyll Island meeting in November 1910, attended by Mr. Davison, Mr. Vanderlip, Mr. Warburg, as well as Senator Aldrich, his secretary and myself. The meeting had to be kept secret lest it be charged that Wall Street was dictating the bill. The draft bill was printed in January 1911 and submitted to leading bankers and economists throughout the country for suggestions and revisions. I suppose that as many as twenty modified drafts were printed during the course of that year as a result of continuous consultations with hundreds of important people.

"Unfortunately, the development of the political situation was unfavorable for the plan. The breach between Colonel Roosevelt and President Taft widened perceptibly during 1911, and it became obvious long before Colonel Roosevelt's formal announcement in February 1912, that he would challenge the President for the party's nomination. To complicate matters still firther, the Democrats won control of the House in the elections of 1910 and in taking over the Banking and Currency Committee assumed a wait-and-see attitude as far as monetary legislation was concerned. For several reasons, the Taft Administration did not actively push for enactment of the bill. For one thing, Senator Aldrich had little regard for Secretary MacVeagh, a lifelong Democrat and free trader. As a result, President Taft's principal advisor on financial matters was not disposed to support Aldrich's plan of reform. In the second place, Taft himself was not much interested in the complex technical questions involved in banking legislation. He was a rather easygoing person, his avoirdupois being such as to make the harder forms of mental labor distasteful. Later, as he lost weight, his mind became correspondingly keener. In the third place, the Payne-Aldrich Tariff of 1909 had put the President in the embarassing position of being obliged to approve a measure he was known to dislike. Since he frankly did not understand the currency question, and since the Aldrich Tariff had displeased him, the President was at best lukewarm about the monetary measure. Senator Aldrich knew that the Tariff had hurt him and told me in personal conversations that his own inclination would have been to liberalize its provisions considerably, but that he could not do so without letting down old associates who had stuck by him through thick and thin. Since personal loyalty of this sort was a cardinal principle with him and indeed the very foundation of his long domination of the Senate, he could not and would not go back on them.

"Of the bankers whose advice he sought, Warburg undoubtedly knew the most about the subject, but the Senator disliked the tenacity with which Warburg would press his points. Often he would shut him off in the middle of a sentence only to reintroduce later the point Warburg had been making as his own. Mr. Davison and Mr. Vanderlip were particularly helpful in arranging contracts with important bankers in different parts of the country and were very adept in the rough-and-tumble debate that accompanied the plan in its various stages. Mr. Davison, by implication, also spoke for Mr. Morgan, who was careful never to intervene personally for fear of hurting the plan's chances. The Senator himself was masterful in his manner of soliciting help and cooperation from people. He used to recount how, upon entering the Senate for the first time, he had asked an old Senator what was the most important prerequisite for Senate leadership and had received the reply 'patience.'

"After it became clear that the Commission's bill was not going to be acted upon, however, the Senator allowed personal bitterness to color his opinion somewhat. His denunciation of the proposed Federal Reserve measure in October 1913 at the Academy of Political Science at Columbia was highly regrettable and can only be attributed to the depth of his disappointment at the failure of what was to have been the crowning achievement of his career. The result of his denunciation was simply to provide fuel for countless subsequent assertions by the sponsors of the Federal Reserve Act that Senator Aldrich can claim little or no credit for that legislation, despite the fact that its provisions reflected virtually all the reforms originally contained in the Aldrich Plan."

Ironically, that speech was the final if unwitting contribution by Senator Aldrich to banking reform, because it convinced a number of hitherto wavering members of the Bryan wing of the Democratic party to vote for the Federal Reserve Act in December. They evidently concluded that any bill criticized so vehemently by Aldrich was, ipso facto, a good thing. And so America got its central bank after all.

PART THREE: HARVEST TIME

From the 1920s to the Post-War Age

With the automobile, radio, and cigarettes well-established, the euphoria of peace took over and it's not surprising that America in the 1920's rushed on with big spurts of economic growth. What is somewhat surprising is that it all ended so quickly; although farm problems, speculative credit and the nation's reluctance to share its growing wealth with the large mass of labor surely had a lot to do with it. Once the depression hit, it knocked the nation off-balance, leaving her broke and despondent until the New Deal patched together a variety of programs to try and get things righted once again. Some ideas and laws coming out of that period were simply hare-brained, and others unconstitutional; but the fact remains that much of it stuck and those efforts to find a middle way dominate in large part the mixed economy that is America today.

Unwillingly, another War came on, young men were sent to fight, government mobilized, goods were rationed, and the industrial power of the nation combined to create a striking economic-military victory. The post-War years witnessed new obstacles in the form of inflationary pressures, periodic recessions, low farm prices, social disorder, and even civil unrest. Sometimes Americans felt the economy wasn't really doing the job, until they reflected a bit more and discovered such imperfections are all part of the human condition after all. What has happened is that the present time is really one of harvest, and the quality of production isn't always up to standard. This means the economy has a lot more work to do, and in approaching the good life yet to come, she must of necessity understand and build on the past that is America's economic history.

THE NATIONAL ASSOCIATION OF MANUFACTURERS AND LABOR RELATIONS IN THE 1920s

By ALLEN M. WAKSTEIN

During World War I American business and labor met the problems of labor relations responsibly. Wartime demands, expediency, mutual advantage, and government policy called for a greater spirit of conciliation, so that the previously hostile relations between labor union and employer were mitigated.[1] There was hope that this new spirit would carry over to the postwar years,[2] but the self-centered interests of each led to conflict. Employers, with a parochial attitude toward labor, moved vigorously in an attempt to deal with the rising "labor problem." Within a few years they won their offensive, and strikes diminished and unions were weakened during the 1920s. Concomitantly, management's power, except in a few industries, increased, and by the end of the depression of 1921 management clearly held a dominant position over labor.

The National Association of Manufacturers (N.A.M.) in the 1920s

---

[1] Gordon S. Watkins, *Labor Problems and Labor Administration in the United States During the World War* (Urbana, 1920), 222-227; Selig Perlman and Philip Taft, *History of Labor in the United States, 1896-1932* (New York, 1935), 403-411.

[2] W. Jett Lauck, *Political and Industrial Democracy, 1776-1926* (New York, 1926), 17; Watkins, *op. cit.*, 223-240. Numerous publications of employer groups and business magazines discussed progressive programs of industrial relations. Examples of these are John Leitch, "Industrial Democracy," *American Industries*, XIX (June 1919), 34; S. S. Smith, "Building and Loan Associations Make Contented Employers," *The Employer*, III (March 1919), 7; William G. Aborn and William L. Shafer, "Representative Shop Committees," *Industrial Management*, LVIII (July 1919), 29-32.

generally was considered to be a major spokesman for the business community. As the largest and the most nationally constituted employer organization, it sought to represent and lead American business in employment relations. Although the N.A.M. was limited in its influence, its size and complex arterial connections with other employer and trade organizations meant that its views and policies reflected those of a significant portion of the business community. An examination of the Association's labor-relations position in the twenties therefore has implications, as this study will suggest, beyond institutional history. What is revealed is the system under which industrial "peace" and the prevalence of management's position was achieved, and what is suggested is the failure of management, once it gained a dominant position, to deal adequately in an area in which it knew its responsibility.

The N.A.M. traditionally was an advisor to its members in regard to labor relations. Until the war its public image was that of adamant anti-unionism, but then it moderated its militancy and emphasized the patriotic need for cooperation.[3] For several months following the armistice it was not clear which policy the Association would follow. At its convention, in the spring of 1919, there were some members who favored a return to the militancy of the early 1900s; but the position that temporarily prevailed was that of encouraging more responsible and constructive approaches, such as had been born of progressive thought in the previous two decades.[4] By fall the conflict on the industrial scene drew the attention of the Association's directors and there was talk of reappraisal of policy. Again, rather than embark on any new activity, the N.A.M. Board decided to maintain the conciliatory approach which the United States Chamber of Commerce and the National Industrial Conference Board were to present at the President's First Industrial Conference.[5] This Conference had been called in the hope that the spirit of wartime cooperation could be expanded upon and extended into the postwar years. When it failed, and as the number of strikes and industrial unrest increased, the N.A.M. Board of

---

[3] Albion Guilford Taylor, *Labor Policies of the National Association of Manufacturers* (Urbana, c. 1928), 35-37; National Association of Manufacturers, *Proceedings,* 1918 (New York, 1918), 22-26, 106; Albert K. Steigerwalt, *The National Association of Manufacturers, 1895-1914: A Study in Business Leadership* (Ann Arbor, 1964), 113-117, 121-128.
[4] N.A.M., *Proceedings,* 1919, 10-42.
[5] N.A.M. Executive Committee of the Board of Directors, "Minutes of Meeting," September 4, 1919, N.A.M. General Files, N.A.M., New York; Board of Directors, "Minutes of Meeting," October 17, 1919, N.A.M. General Files.

Directors began to reassess its position and to seek out the means by which the labor situation could be handled.[6]

The policy that emerged, based upon already established employer philosophy, can be summarized by the term "open shop." Specifically the term meant that the employer was to be free to conduct his business affairs, to make decisions, and to determine—unencumbered by outside pressures—the relationship between management and labor. Without being explicit, employers had followed these policies throughout the nineteenth century. In the past, only when union pressure threatened the norm did individual employers react by formalizing their thoughts under the label "open shop" and by offering it to their employees as an alternative to the union closed shop. In 1902, for example, after unionism had made considerable gains for five years, employers felt their position threatened and coalesced first into local associations and then into a national organization led by the N.A.M., in order to reestablish more effectively open-shop conditions.[7] During the Progressive era and the war that followed, unionism once more regained strength and, in seeking to utilize it, again gave the business community cause for concern. Holding that their sacred tenets were being violated by union practices, particularly by that of the closed-shop rule, business saw the need for group effort to counteract them.[8] In 1920, following the spontaneous formation of numerous local open-shop associations, the N.A.M. sought to coordinate these efforts and, through national leadership, to fulfill its labor-relations principles. Toward this goal it appointed from its membership an Open Shop Committee and hired a professional staff to manage an Open Shop Department. The latter soon became the compiler of, and the clearing house for, information concerning the evils of unionism and the advantages of the open shop.[9]

Thus armed, the N.A.M. took up the struggle. It labeled trade-unions "un-American" since, it held, the closed-shop rule—a major goal of the unions—was a denial of the inherent American principles of *laissez-faire*, free competition, and the right of private property. Union at-

---

[6] N.A.M. Board of Directors, "Minutes of Meeting," November 6, 1919; notation by Secretary of N.A.M., in N.A.M. Board of Directors "Minutes of Meeting [Book]," December 11, 1919.
[7] Clarence E. Bonnett, *History of Employer Associations in the United States* (New York, c. 1956), 46, 51, 53, 102-166; N.A.M., *Proceedings*, 1903, 57-62; Philiip S. Foner, *History of the Labor Movement in the United States* (4 vols., New York, 1964), III, 36-39; Steigerwalt, *op. cit.*, 108-113, 121-123.
[8] Perlman and Taft, *op. cit.*, 435-460; Watkins, *op. cit.*, 87-89.
[9] N.A.M. Board of Directors, "Minutes of Meeting," July 7, 1920; Open Shop Committee, "Report," in *ibid.*, October 22, 1920.

tempts to influence wages, restrict production, and determine working conditions were considered to be a violation of "natural law." The open shop, to the contrary, which did permit economic laws of supply and demand to operate, was entitled the "American Plan."[10]

Besides these theoretical arguments, the N.A.M.'s campaign considered practical elements. Unions, supported by the closed shop, hindered the nation's prosperity, destroyed the incentive of business, and rendered the workingman dependent and slothful. Studies were undertaken to show that a community's economic development, costs of public and private construction, and wholesale prices of building materials, clothing, and fuel were all adversely affected by a preponderance of closed-shop unionism. In contrast, it was argued, the open shop had been the system under which business had built a great industrial society and was the cornerstone for continuing America's prosperity. The system achieved the greatest material success for the community and also provided for the best interests of both employee and employer.[11]

The N.A.M. supposedly was "not opposed to organization of labor as such . . .,"[12] but in denying the legitimacy of the regular functions of a union it relegated such organization to a social role or to being a mere conveyer of management's decisions. By being unwilling to bargain with union representatives, it rendered meaningless the right of labor to organize. It refused the challenge to approve publicly of employers dealing with union leaders in matters affecting union employees, even in return for pro-labor support of the open-shop idea.[13] This attitude, along with other responses to unionism, clearly indicates that the open-shop drive sought, if not to obliterate unions, at least to remold them in the employers' image of what they should be.

Since open shop meant that employers would not deal with unions, how then did they assess the status of union members? The position

---

[10] The best single source for capturing the N.A.M.'s view toward unions and the open shop is N.A.M., *Open Shop Encyclopedia for Debaters,* (New York, 1921,) *passim.* Further explanation of the Association's position can be found in the *Open Shop Bulletin* and the *Open Shop Newsletter,* published by the N.A.M. from November 1, 1920 to March 28, 1932; and in pamphlets and reports of the Open Shop Department, Open Shop Department Files, N.A.M.

[11] N.A.M., Open Shop Department, *Evidence in the Case of the Open Shop* (New York, 1923); *ibid., Evidence for the Open Shop* (New York, 1927); copies of studies in N.A.M., Open Shop Department Files and reports of Activities in Open Shop Department, "Report," April 1-September 30, 1926 and *ibid.,* June 1 to Sept. 30, 1927, Open Shop Department Files, N.A.M.

[12] N.A.M., *Proceedings,* 1920, back cover and in other years.

[13] William Frew Long to James A. Emery, September 19, 1923, N.A.M., Emery Collection, N.A.M., New York.

of the N.A.M., as stated in its convention-approved Declaration of Labor Principles, was representative of what most open-shop advocates believed:[14] "No person should be refused employment or in any way discriminated against on account of membership or non-membership in any labor organization. . . ."[15] The major intent was to protect non-union members. To be fair, however, employers had to declare that they also would not discriminate against union members. Having condemned unions for an unfair practice, they could not have exposed themselves to a similar charge. It is also noticeable that the word "should" is used in preference to stronger language, allowing the actual implementation of the principle to be open to interpretation. A belligerent company could read militancy into any given set of statements, while the benevolent one could use the same principles cooperatively.

That employers, in practice, were not uniformly agreed as to whether or not an open shop could hire union members is evident. John E. Edgerton, speaking as President of the N.A.M., chided those who were not abiding by the "true" definition, those who claimed to be open shop but who, at the same time, hired only non-unionists. This practice, he said, was a violation of the principle. Although an employer could rightly refuse to hire union members, he could not then label his company "open shop." Such action "reflects a misinterpretation of the American Plan or the Open Shop Principle as commonly understood and practiced, and it gives comfort to those who falsely claim that the Open Shop movement is a subtle attack upon the right of employees to organize."[16]

Although vigorous in disseminating information encouraging the open shop, the N.A.M. promulgated neither conciliatory nor blatantly militant labor-relations principles. Beneath its seemingly monolithic and moderate image there simmered an abundance of discord, with members organizing themselves into various factions vying for dominance. The role of the executive office necessarily became one of moderator, between those who wanted more militancy and the expansion of managerial prerogatives, and those who wanted less aggression and a greater voice for labor in industry. The degree of success of each faction's in-

---

[14] A series of organizations who advocated nondiscrimination as the principle for the open shop are cited in Citizens' Alliance of Duluth, *Citizens' Alliance Bulletin*, III (March 1921). It states that no open-shop association had declared in favor of discrimination aaginst union men.

[15] N.A.M., *Proceedings*, 1920, back cover, and in other years.

[16] "Manufacturers' Head Deprecates Employers' Twist of 'Open Shop,'" *The Employer*, VI (August 1921), 7.

fluence was determined to a large extent by general economic and social conditions.

Soon after the formation of the Open Shop Department, the N.A.M. members who believed that the official policy of the open shop was not vigorous enough to counteract the threat of union power, and who desired to drum up support for an autocratic managerial position, presented to the Association's Executive Committee a series of amendments to the Declaration of Labor Principles.[17] If the proposed amendments had been adopted, the N.A.M.'s policy would have changed to closed-nonunion shop. Given the turbulence of labor conditions in the immediate postwar years and the negative label attached to labor during the "Red Scare," the N.A.M. could have justified assuming such an extreme role. Fearful of alienating people who traditionally had followed policies of non-discrimination, however, the Association took no action. In order to continue to project a favorable public image it dared not advocate a position that could be accused of being no better than that of the union. It was easier and more to its own self-interest to avoid having to deal with consequences wrought by change.

Despite the fact that the N.A.M. did not subscribe to the more militant approach, its governing body—the Board of Directors—did nothing to curtail individual action along these lines by any of its members. The Board, in fact, condoned closed-nonunion practices when circumstances seemed to warrant it. In the coal industry, where it believed the unions were very negative and militant, it contended that toleration of union members was bound to lead to agitation and disruption of normal operating conditions. Consequently, it stated that discrimination against unionists was in this case understandable. It also saw fit to advocate a policy of not hiring unionists who were particularly troublesome.[18] On the issue of "yellow dog" contracts, which in essence meant a closed-nonunion shop, it did not take a position. When a general manager requested instructions from the Executive Committee as to such contracts, the Committee stated that it "would not assume or presume to assume any attitude in this matter of contractual relations between employer and employe.'"[19] Although the N.A.M., in theory and in its pro-

---

[17] N.A.M. Executive Committee, "Minutes of Meeting," November 19, 1920.
[18] N.A.M., *Proceedings*, 1922, 21, 24-28; Fuel Supply Committee, Reports and mimeographed material, N.A.M., General Files; Noel Sargent, Manuscript of an Address to Pennsylvania Trade Secretaries Association [1922?], Open Shop Department Files, N.A.M.
[19] N.A.M. Executive Committee, "Minutes of Meeting," April 27, 1921.

jected image, assumed a somewhat middle-of-the-road approach, actually the more belligerent attitudes, as always in periods of industrial strife, dominated its policy.

Attempts by some members to obtain a more conciliatory or progressive handling of labor-management relations were disregarded, but nevertheless did exist. One such group within the Association which was concerned with labor matters, the Industrial Betterment Health and Safety Committee, included in its 1921 annual report the recommendation that, "The National Association of Manufacturers should lend its powerful influence towards the extension of industrial representation. It should clearly demonstrate that industrial representation does not involve the disruption or destruction of present labor organizations." The Executive Committee responded by deleting the paragraph,[20] and the committee which issued the recommendation failed to be reconstituted at the subsequent annual convention.[21] This official rejection of a more enlightened approach to labor relations is not surprising given the timing of the request. In the midst of the aggressive phase of the open-shop movement, leadership for the expansion of management's power, not labor's, was demanded of the N.A.M.

From 1920 through the summer of 1922, the N.A.M. provided aggressive leadership in promoting the open-shop idea. Reacting to the strength that unions had gained during the war and to their postwar militancy, the open shop had been used largely to weaken the union position. As with most of the ideas inherited from the Progressive era, the hope of bringing about "industrial democracy" also was shelved during the early twenties. Not until late 1922, when the "labor problem" had diminished, when the threat of union power had become less pressing, and when the public attitude seemed to be changing, could progressive ideas on labor relations again begin to receive a hearing. Now that the weight of union power had been lifted, belligerent employers, including members of the N.A.M., had become complacent. This allowed for the more progressive membership of the N.A.M. to advance its ideas in the hope that the assembled Association would adopt a more moderate and more inclusive open-shop position.

A beginning in this direction may be seen at a conference of representatives from prominent local employer associations which the Open Shop Committee sponsored in October 1922. The usual diatribes against

---
[20] *Ibid.*
[21] N.A.M., *Proceedings*, 1922, 9, *passim.*

unions and reaffirmations of open-shop principles were absent. Instead, emphasis was placed upon the need to expand the horizons of the open shop so as to offer a meaningful alternative to unionism, to present a better public image, and to initiate programs that would both utilize and benefit by current personnel-management techniques. The conference recommended to the Open Shop Department that the N.A.M. enlarge and develop its approach by educating employers, employees, and the public about their mutual rights, duties, and obligations, and by alerting employers about "the need for the most enlightened, progressive, and statesmanlike conduct of their activities. . . ." It passed a resolution calling for the dissemination of information "for the purpose of impressing on employers a larger realization of the duties and responsibilities of management."

The Open Shop Committee, at the same time, recommended that the N.A.M. establish a Department of Industrial Relations, only one phase of whose work would be the open shop, and that any further conferences should deal not only with the open shop, "but with all constructive phases of industrial relations."[22] Subsequently, the Association established such a Committee[23] which, having been instructed to gather and disseminate information on these programs of industrial relations, proceeded to solicit data from those who had in operation plans that sought to bring workers and management into a closer relationship. Among the items it concerned itself with were employee representation, profit sharing, group insurance, housing assistance and pension plans.[24]

But it is a long way between passive acceptance of new concepts and the active decision to support them vigorously; and in this case the former dominated the N.A.M. Despite growing sympathy toward progressive ideas, they were not meaningfully incorporated as the logical extension of the open shop. Instead a quiet struggle ensued. In 1923 the Industrial Relations Committee, now in existence for about a year as complement to the Open Shop Committee, submitted to the Board of Directors a confidential report critical of the "previous and present activities" and recommending a new tack. Although reaffirming its

---

[22] Open Shop Conference, "Minutes of Meeting," October 10, 1922, Open Shop Department Files, N.A.M.
[23] N.A.M., *Proceedings*, 1923, 52-55.
[24] Noel Sargent to members of N.A.M., February 18, 1924, Open Shop Department Files, N.A.M.; Industrial Relations Committee, "Minutes of Meeting," Jan. 25, 1925, April 4, 1924, Open Shop Department Files, N.A.M.; N.A.M. Board of Directors, "Minutes of Meeting," February 19-20, 1924, N.A.M., *Proceedings*, 1924, 88-90.

belief in "the true Open Shop principle," it contended "that an aggressive, militant and public campaign [as fought by the Open Shop Committee], is at variance with the generally understood and accepted principles of Industrial Relations." It concluded that to stress the need for cooperation between management and labor in light of the present militant tactics was to expose the Association to the public as "carrying an olive branch in one hand and a club in the other...." Explicit in its report was the entreaty that the N.A.M. reconcile the conflict by replacing the Open Shop Committee's stratagem of "Let's Fight" with the more diplomatic approach of "Get Together."[25] This attempt by the progressive faction within the Association at least to influence, if not to dominate, the N.A.M.'s labor policy was unsuccessful. Through 1924, the Industrial Relations Committee solicited information from members and discussed an expansion of programs, and for several years it continued to coexist with its adversary, the Open Shop Committee.[26]

At the same time the public and business community were becoming interested in many concepts of welfare capitalism and of industrial democracy that had been brewing in progressive thought during the previous quarter of a century. They were beginning to perceive these concepts as constructive, just, and economically wise. Receptive to public opinion, the N.A.M. also began to consider them as supplements to the open shop. It knew that more than militancy was expected of it, that public views were like a pendulum, now swinging against unions because of the radical image they projected, but all too ready to swing back against management if it inaugurated radical practices or took advantage of monetary conditions.[27] The Industrial Relations Committee was a reminder that the employer could benefit from the kinds of programs that would "tie men to management" and increase the "loyalty" of the employee.[28]

Against this background and out of the conflict between the Industrial Relations and Open Shop Committees there finally emerged, in 1926, the Association's decision to attempt a change in its formal labor policy. At the Board of Director's meeting in June the two committees

---

[25] N.A.M. Board of Directors, "Minutes of Meeting," July 12, 1923.
[26] The name of the Committee was changed from Industrial Relations to Employment Relations following the 1924 annual meeting. N.A.M. Executive Committee, "Minutes of Meeting," June 26, 1924. The change was only nominal since the composition and the function of the committee remained essentially the same. To avoid confusion the label of "Industrial Relations" will be used throughout.
[27] "Memorandum on Open Shop Movement," [1923], Open Shop Department Files, N.A.M.
[28] N.A.M. Board of Directors, "Minutes of Meeting," February 19-20, 1924.

were "directed to give consideration to revision of the Declaration of Principles, and to report at the next meeting specifically the Committees' recommendations as to changes to be made."[29] Subsequently, the chairman of the Industrial Relations Committee announced to the Board that "the present 'declaration' tends to give the impression that the National Association of Manufacturers is opposed to all labor organizations."[30] He presented the suggested revisions in the old format, but one lacking the strong, militant, and authoritarian language of the original.

The revisions were conciliatory and demonstrated recognition of a role for labor beyond that of merely contributing its skill and energy. Such phrases as "unalterably opposed" or "disapproves absolutely," which had been used earlier by the Association to describe its rejection of certain labor practices, were replaced by the milder phrase, "particularly disapproves." Excluded entirely in the new document was any statement of managerial prerogatives, as well as the statement that "employers must be unmolested and unhampered in the management of their business. . . ." While the original document had allowed the worker only two rights, the rights to work or not, and had seen the open shop as protecting the worker's freedom from interference by his colleagues and from outside pressure, it had made no reference to worker participation and consultation in plant affairs or to a cooperative mutual interest between employer and employee. The revisions embodied a completely new tone as exemplified by the following principles for which there had been no precedent:

>   2. The interests of employer and employe can best be promoted by free discussion of employment problems and by cooperative effort.
>   3. Adequate provision should exist in the individual plant for the administration of justice, the discussion of employment problems and the promotion of cooperative effort.[31]

The Board of Directors considered the proposed changes[32] and later approved the appointment of a special committee to study them further.[33] At the time of the Annual Convention, in the fall of 1927, no

---

[29] *Ibid.*, June 15, 1926.
[30] *Ibid.*, September 14, 1926.
[31] The "Proposed N.A.M. Labor Principles" were presented at the N.A.M. Board of Directors meeting on September 14, 1926. A copy of the recommendations is in the Open Shop Department Files, N.A.M.
[32] N.A.M. Board of Directors, "Minutes of Meeting," September 15, 1926.
[33] N.A.M., *Proceedings*, 1927, 59.

conclusion or decision had been reached."³⁴ The attempt of the Industrial Relations Committee to revise the Declaration of Labor Principles apparently failed, in view of the facts that no consideration was given the matter after the 1927 annual meeting and that the long established principles remained the N.A.M.'s avowed position.

The new recommendations had reflected the growing awareness on the part of some members that the previous militancy and the espousal of essentially negative principles needed to be replaced by policies more positive and more in keeping with progressive ideology. It was not enough merely to establish an open shop and then, while still emphasizing it, to give minor attention to constructive alternative programs. Acceptance of the revised Declaration of Labor Principles would have represented a commitment on the part of the N.A.M. to encourage business to take a decisive step in the direction of a new system of industrial relations. That some administrative officers of the Association did seek to work along lines suggested by the new declaration became evident in the following few years; but efforts in this direction and, more specifically, effectiveness in influencing the business community along "right lines," were hampered by the failure of the Association to unite vigorously behind the "new" concept.

The N.A.M. was not unaware, as exemplified by the attitude of the Industrial Relations Committee, of the responsibilities inherent in its position of power. Throughout the twenties this Committee and its predecessor, the Committee on Industrial Betterment, Health and Safety, had advocated that the Association enlarge its labor policy so as to encourage management, under the open-shop system, to lead in establishing welfare practices and communications procedures. Furthermore, N.A.M. officials throughout the period acknowledged the duties and obligations that the open shop imposed upon those who practiced it. The Open Shop Committee discussed some of them in its 1923 annual report, where it held that employers were "fully alive to the obligations they owe to the community and the workers as managers and guardians of industrial production."[35] At a conference hosted by the Committee that year Walter Drew, who at times served as legal council for the Open Shop Department, stressed that the "real fate of the open shop rests with employers to see that they don't abuse their responsibilities."[36]

---

[34] *Ibid.*, 100.
[35] *Ibid.*, 1923, 158.
[36] Open Shop Conference, "Minutes of Meeting," February 14, 1923, Open Shop Department Files, N.A.M.

The Industrial Relations Committee, believing that employers were not yet fully aware of their role, saw fit to warn manufacturers and associations that if they neglected their obligations "to present the true facts and to shape their policies and actions in accordance with right principles and present-day needs they will some day, not far distant, find it too late for either preventive medicine or surgery."[37]

At its 1923 meeting the Association devoted an entire session to discussion of the open shop. President John Edgerton, in his major address before the session, called upon manufacturers to begin thinking "more and more of their obligations and to insist less upon their rights." The fulfillment of these obligations would be the means of solving the industrial problems of society. "The responsibilities of industrial managers," he said, "imposed upon them a vast duty. . . ." It required them to give their workers a "square deal in industry," and to be "liberal" in dealing with them.[38] The assembly concluded with the adoption of a series of resolutions entitled "Private Employment Relations and the Open Shop." After defining the open shop and summarizing the basis for it, the Association declared that "an obligation or trusteeship to his employees and to the public rests upon the employer."[39]

This obligation was not an abstraction since, in the 1920s, the dominant position of management found the worker in a dependent position. The open-shop movement, successful in achieving public acceptance, contributed to the curtailment of the number and effectiveness of unions.[40] This situation left the employee under employer tutelage and, if not supplemented, would provide the twentieth century with an archaic and inadequate system of industrial relations. Noel Sargent, manager of the Open Shop Department, was aware of the power and responsibility held by employers. The open shop, he believed, could be the means of implementing the necessary industrial-relations and welfare-capitalism programs by which business could meet its obligations.[41]

Just as the N.A.M. response to this need was insufficient, so was the response of management as a whole. Although employee-representation plans were established to take care of an increase of approximately one

---

[37] N.A.M., *Proceedings*, 1925, 53.
[38] N.A.M., *Proceedings*, 1923, 152, 155; see also Edgerton's comments in *ibid.*, 1924, 92-93.
[39] *Ibid.*, 200.
[40] Leo Wolman, *Ebb and Flow in Trade Unionism* (New York, 1936), 16, 17, 35-37, 172-191; "How Manufacturing Industries Operate," [1928?] in Open Shop Department Files, N.A.M.
[41] Interview with Noel Sargent, December 19, 1960; Australia Industrial Delegation to the United States, *Report* (Canberra, Australia, 1928?), 68.

million workers,"⁴² such plans did not compensate for the million and a half decline in union membership."⁴³ Moreover, a much larger number of workers—the vast majority—had no means by which they could cooperatively discuss items of mutual concern with management. The introduction of personnel programs was perhaps more successful, with 12 percent of the wage earners in manufacturing being employed in establishments where there were departments concerned with such matters.⁴⁴ In addition a wide variety of welfare plans were instituted, but here again for only a small minority of workers. Considering what little progress management made in industrial relations during this period, one may conclude that it failed to provide adequately for the workers' economic, psychological, and political health.

Possibly much of the business community-at-large was oblivious of the need for employers to conduct their affairs with workers in a manner different from that of the previous half century. The generation of the 1920s existed in a milieu which placed primary emphasis on the notions that business must remain unencumbered in making decisions and that profits must be maximized. The N.A.M., however, cannot shield itself with a similar alibi. As has been shown, it *was* aware of the responsibility of management to provide an alternative to trade-union collective bargaining. It spoke in terms of management and workers having a mutual interest, of employer responsibility for the welfare of the employee, and of the benefits to the economy that would be provided by effective communications between them.⁴⁵ From 1923 on, factions within the Association pleaded with it to assume more determined leadership for counteracting the militancy and autocracy of the traditional open-shop concept and for encouraging more progressive programs by management. If the N.A.M. had given its primary support to industrial cooperation programs with the same intensity that it had earlier fought for the open-shop idea, a more meaningful industrial-relations system could have developed. Instead, with the Association's passivity, employer dominance became even more firmly entrenched.

---

⁴² National Industrial Conference Board, *Collective Bargaining Through Employee Representation* (New York, 1933), 16-17.
⁴³ Wolman, *op. cit.*, 16, 17.
⁴⁴ National Industrial Conference Board, *Industrial Relations Program in Small Plants* (New York, 1929), 20; Don D. Lescohier and Elizabeth Brandeis, *History of Labor in the United States, 1896-1932* (New York, 1935), 323-335.
⁴⁵ That there was mutual interest between employer and employee and that the open shop served society's economic interest are implicit and explicit in many N.A.M. statements. e.g., N.A.M., *Proceedings*, 1924, 54-58; *ibid.*, 1925, 77; *ibid.*, 1926, 113, 138.

Anticipation of the disproportionate difficulties brought about by the process of changing established policy in a large heterogeneous body prevented the adoption of progressive ideas. The Association had a long tradition of opposing labor unions, and its alternative proposal of the open shop involved a set of principles that supposedly advanced nondiscriminatory employment practices and freedom to work, to make contracts, and to conduct one's business. Once these goals had been achieved most Association members were satisfied and, therefore, apathetic about pushing on to assume the responsibilities that accompanied more progressive programs. Rather than providing, therefore, the strong and aggressive leadership that would have encouraged the wide initiation of industrial-relations plans, welfare programs, and a more equitable distribution of profits, the N.A.M. accepted the victory of the open shop as paramount and worked only in token fashion for the introduction of a meaningful alternative to unionism. Efforts to distribute information on new approaches meant little in the face of the negative and militant position the Association projected. And unwillingness to adopt the proposed revisions to its traditional Declaration of Labor Principles was symbolic of its failure.

THE HARLAN COUNTY COAL STRIKE OF 1931

By TONY BUBKA

The Plight of the Miner

Harlan County was still a primitive, backwoods region in 1910. Until the advent of the railroad in 1911, travel was either by horseback or buggy, and hard-surfaced roads were unknown.[1] For two decades thereafter the County's coal industry expanded. In 1911, for example, there were only six mines operating. But by the end of 1931, with new mines continually opening, seventy-one had begun operation.[2] How Harlan County became coal-conscious is reflected in these figures: In 1911 there were 169 miners digging in the pits of Harlan County; by 1923, 9,260 miners were employed, and this number kept increasing until it reached a peak in 1930.[3]

The World War I years created limitless need for coal, and consequently huge profits awaited those who undertook to exploit coal developments. Coal output in 1918 surpassed all previous levels of production when 346,540,000 tons were mined. Production figures for the years between 1913 and 1918 are equally impressive. In that brief five-year span, 2,960,938 tons of coal poured out of the pits; this was equivalent to approximately 33 percent of all the coal mined in the United States since 1807.[4] Much of it came from new mines in the southern coal fields which were opened to supply the needs of expanding industry.[5]

---

[1] *Harlan Daily Enterprise* (Kentucky), April 25, 1932, 1.
[2] U. S. Congress. Senate. Committee on Manufacturers, Hearings on S. Res. 178, *Investigation of Conditions in the Coal Fields of Harlan and Bell Counties. Kentucky*, 72nd Cong., 1st sess., (Washington Printing Office, 1932) Chart, 287.
[3] Homer L. Morris, *The Plight of the Bituminous Coal Miner* (Philadelphia, 1934), 21.
[4] David J. McDonald and Edward A. Lynch, *Coal and Unionism* (Maryland, 1939), 136.
[5] *Ibid.*, 181.

Over-expansion of the coal industry was stimulated by factors other than the market created by the First World War. At first the wartime Fuel Administration set a ceiling price of $2.58 a ton for bituminous coal. The maximum limit brought about a temporary stability in the coal industry and also provided a tidy profit for the coal producers. But in 1920 Italy and France were in dire need of coal. The Fuel Administration cancelled its wartime limitation on prices, and the profits on coal shipped to European countries reached their peak, while coal shipped to the Great Lakes brought $10 a ton. The operators rushed into the field to grab up these profits.[6] Kentucky's coal output was temporarily stimulated by the 1926 general strike in England.[7] This general strike lasted a few days, but the coal mine stoppages began in May, varying in intensity in different localities, and the disputes were not finally terminated till November and December.

Still other factors stimulated coal expansion in the South. Wage discrepancies between the union and the non-union mines tended to encourage new operations in the southern coal fields where there was no unionization. Coal operator Howard N. Eavenson, testifying before a Senate Committee, asserted that periodic strikes in the northern fields, which grew out of the miners' struggle for greater bargaining power, was one of the chief reasons for the development of Harlan's coal fields.[8] An investigator for the La Follette Civil Liberties Committee made this point more emphatic by declaring that the sole reason for opening Harlan's coal pits was to supply coal that would be mined by non-union labor and thus, by selling coal cheaper, to compete favorably with the union production price scale of the northern fields. In short, the tapping of Harlan coal veins by non-unionized labor was to serve as a cudgel; it would force union labor's submission to the operators' terms.[9]

The Harlan County operators' attitude toward labor is explained by Howard N. Eavenson. From the beginning of the country's coal history, he asserted, operations were non-union, and during the World War I years Harlan's owners negotiated contracts with the United Mine Workers of America (U.M.W.) only because of the coercive tactics of the Fuel Administration. These contracts were terminated after the war

---

[6] Malcolm Ross, *Machine Age in the Hills* (New York, 1933), 51.
[7] *Ibid.*, 54
[8] U. S. Congress. Senate Committee on Mines and Mining. Hearings on S. 2935, *To Create a Bituminous Coal Commission and for Other Purposes,* 72d Cong., 1st sess. (Washington: Government Printing Office, 1932), 497.
[9] U. S. National Archives. Hearings on S. Res. 266, *Violations of Free Speech and Rights of Labor,* 74th Cong., 1st. sess., Harlan County, Miscellaneous notes folder #4.

ended.[10] But before they were, because of Fuel Administration pressure, 70 percent of the soft coal output was mined by 1919 according to union-wage contracts. During the post-war years, however, union organization dwindled, so much so that by 1930 the soft coal production by non-union mines amounted to 80 percent of the total.[11]

The crippling effects of a hastily expanded industry became noticeable a decade after World War I. An overgrown industry had stagnated owing to many factors, notable among them several modern mining innovations which increased the rate of production; the digging of countless new pits which tended to over-supply the market with coal; improvements in technology by which more energy was derived from a ton of coal; and the relentless competition coming from other energy-producing markets.

The depression was still another disturbing factor which affected the coal industry. Its consequences for the coal fields are illustrated by the national production figures for this period. The total coal production in the bituminous mines for 1930 was 467,000,000 tons, but in 1931 tonnage declined to 360,943,000, a drop to the low production record set in 1909.[12]

Because of these various figures, contributing to a dysfunctional coal industry, unemployment increased in the fields. The soft coal operators employed 704,793 wage earners in 1924; this number was reduced to 294,000 nine years later.[13] The Depression and declining production notwithstanding, the soft coal operators did not unite among themselves to foster fair prices, to limit expansion, or to set up a framework of codes which would govern regulation of the industry. If the entire coal industry was disorganized, these operators seemed particularly so.

As new mines had opened in Harlan County during the 1915-1918 coal boom, a corresponding increase in population occurred. To recruit labor for these mines, a group of labor agents were sent out to scout the surrounding counties. They offered seemingly incredible inducements; in particular a high-wage appeal that guaranteed more earnings for one month of work than a mountaineer could possibly earn in a year's labor in the hills. They also offered to assume the transportation cost to the mines, and held out the lure of decent living quarters: homes

---

[10] Hearings on S. Res. 179, 203.
[11] Morris, *op. cit.*, 13-14.
[12] U. S. Congress. Senate. Committee on Manufactures, Hearings on S. 174 and S. 162, *Unemployment Relief*, 72nd Cong., 1st sess. (Washington: Government Printing Office, 1932), 225.
[13] Norman M. Thomas, *Human Exploitation in the United States* (New York, 1934), 97.

with electric lights and running water.[14] High wages, however, were the greatest appeal. Before the advent of the railroad in the mountain counties (and the Louisville and Nashville began to extend its operations in 1911), there was scarcely a farmer who earned $100 a year.[15] The mountaineer entered a machine-dominated world when the railroad came and when World War I sent coal prices soaring. He tasted a new way of life, and when the boom passed there was nothing to go home to.

The data from Morris' field investigations in 1932, based on interviews with unemployed miners, reveal that their lack of vocational training and their educational shortcomings handicapped them in seeking employment in other industries. Of the miners questioned, 78.2 percent never completed any education beyond the primary-school level.[16] The Committee for Kentucky report notes that, in the national scale of illiteracy, Kentucky was ranked 47th.[17]

Kentucky's coal fields are situated in rural areas, so that the miners were detached from industrial centers. Moreover, the only major source of employment in eastern Kentucky, other than coal mining, was farming and small-scale lumbering.[18] Therefore, when the Harlan disorders occurred, geography, depression, and limited occupational opportunities made it impossible for unemployed miners to transfer to other industries.[19]

Nor did Kentucky's limitations on public relief help the situation. Indeed, when unemployment became widespread in the Harlan region after 1929, Harlan's unemployed miners suffered more than did their northern counterparts. Many communities simply lacked tax resources to support a public welfare program.[20] They did their best in assisting the destitute miners. And the majority of coal operators also helped, at least to the extent of providing the unemployed miner a dollar-a-day credit for food.[21]

The significance of inadequate relief, and its relationship to the Harlan strike, cannot be overemphasized. Being impoverished, the miners gradually became willing to follow the leadership of any or-

---

[14] Morris, *op. cit.*, 60-61.
[15] Hearings on S. Res. 178, 152.
[16] Morris, *op. cit.*, 76-77.
[17] John Gunther, *Inside U.S.A.* (New York, 1947), 640-641.
[18] Hearings on S. Res. 178, 187.
[19] Morris, *op. cit.*, 22.
[20] Grace Abbott, "Improvement in Rural Public Relief: The Lesson of the Coal-Mining Communities," *Social Service Review*, VI (June, 1932), 205.
[21] Hearings on S. Res. 178, 183.

ganization that would help them in their plight. The promise of food and relief was a principal factor in encouraging membership in the left-wing unions that entered the Harlan region to capitalize on this unrest.

The "Evarts Battle"

With the Depression in full swing, Harlan County operators decided to minimize their production costs in order to remain in business. In February 1931, the Black Mountain Coal Corporation issued notice of a 10 percent wage reduction to all of its employees. Rather than accept these terms, the miners chose to walk out of the pits.[22] Their action was the first in a series of events that helped to precipitate the "Battle of Evarts." The walkout was not, it must be noted, prompted by outside agitation. Indeed, the U.M.W. did not endorse, nor act in any way to stimulate, the strike.[23]

Some of the older miners who had previously belonged to the United Mine Workers began to have visions of the good old U.M.W.A. days when wages were high. The unorganized miners, happy to better their conditions, talked of unionization and held meetings toward this end in various sections of the country. The miners had resorted to this 1931 strike to dramatize their plight; and now they staged marches, made speeches, and voiced their grievances in militant, unyielding terms.

The restlessness of Harlan County miners was conveyed to the U.M.W., whose first official action occurred on February 15, when William Turnblazer, U.M.W.A. President of District #19, circulated thousands of leaflets urging the miners to join the union. This literature outlined the miners' grievances, denounced the industrial overseers, and pointed out that the solution to the miners' problems lay in union organization.[24]

Turnblazer scheduled a general mass union meeting at the Gaines Theatre in Pineville, Kentucky, on Sunday, March 1, 1931. More than 2,000 miners attended, filling the theatre to capacity; and many more, unable to get in, gathered outside. Philip Murray, vice-president of the U.M.W.A., addressed the crowd assembled there, and appealed for unionization as the key to better conditions. When the meeting was concluded, union agents began to recruit miners into the union.[25]

---

[22] Louis Stark, "Harlan War Traced to Pay-Cut Revolt," *The New York Times*, September 29, 1931.
[23] "Fine Bedfellows," *United Mine Workers Journal*, XLII (June 1, 1931), 7.
[24] "Appeal Made to Kentucky and Tennessee Miners to Join United Mine Workers of America At Once," *United Mine Workers Journal*, XLII (February 15, 1931), 8-9.
[25] "Union is Reborn in Two Kentucky Counties After Big Meeting at Pineville; Philip Murray Talks," *Ibid.*, XLII (March 15, 1931), 12.

U.M.W.A. policy at this time was to organize, and not to sanction or to engage in, strikes. Its unwillingness to do either was bitterly opposed by radical groups who claimed that the U.M.W.A. had betrayed the miners. But the union seemed to have good cause, for its membership strength suggests U.M.W.A. impotence. For instance, of the nation's total coal output in 1930, only 20 percent was mined under union contracts.

Following this Pineville meeting, the miners flocked to the union. The U.M.W.A. drive developed into a threatening movement, and operators retaliated by dismissing union organizers and those sympathetic to union demands. The Black Mountain Coal Corporation, for example, discharged a number of U.M.W.A. agitators and other mines in the county followed suit. Before long, about 500 of the most militant miners were without work.[36]

Miner-operator relations continued to be unsatisfactory, with ill-feeling aggravated by the operators' policy of forcibly evicting the discharged miners from company-owned homes. Justifying such evictions, one operator explained: "It is a nasty business and I hate to do it —but I can see no other way of dealing with the problem. I must have these houses for other employees who will work under my conditions, accept my wage scales and pay their rent."[37]

The miners evicted by the Black Mountain Company began to gather at the independent town of Evarts. Picket lines were formed there, and the road leading to Black Mountain Camp was patrolled. On April 1 a miner named Carpenter, who continued to work at the Black Mountain Camp in spite of several warnings, came to Evarts. Several unemployed miners seized him, took him out to a town lot, forced him into a kneeling position, and then lashed him unmercifully.[38]

Violence increased as the strike progressed. A dynamite explosion wrecked the mine shaft structure of the Bergen Coal Company (located at Shields) on April 20.[39] The month ended with a series of store robberies. Seven shops were burglarized and some of them were completely looted of their stock.[40]

The strike situation became further aggravated when the operators decided to bring in strike-breakers. On April 27 the Black Mountain

---

[36] Hearings on S. 2935, 142.
[37] Morris, op. cit., 124.
[38] *The New York Times*, November 27, 1931, 6.
[39] *Harlan Daily Enterprise* (Kentucky), April 20, 1931.
[40] Ibid.

Company issued a lockout order designed to evict the striking miners, so that the company-owned houses would be made available for those willing to work under its terms. Many of the families were ordered to vacate.[31]

The influx of strike-breakers inflamed the situation. And, on April 28, gun fire broke out and shots were fired upon Black Mountain Company deputies from the hills, when these deputies attempted to bring miners in to work the pits.[32] Further trouble occurred the next day when sixteen unoccupied company dwellings at the Ellis Knob Coal town were consumed by fire.[33] Taking notice of the increasing acts of violence, Judge D. C. Jones of Harlan declared, on May 4, that "a state of lawlessness" now prevailed in the county, and requested a special grand jury to assess the state of unrest.[34]

On the night of May 4, while preparations for a grand jury inquiry were taking shape, between 300 and 400 striking miners (non-union miners were excluded) gathered at the Evarts Theatre to form a plan of action.[35] Incendiary speeches characterized the meeting. William Hightower, Evarts local U.M.W.A. president, exhorted the group: "Come to Evarts with your rifles and shotguns and bring a good head. We'll win out if we have to wade in blood up to our necks. We won't have those Black Mountain thugs to contend with much longer."[36]

On the morning of May 5, a procession of miners—100 to 200 men, townsfolk judged—poured into Evarts. This group carried assorted firearms, including shotguns, rifles, and pistols. They scattered along the highway and concealed themselves behind any object which provided protection. Numerous and contradictory stories relate what happened then. It was reported that Jim Daniels, a deputy sheriff employed at the Black Mountain Camp, relayed a message to the miners "that he was coming down to clean up the whole damn town."[37] We do know that three cars conveying a force of ten deputies arrived in Evarts, and that gunfire broke out all around the cars when they crossed the railroad junction.[38]

---

[31] American Civil Liberties Union, "The Kentucky Miners Struggle," 6.
[32] *Harlan Daily Enterprise*, April 28, 1931.
[33] *Ibid.*, May 1, 1931.
[34] *Ibid.*, May 4, 1931.
[35] *Kentucky Reports—Reports of Civil and Criminal Cases Decided by Courts of Appeals of Kentucky: Jones v. Commonwealth*, vol. 249 (Frankfort: April 12, 1933), 505-506.
[36] John F. Day, *Bloody Ground* (New York, 1941), 300.
[37] American Civil Liberties Union, *op. cit.*, 6.
[38] Jones *v.* Commonwealth, Hightower *v.* Commonwealth, 249, Ky., 506.

George Dawn, another of the deputies in the squad, declared that when the miners began shooting, Daniels replied with machine gun fire, and that when he, Daniels, reached the road bank a shotgun discharge tore his head from his body. Thousands of shots were exchanged, and when the fury of the skirmish subsided four persons lay dead."[39] The brief encounter brought other repercussions to strike-torn Evarts. Schools suspended classes, and all mining operations were halted.

The "Reign of Terror"

The "Battle of Evarts" was the climax of those disorders growing out of the miners' desire to organize. It helped catalyze an emotional response of such proportions that county officials were unable to restrain the insurrectionary mood; and troops were ordered to the strike area. Coal operator R. W. Creech of the Wallins Creek Company plainly summed up the employer attitude: "They'll bring a union in here over my dead body. I would rather close this mine forever than work with a union."[40]

Such intransigence meant that operators and union miners were on a collision course. Only chaotic and ineffectual unionization drives deterred a head-on crash. The initial drive was carried by the U.M.W.A. But the union's strength had waned to the point where it was unable to make any progress. Consequently, rival, more radical, union groups entered Kentucky.

About this time, the Industrial Workers of the World dispatched representatives, who began to set up locals in Harlan County. Believing that, "the working class and the employing class have nothing in common," the Wobblies had the destruction of capitalism as their ultimate goal.[41] Their activities, as could be predicted, were suppressed by county authorities. Moreover, the Wobblies were handicapped by lack of funds, and could not compete with Communist organizers.[42]

Shortly after the "Battle of Evarts," the National Miners Union (N.M.U.), an organization directed by the Communists, sent agents into Harlan County to discredit the U.M.W.A. and to capitalize on its failure to organize the miners. The N.M.U.'s organizing drive was supported by other radical groups, especially by the International Labor

---

[39] *Harlan Daily Enterprise* (Kentucky), May 5, 1931.
[40] J. C. Byars, Jr., "Harlan County: Act of God?" *The Nation,* CXXIV (June 15, 1932), 673.
[41] *Industrial Worker* (Chicago), January 19, 1932.
[42] Selig Perlman and Philip Taft, *History of Labor in the United States* (New York, 1935), IV, 610-611.

Defense (I.L.D.), a left-wing agency set up in Chicago in 1925 in order to fight sedition and criminal syndicalism statutes. The guiding spirit of the I.L.D. was J. Louis Englahl, a Communist Party member and former *Daily Worker* editor.[43] The Federated Press, a Party wire service, also helped by sending special correspondents into the strike zone. Established in 1918, its purpose was to supply news that supported the Communist viewpoint.[44]

The N.M.U. was the largest affiliate of the Trade Union Unity League (T.U.U.L.), the central agency of all Communist unions in the United States. The League had as its objective (according to the central council of the Red International of Labor Unions, at a Moscow convention held on February 15, 1930) the mobilization of "the masses of workers in order to smash the offensive of the capitalists."[45] Its salient aim was trade-union control, which was essential because the expected revolution was considered possible only with working-class support.[46] T.U.U.L. subsidiaries, according to the House Report on Communist Propaganda, assumed organizational titles similar to those of legitimate unions in order to conceal their identity.[47]

In July 1932, the N.M.U. held a national convention in Pittsburgh. Its purpose was to adopt strike action tactics to be carried on throughout the national coal fields. Twenty-eight miners who represented the various coal camps in Harlan County attended the conference.[48] The delegates adopted the following resolution: ". . . Our answer . . . is the organization of the . . . miners, together with all workers toward the breaking of the rule of the capitalist class, the confiscation of the coal mines, and all factories, . . . which can only be achieved through the establishment of the rule of the workers, along the path of the workers in the Soviet Union. . . ."[49]

Supporting the N.M.U., the Communist Party, on July 29, affirmed: "In those sections [of Kentucky] where the Party is now being built for the first time the Politbureau must give direct attention. In the mine

---

[43] U. S. Congress. House. Special Committee to Investigate Communist Activities in the United States. House Report 2290 on H. Res. 220, *Investigation of Communist Propaganda*, 71st Cong., 3d sess. (Washington: Government Printing Office, 1931), 57.
[44] *Ibid.*, 58.
[45] *Ibid.*, 24.
[46] According to the House Report, the T.U.U.L. was a branch of the Red International Labor Unions. *Ibid.*, 82.
[47] *Ibid.*, 16, 23.
[48] Sterling D. Spero and Jacob Aronoff, "War in the Kentucky Mountains," *The American Mercury* XXV (Feb., 1932), 231.
[49] *Daily Worker*, (New York), July 25, 1931.

fields of Kentucky ... we must aim to develop the strike movement as rapidly as possible...."⁵⁰

N.M.U. agitation continued throughout 1931, and the union movement encountered bitter resistance from the operators, who resorted to extra-legal devices to subdue it. The months following the "Evarts Battle" were characterized by violence so intense that writers who reported upon events in the strike area declared "a reign of terror" existed in Harlan County. A small-scale war was waged against unionism in which shootings were an almost daily occurrence. The tactics used against the miners and their sympathizers included evictions, black-listing, raids on miners' homes, and arrests of miners on charges of criminal syndicalism. Intimidation and wholesale arrests were the mildest aspect; dynamiting, beatings, the shooting of reporters, the deportation of undesirable visitors, and the closing down of miners' relief kitchens were the more tangible ways in which the union drive was thwarted.

On November 6, a group of writers led by Theodore Dreiser entered Harlan County to conduct a self-appointed inquiry into the situation. The basic purpose of the Dreiser Commission was to investigate, and to learn whether civil rights and liberties were being denied in the strike area. But its importance lay in the Commission's effect on events in Kentucky. Major Chescheir, a National Guard officer who was sent to the strike zone, testified that Commission activities gave the N.M.U. a new lease on life, for these tended to stimulate and encourage the N.M.U. organization drive.⁵¹ Some credence may be given to his testimony, for we learn that George Maurer, an I.L.D. representative, spoke at a public meeting sponsored by the Commission, where he declared:

> ... We are for the working class and against all enemies of the working class, the coal operators, the county officials, the State officials that are the tools of the bosses....
> We will try to have, I hope, something in America like they have in the Soviet Union....⁵²

The Evarts Battle Conspiracy Trials

On May 6 the special grand jury called by Judge Jones began its inquiry. The grand jury was kept busy turning out indictments and 335 criminal cases were listed for trial when the fall court session convened, a number setting a new record in Harlan County judicial history.⁵³

---
⁵⁰ *Ibid.*, August 3, 1931.
⁵¹ Hearings on S. Res. 178, 151.
⁵² *Ibid.*, 142-144.
⁵³ *Harlan Daily Enterprise*, August 10, 1931.

Charges of Communism were rife throughout the Evarts conspiracy trial proceedings. But there is no evidence that the Communists were responsible for the disorders which culminated in the battle of Evarts. According to Daniel M. Carrell, a Kentucky National Guard Colonel who entered Harlan County with troops immediately after the battle, no trace of the radical element (which included the I.W.W. and the Communists) was noticeable. Radical groups did not begin their operations until it became evident that the U.M.W.A. efforts to unionize the mines were futile.[54] Some of the miners on trial were U.M.W.A. officials. Yet the Congressional Report on *Communist Propaganda* concluded that the A.F.L., including the U.M.W.A., had ". . . patriotically borne the greater brunt of the communist attack in this country, and still constitutes our one great bulwark of defense against potential dangers of communism."[55]

Harlan County officials, however, accused radicals and Communists of being responsible for the lawlessness. What is important is whether such accusations influenced the outcome of the Evarts murder trials, and whether in fact they were so designed. That is, were those on trial identified in the minds of the jury and the public with "communism"? The miners on trial were the key figures in the union drive; to associate them with radicalism would tend to discredit any attempt at labor organization. The Communist issue and its relevance to the trials can be demonstrated by a study of the August 1931 *Harlan Daily Enterprise.* Throughout the conspiracy proceedings, this organ of the coal operators barraged its readers with anti-Communist articles. Significantly, all of these articles were long, which is unusual for a country newspaper. The Harlan trials, on a prosecution motion, were transferred to other counties on August 24. The Mt. Sterling Court took jurisdiction of the Evarts cases. Other cases were to be tried at Winchester, Kentucky. After this date the *Harlan Daily Enterprise* abruptly discontinued its publication of anti-Red articles, or at least did so for the time being.

The atmosphere at the Mt. Sterling trials is best revealed in a *Mt. Sterling Gazette* editorial shortly before the court opened: "It is useless to send men and women of the stripe of the Harlan agitators to the penitentiary. They would be safer in a pine box six feet under ground."[56]

On December 10 the first of these trials ended; W. B. Jones was convicted of murder, and the punishment fixed by the jury was life im-

---
[54] Hearings on S. Res. 178, 121.
[55] House Report 2290 on H. Res. 220, 82.
[56] American Civil Liberties Union, *op. cit.,* 14.

prisonment.⁵⁷ Court sentiments during his trial are reflected in prosecuting attorney W. C. Hamilton's closing address to the jury: "In Russia they will read the fate of this man . . . and if you turn him loose there will be celebrations in thousands of places and in Moscow the red flag will be raised higher."⁵⁸ *The Knoxville News-Sentinel,* which covered the trials, pointed out that a good proportion of Hamilton's address was devoted to denouncing radicals, and that no evidence was introduced to indicate that Jones was affiliated with any radical organization.⁵⁹ William Hightower, the next miner to be tried, also was found guilty of the charge of conspiracy to murder and, like Jones, was sentenced to life imprisonment.⁶⁰

Lacking funds to pay the expenses of its witnesses, the defense introduced a motion to have the remaining Evarts cases returned to Harlan County. On January 14, 1932, Judge Prewitt, in spite of protests from the prosecution, ordered the remaining twenty-eight cases to be sent back to the Harlan circuit court.⁶¹

Fifteen trials were ultimately held; seven ended in convictions and life sentences and three in acquittals. There were five hung juries. Herbert Mahler, secretary-treasurer of the Kentucky Miners Defense and defense strategist of the I.W.W., worked unceasingly on the cases from the beginning, and his efforts resulted in the release, in 1935, of three of the seven who were imprisoned. Mahler continued to fight until the last man was pardoned in 1941.⁶²

The Jones-Hightower cases were reviewed by Kentucky's highest judicial body, the Court of Appeals. In its opinion the court declared that Jones and Hightower did unlawfully and intentionally conspire to murder Jim Daniels, and that the miners gathered at Evarts to prevent the law officers from fulfilling their duty. Many defense witnesses testified that the deputies fired the first shot; but the prosecution's evidence indicated that the miners opened fire. For the court, however, the question about which side fired first had no real bearing on the issue since, "The man who seeks his adversary for the purpose of shooting him cannot be acquitted on the grounds of self-defense, if, when

---

⁵⁷ *The New York Times,* December 11, 1931.
⁵⁸ American Civil Liberties Union, *op. cit.,* 15.
⁵⁹ American Civil Liberties Union, *op. cit.,* 15.
⁶⁰ *The New York Times,* January 15, 1932.
⁶¹ *Mt. Sterling Advocate* (Kentucky), January 14, 1932.
⁶² Loyal Compton, "He Won a Fight for the Right. Now What to Do?" Louisville *Courier-Journal,* February 12, 1941, magazine section.

he finds his adversary, the adversary is quicker than he and makes the first shot."[63]

The obstacles which the miners faced in the Evarts trials are indicated by Daniel B. Smith, who had served as defense counsel for six of the defendants. In a letter to Governor Ruby Laffoon, dated February 5, 1934, Smith, who was now state attorney for Harlan County, reminisced:

> In the cases that I practiced, and those that were tried in Harlan County, I can safely say that the defendants never had a fair chance in any of the trials.
> Handicapped by some court officials who were in my opinion over-zealous, and by lack of financial help, under the disapproval of powerful interests, these men were at a great disadvantage.[64]

The National Miners Union Strike

The N.M.U. had a short life, one marked by militant attempts to gain recognition as a bona fide labor union. It had been conceived on April 1, 1928, at a Pittsburgh National Mine Conference sponsored by the Communists and designed to displace the U.M.W.A.[65]

Benjamin Gitlow, an important Communist official who renounced the Party, claimed that the World Congress of the Profintern ordered a change of strategy. Communists were instructed to cease "cooperating" ("boring from within") with legitimate trade unions and to start a drive for dual unionism. Destruction of the U.M.W.A., it was hoped, would prompt the miners to enlist in the Communist trade union. To finance this new movement, $29,329.46 was given to Gitlow and connections established with Moscow for further financial aid.[66]

With these funds, an organizing drive was mounted. For Communists, the miners seemed a perfect choice, a typical example of the exploited classes. Apparently the N.M.U., with its obviously revolutionary nature and demands, appealed to some of the Kentucky miners. For by December the zealous union organizers were convinced that they had sufficient adherents to strike every mine in that territory. They called a district convention which met at Pineville on December 13, 1931, and some 250 delegates in attendance voted to strike the southeastern Kentucky and Tennessee coal mines on January 1.[67]

---

[63] Jones v. Commonwealth, Hightower v. Commonwealth, 249 Ky., 505-511.
[64] Hearings on S. Res. 266, 4309-4310.
[65] Hearings on S. Res. 178, 28.
[66] Benjamin Gitlow, *I Confess* (New York, 1940), 387-388.
[67] *The New York Times*, December 14, 1931

This N.M.U.-sponsored "strike against starvation wages" initially met with a disappointing response, and only a few mines in Kentucky were closed down. Thirty-five Harlan County pits continued working, while seven were idle. In Bell County the response was even less encouraging for the union; only two mines were closed with approximately 120 miners going out on strike. Lack of coal orders caused a few Knox County mines to be idle, and their owners expressed their intent to continue operations under non-union wage contracts when the pits were re-opened.[68]

Arrests of the principal N.M.U. leaders became an important method of stemming the strike, and a number were jailed on the charge of violating Kentucky's criminal syndicalist law. On January 6, county officials also arrested on the same charge Allan Taub, an International Labor Defense lawyer sent down to Pineville as defense counsel.

The N.M.U. continued to agitate for broadening the strike movement, and scheduled a rally for January 16 at Harlan Town. This demonstration never materialized because precautions were taken to prevent it by sheriff's deputies who kept a sharp lookout for union organizers seeking to work in the country. A more positive effort to check the new strike offensive occurred the following day when Joe Weber and Bill Duncan, active N.M.U. members, were apprehended in Tennessee, placed in custody of Harlan deputies, deported from the county, and then beaten.[69]

A New York relief group—the "Independent Miners' Committee"—composed predominantly of writers sought admission to Pineville. One of their objectives was "to defend the rights of the Workers International Relief to distribute relief"; another was to circulate subversive literature and to make a lecture circuit of the coal towns. A few hours after their arrival in Pineville, the entire committee was placed under arrest on grounds of being public nuisances. The misdemeanor warrants against them were cancelled and, instructed that their presence was not desirable, they were convoyed out of the state. Their expulsion was carried out in the best Harlan manner, including the usual beatings—given to Allan Taub and Waldo Frank, chairman of the committee, both of whom were dragged from their car at the state line.[70]

The ejection of the writers' committee had a deeper significance than

---

[68] *Ibid.*, January 2, 1932, See also, Hearings on S. Res. 178, 37-42.
[69] American Civil Liberties Union, *op. cit.*, 19.
[70] *The New York Times*, February 11, 1932.

according brutal treatment to two "undesirable" visitors. Their automobile escort was composed of many prominent Pineville citizens. It symbolized the mass participation of the respectable element in the community who, believing their labor problems local in nature, had combined with county authorities against outside interference.

The strike agitation extended to neighboring counties. About the time of the ejection of the writers' committee youthful Harry Simms, a key N.M.U. organizer, entered Knox County to speak at a miners' demonstration there. But he was intercepted and shot to death by a deputy sheriff.[71] Charging that it was a case of deliberate murder, New York City's Communists attempted to make a martyr of Simms, a kind of John Reed of the great American class struggle.[72] The *Daily Worker* printed a thumb-nail sketch of the nineteen-year-old hero: "He was sent to the South by the Young Communist League as District Organizer . . . arrested there repeatedly . . . released by the Young Communist League at the request of National Miners Union to organize youth sections in Kentucky and Tennessee."[73] Simms' body was shipped to New York City, in time for a mass demonstration there amid calls of revenge for the fallen comrade. His coffin became the center piece for a mute, rain-drenched procession, dotted with red flags and the glittering scarlet neckerchiefs of uniformed teen-age Communists, that marched to the Lower East Side.[74] When Simms' body was transferred to the Bronx Coliseum, an estimated 10,000 people gathered there for what the *Daily Worker* termed "one of the most impressive working class demonstrations ever held in New York City." Vindictive oratory and fists clenched in salute marked the occasion. Simms' father declared: "He died on the battle field of the class struggle." William Z. Foster, the principal speaker, proclaimed: "One thing we must do is to make the capitalist class pay dearly for the murder of Simms . . . it will not be long before they will face workers' courts in America." Party functionary Israel Amter concluded: "His body goes to the soil, but his spirit goes to the masses."[75]

In March 1932, the N.M.U., aware of its shortcomings and mistakes, formally declared the strike a failure and retreated from the Kentucky coal fields. Party leader Jack Stachel, in reviewing the debacle, offered

---

[71] American Civil Liberties Union, *op. cit.*, 20.
[72] Ross, *op. cit.*, 180-181.
[73] *Daily Worker* (New York), February 13, 1932.
[74] *The New York Times*, February 18, 1932.
[75] *Daily Worker* (New York), February 19, 1932.

a long list of reasons why the venture failed. What did the strike accomplish, he asked? Communists, according to him, "succeeded in recruiting dozens of miners into the Party and for the first time establish [ed] our organization among the miners in Kentucky."[76]

In retrospect, one of the greatest blunders was not the N.M.U.'s failure to educate the miners in the importance of the historic class struggle, but rather the Communists' failure to understand the background, nature, and character of the Kentucky miners. Essentially mountaineers, they were people who preferred to be left alone. Their character had been moulded by endless years of seclusion, continual conflict with the "law," generations of family feuding and moonshining, all of which combined to produce the psychology of the Kentucky miner. From a religious and behaviourial viewpoint, the mountaineer miners formed a homogenous class. The N.M.U. leaders ignored their deep-rooted religious inheritance. They had selected several promising local miners and sent them north to be briefed and instructed in revolutionary practices. But these novitiates learned that the N.M.U. was Communist-inspired; and, given their view of communism, they believed they now belonged to a union embracing atheism and treason to their country. The would-be converts, upon return, became the strongest anti-N.M.U. propagandists in the strike area. They were encouraged by county authorities to reveal what they had learned about the N.M.U. while in the north, and their penitent revelations were printed in newspapers and in broadsheets entitled "Miners Expose Reds," which were widely distributed.

The anti-religious sentiments unwisely expressed by some N.M.U. representatives confirmed their opinion and helped undermine the union. Trial testimony of Doris Parks, W.I.R. secretary, serves as an illustration. She was arrested in February and jailed on a criminal syndicalism charge. During a court trial she replied to attorney Smith's queries as follows:

> "Do you believe in any form of religion?"
> "I believe in the religion of workers...."
> "Does your party propose to take over private property for their own use?"
> "They intend to take over everything for their own use...."
> "Do you believe the Bible and that Christ was crucified?"
> "I affirmed, didn't I, that I believe only in the working class and their

---

[76] Jack Stachel, "Lesson of Two Recent Strikes," *The Communist*, XI (June, 1932), 529-535.

right to organize and to teach them they can be led out of this oppression by the Communist Party."[77]

Nor did prison silence her. "When that day comes," she boasted, "we will not be satisfied with the crumbs—with beans and potatoes . . . . our victory will be brighter than the sun in the sky."[78]

The N.M.U. withdrew from the Kentucky fields, and for the remainder of its history the union's activities centered in the tri-state area of Pennsylvania, Ohio, and West Virginia. But it made no advances in these northern coal fields, and along with the entire family of T.U.U.L. Party-line locals, the N.M.U. was dissolved in 1935. The dissolution was ordered by a New York T.U.U.L. convention directive: "In line with this policy of a unification of its unions with the A.F. of L., the Trade Union Unity League . . . had no further need of continuing in its present organizational form . . ."[79] The T.U.U.L. now reverted to its original position of "boring from within."

The Harlan County coal strike of 1931 had its foundation in a combination of many factors: a wildcat walkout, a lockout, unemployment, undesirable working conditions, and a disorganized coal industry. The union struggle developed into a local war when the Communists invaded the area to exploit the miserable conditions that prevailed. The miners, in the long run, were adversely affected, since their struggle delayed the coming of unionism. Hence the coal operators had successfully defeated all attempts to unionize the mines, and industrial peace prevailed in Harlan County until 1933. With the passage of the National Industrial Recovery Act in that year, the U.M.W.A. once again launched an organizing drive; but not until about 1939 did it achieve recognition in Harlan County.

---

[77] Morris, *op. cit.*, 134-137.
[78] Ross, *op. cit.*, 183.
[79] William Z. Foster, *From Bryan to Stalin* (New York, 1937), 274.

Forty Years Ago: The Great Depression Comes To Arkansas

By GAIL S. MURRAY

THE STOCK MARKET CRASH OF OCTOBER 29, 1929, MARKED for most of America the beginning of long years of unemployment, deprivation, and hopelessness. On that day a record-breaking 16,410,131 shares were traded on the New York Stock Market, and billions of dollars in open-market values were wiped out in a single day.[1] By November 13, the market had reached the lowest point in its history and industrial stocks had fallen 228 points from the September figures.[2]

The impact of this stock market failure, however, was slow to reach the more agricultural sections of the country, including Arkansas. Grain and cotton prices fell on the Chicago market, of course, but low agricultural prices were nothing new to farmers. Since 1920 the southern farmer particularly had known acute hard times. There had been a steady increase in the number of tenant farmers in the South and an accompanying decline in the number of farm owners.[3] In Arkansas, the average wage per day of a farm laborer as of July 1929 was $1.65, about one-half the average wage of non-southern states.[4]

---

[1]David A. Shannon, *The Great Depression* (Englewood Cliffs, N. J., 1960), 4.
[2]Edward H. Merrill, *Responses to Economic Collapse: The Great Depression of the 1930's* (Boston, 1964), 7.
[3]John Samuel Ezell, *The South Since 1865* (New York, 1963), 428.
[4]Clarence Heer, *Income and Wages in the South* (Chapel Hill, 1930), 15.

In addition to the low-wage problem in the 1920's, farm credit was discouragingly tight. Due to poor management and lax banking laws, Arkansas experienced a disproportionate number of state bank failures during the 'twenties.[5]

As if these factors were not sufficient to squeeze the Arkansas farmer in a financial vise, the winter of 1926-27 was one of unprecedented cold weather in the state. This in turn was followed by some of the worst spring flooding Arkansas had ever experienced.[6] Many farmers survived their flood losses only through Red Cross disaster relief. Approximately $53,000 had to be appropriated from the state's general revenue fund to provide aid for the flood victims, and in addition, the federal government loaned $1,804,000 to the Arkansas state highway fund for reconstruction of flood damaged roads and bridges.[7]

Responding to the plight of many citizens, Governor Harvey Parnell in his inaugural address of 1929 urged a revised tax structure to eliminate "inequalities" and mentioned the possibility of getting federal loans to improve agricultural conditions. He also announced plans to concentrate less on state highways and more on development of county roads that would benefit farmers taking goods to market.[8] But nature had not yet spent its fury: several tornadoes ripped through Arkansas in the spring of 1929, killing dozens of persons and destroying hundreds of thousands of dollars worth of property.[9]

This was more than the state's finances could endure. In an attempt to acquire needed funds to meet this most recent disaster, the state sold bonds throughout the country. Represented by the firms of Halsey, Stuart, and Company, Inc., of New York City, the state claimed that its assets

---

[5]Emory Q. Hawk, *Economic History of the South* (New York, 1934), 532-33.
[6]"Stricken Arkansas," *Outlook*, LVII (Jan. 21, 1931), 85.
[7]*Arkansas Acts* (1929), 39-40, 80-81, 881.
[8]Harvey Parnell, *Inaugural Address* (Little Rock, 1929), 5, 7, 11-12.
[9]*Arkansas Gazette, State Centennial Edition, 1836-1936* (Little Rock, 1936), 173.

totaled $1,219,441,326 in real estate. Actually, however, the official assessed value in mid-1929 was only about one-half this amount. They had quoted the 1928 value! When this fact later came to light, bondholders were understandably outraged.[10]

As a result of this accumulation of difficulties during the 'twenties, Arkansas found herself ranked forty-sixth in the country in per capita income in 1929. At the same time she ranked *first* in per capita *indebtedness*.[11]

It is no wonder, then, that the New York stock market crash of October 29, 1929, was not unduly noticed in Arkansas. The sinking market prices did not even make front-page news in the *Arkansas Gazette* until Friday, October 25, when the news of "Black Thursday" was revealed: 12,894,650 shares had been traded, making that "the most disastrous day in Wall Street history."[12] But like most Americans, Arkansans probably tended to believe this was only a temporary crisis, and they accepted President Herbert Hoover's statements that business was on a sound basis. The worst day of the crash, so-called "Tragic Tuesday," (October 29) was accorded only minor space in the *Gazette*. The following day the lead item was *not* the almost eleven million shares that had been traded on the New York market, but an article on "Halloween Goblins and Spooks!"

By November 11 the news services were announcing that the "crisis in the stock market is passed," and recovery was declared imminent. One securities company in Little Rock advertised: "We believe that the investor who purchases securities at this time with the discrimination that is always a condition of prudent investing may do so with the utmost confidence."[13] On November 17, however, sixty-six

---

[10]John T. Flynn, "Other People's Money," *New Republic*, LXXVII (Jan. 24, 1934), 303.
[11]Lee Reaves, "Highway Bond Refunding in Arkansas," *Arkansas Historical Quarterly*, II (Dec., 1943), 318.
[12]Little Rock *Arkansas Gazette*, Oct. 25, 1929, p. 1. (Hereafter cited *Gazette*.)
[13]*Gazette*, Nov. 11, 1929, p. 11.

Arkansas banks closed their doors; forty-six others soon followed.[14]

When the Great Depression began Arkansas was still attempting to recover from the general depression and the natural blows it had suffered for a decade. In late November of 1929 Governor Parnell telegraphed President Hoover the plans for highway construction in Arkansas and suggested that an increase in federal funds would be helpful. In another telegram he asked that the Federal Farm Board increase its loans to Arkansas cotton farmers.[15] Obviously, then, conditions in Arkansas were far from prosperous as the year 1929 drew to a close.

THE DROUGHT, 1930-1931

Because of the floods and tornadoes of 1927 and 1929 Arkansas farmers were eagerly anticipating large spring plantings and prosperous fall harvests in 1930. Mother Nature had different ideas! The worst drought ever to hit the South and Southwest occurred during the summer of 1930. Weather statistics show that the drought was 16 per cent more severe in Arkansas than in any other state.[16] On August 2, 1930, the *Arkansas Gazette* announced Little Rock's seventy-first rainless day. Few if any crops were harvested that fall.

Most farmers, however, did not realize the severity of the drought until late December when their supplies of produce, canned goods, and feed began to run out.[17] When these same farmers attempted to borrow money to see them through the winter, bankers and businessmen were unable to respond to their requests. Too many banks had already closed. The continuing financial crisis made loans almost impossible to obtain.

Congressmen from the drought-stricken states attempted

---

[14] Charles Morrow Wilson, "Famine in Arkansas," *Outlook*, LVII (Apr. 29, 1931), 596.

[15] *Gazette*, Nov. 26, 1929, p. 1.

[16] "Stricken Arkansas," *Outlook*, LVII, 85.

[17] "Drought: Field Report from Five of the States Most Seriously Affected," *New Republic*, LXVI (Feb. 25, 1931), 40.

to push federal relief programs through Congress, but these were met with Administration opposition. In October, President Hoover rebuked those who wanted a special session to deal with relief measures and reaffirmed his belief that the nation's voluntary organizations and community service projects could take care of the distressed. In December, however, the President acquiesced and signed a bill authorizing $45,000,000 in federal funds for loans to drought-stricken farmers; Arkansas received about $5,000,000 from this appropriation.[18]

Unfortunately this money was not as effective in relieving the poverty of the farmers as was expected. These funds could not be used in loans for food and clothing so necessary to survive the winter. Instead, the money was designated solely for the purchase of seed and equipment to be used in spring planting. In order to qualify for federal funds a farmer was required to mortgage his *anticipated* fall crop. However, the most destitute Arkansans already had mortgages from the previous year which they had been unable to redeem because of the drought and subsequent crop failures. Yet this new federal aid was unavailable to them. In Lee County alone, only twenty-two of the 5500 farmers (most of whom were destitute) could qualify for a government loan. It seemed that those who needed aid least were the only ones the government considered "good risks."[19]

It was obvious to the legislators from the drought area that a much broader program of federal aid was needed for their constituents, and thus a group of Democrats in both houses drew up additional relief legislation.[20] This bill called for an additional $15,000,000 to be used specifically for *food* loans in the drought areas. In a moving speech before the Senate, Arkansas's Senator Joe T. Robinson told of the vast suffering that he had witnessed in his state. He described conditions in three Arkansas counties with a total

---

[18]*Gazette*, Dec. 21, 1930, p. 1.
[19]"Drought: Field Report," *New Republic*, LXVI, 40-41.
[20]*Gazette*, Dec. 21, 1930, p. 1.

population of 100,000 people in which not one bank was open for business. He told of one Red Cross worker in Mississippi County who had received 1200 calls for relief in just two days.[21]

Once again Arkansas tried to secure additional funds by issuing bonds to bolster its faltering financial structure. As in 1929 the state's assets were overvalued by about one-half. Although the state managed to raise $18,000,000 this way, it was at the cost of perpetuating false assumptions about the soundness of the Arkansas economy. Within a few years Arkansas was to find itself the defendant in several law suits stemming from these overvalued bonds.[22]

As Arkansans suffered through this long winter, an event occurred that dramatized their needs and focused country-wide attention on the drought-stricken area. On Saturday, the third of January, 1930, a group of some three hundred farmers from Lonoke County congregated in England, Arkansas, and demanded that local merchants give them food for their hungry families. The men were orderly but determined. Apparently they were angry with local Red Cross officials who had told them that morning that the Red Cross had run out of application blanks, and no aid could be distributed until more blanks were obtained. Coming only a few days after the local bank and trust company had closed, destroying all the savings of many of these farmers, the refusal of Red Cross aid was too much to bear. Fearing a riot, some of the merchants made arrangements to be reimbursed through the Red Cross office in Kansas City; then they began distributing food to the agitated farmers. For the first time in Lonoke County, a bread line was formed, and over 150 loaves of bread were distributed to the needy families. Soon the farmers dispersed.[23]

The desperation of these men, coupled with the harsh realities of the situation, gave the "England Riot" nation-wide publicity, much of which was exaggerated. Some U. S.

---

[21]"Stricken Arkansas," *Outlook*, LVII, 85.
[22]Flynn, "Other People's Money," *New Republic*, LXXXVII, 308.
[23]*Gazette*, Jan. 4, 1931, pp. 1, 2.

Congressmen claimed that the demonstration was "staged" by the proponents of the food-relief bill pending in Congress. Others claimed the "riot" was Communist inspired, and there was a great danger that similar incidents would occur throughout the drought area.[24] Governor Parnell further confused the issue by stating in a telegram to the *Baltimore Sun* that the incident was of little significance and that Arkansans could handle their own relief problems. He soon reversed himself and sent a telegram to Congress saying: "I hasten to correct any impression that Arkansas does not need assistance regardless of what eastern papers may say."[25]

There were also conflicting reports by those present during the episode. George Morris, a lawyer who spoke to the angry crowd and tried to convince them to disperse, said that almost all of the men were hard-working, honest farmers who were simply desperate for food. On the other hand, Albert Walls, Red Cross official, claimed that at least 25 per cent of those present were imposters, and not really in need of food.[26] According to the leader and spokesman for the farmers, the group that originally drove into England numbered only forty-seven, but some 300 to 500 persons collected after he and his group arrived; this, he said, was the "mob" referred to in the press. He denied newspaper accounts that said half the mob was armed; none of his men had weapons, he said.[27]

In any case, the "riot" did get results. England merchants sent appeals to Red Cross headquarters on behalf of local farmers, and efforts were begun to get highway construction work for many of the men. Prominent Arkansans sent twenty-six telegrams to Congressman Tillman B. Parks describing the grave need for federal relief funds in the state. Arkansas Senators Thaddeus Caraway and Robinson

---

[24]*Ibid.*, Jan. 13, 1931, p. 1.
[25]*Ibid.*, Jan. 9, 1931, p. 1.
[26]*Ibid.*, Jan. 4, 1931, p. 1.
[27]Lement Harris, "An Arkansas Farmer Speaks," *New Republic*, LXVII (May 27, 1931), 40-41.

continued to press for adoption of the $15,000,000 food-loan relief bill. Administration opposition to this bill was strong; Hoover considered the proposal a "dole" and wholly un-American.[28] Although this measure was defeated in January, a similar bill did pass and was signed by the President on February 14, 1931.

January through March marked the period of greatest suffering in the drought-stricken areas. It is difficult to comprehend the pitiful conditions under which many families lived that winter. One Red Cross field worker commented that: "Unlike a spectacular flood or cyclone . . . starvation has crept up so slowly that people are unaware of the dire need and destitution. . . . People have no food and no clothing. . . ."[29] One farmer reported that by the spring of 1931 his children had been without new clothes for three years, and his eldest boy had to be kept out of school to help with whatever spring planting could be done. The family's supply of canned fruits and vegetables was used up by early winter; most of the time after that they ate beans cooked with lard. Because this farmer had borrowed against his 1930 crop and then lost it in the drought, he was ineligible for a federal loan. He was, therefore, completely dependent on Red Cross food relief for survival.[30]

In February 1931 the Arkansas Red Cross reported that 519,000 people were receiving food allowances, and still others were being fed through the public school hot-lunch programs. The goal of the Red Cross was to provide needy families with a $2.00 per week food allowance with an additional fifty cents available for each child. The maximum any family could receive was $4.50 per week.[31] However, the actual average ration in Arkansas was closer to $1.20 per person per *month*. For a family of five, this $6.00 per month allowance usually consisted of 25 pounds of meal, 48 pounds of flour, 25 pounds of pinto beans, 2 gallons of

---

[28]*Gazette*, Jan. 4, 8, 9, 1931.
[29]*Ibid.*, Jan. 9, 1931, p. 1.
[30]Harris, "An Arkansas Farmer Speaks," *New Republic*, LXVII, 41.
[31]"Arkansas's Fight for life," *Literary Digest*, CVIII (Feb. 28, 1931), 6.

molasses, 10 pounds of lard, 1½ bushels of potatoes, 10 pounds of salt, one package of soda, and one can of baking powder.[32] Obviously absent were such staples as meat, milk, fruit, and sugar.

The National Red Cross sent forty relief experts into Arkansas, but most of the work was done by a staff of over 6,000 volunteers.[33] Even so there were not enough workers to meet the overwhelming task. And certainly there was not enough money in the Red Cross treasury to care adequately for all the drought victims. In January 1931 a huge nationwide campaign for Red Cross funds was begun; the goal was $10,000,000.

Perhaps the most pitiful conditions in the drought-stricken area were found in the "plantation" counties of eastern Arkansas. There the Negro and white sharecroppers were already living a hand-to-mouth existence *before* the drought and the Depression. They worked for fifty or seventy-five cents a day and often bought all supplies at the "plantation commissary" where prices were high. They borrowed money on their crops from the plantation owners, often at interest rates as high as 25 per cent. At the end of the year the balanced statements almost always showed the sharecroppers in the red. When the effects of the Depression began to be felt in 1930, the cropper discovered that the plantation owner would no longer "tide him over" through the winter. Neither could a sharecropper get credit at local banks or qualify for a government loan. Even the Red Cross sometimes turned down his application for relief, considering him the responsibility of his plantation owner.[34]

As early as mid-January 1931 the Red Cross was feeding half the farming population of St. Francis County. Their funds became so depleted that the food allowance was reduced to a dollar per person per *month*.[35] Conditions were

---

[32]Wilson, "Famine in Arkansas," *Outlook*, LVII, 596
[33]"Arkansas's Fight for Life," *Literary Digest*, CVIII, 6.
[34]Lucien Koch, "War in Arkansas," *New Republic*, LXXXII (Mar. 27, 1935), 183.
[35]*Gazette*, Jan. 12, 1931, p. 2.

much the same throughout the agricultural areas of Arkansas.

Gradually the plight of the drought-stricken farmer became common knowledge throughout the country, partly because of the Congressional speeches by Senators Robinson and Caraway, and partly due to the efforts of the national news media. Such influential national publications as *The New York Times, The New Republic, Outlook,* and *Colliers* sent reporters to Arkansas to survey conditions. Their stories were usually heart-felt pleas for aid to a proud but needy people who wanted work, not charity.

The result of all this publicity was not exactly what most Arkansans had hoped. No real remedies were proposed; no work-relief projects initiated. Instead, charity poured into the state. Thirty-two states sent a total of 530 carloads of foodstuffs; the railroads provided free transportation.[36] The United Drug Company, with 152 branches throughout the state, authorized its stores to supply medicine without charge to needy persons upon authorization from their physician.[37]

In February Will Rogers made a good-will tour of the state to raise funds for relief work. He paid all his own expenses during the four-day appearance and raised over $39,000 which he turned over to the Red Cross in Arkansas.[38]

The Columbia Broadcasting System produced a program at KLRA Radio in Little Rock featuring several drought sufferers from around the state. The program was introduced by Will Rogers, and then several farmers and young people told of their efforts to get work, find food, and eke out some sort of existence. The program was aired over seventy-six CBS affiliate stations across the country. As a result, more voluntary funds poured into the state.[39]

Seeking to prevent a repetition of the suffering that

---

[36]Wilson, "Famine in Arkansas," *Outlook,* LVII, 597.
[37]*Gazette,* Feb. 15, 1931, p. 1.
[38]Floyd Sharp, *Traveling Recovery Road: The Story of Relief, Work-Relief and Rehabilitation in Arkansas* (Little Rock, 1936), 17.
[39]*Gazette,* Jan. 28, Feb. 1, p. 1.

the drought had caused, the state extension service in cooperation with the University of Arkansas launched a "self-sufficiency" farming campaign. The purpose of this program was to educate farmers in diversified farming and to make each farm family more self-sufficient. This program included such recommendations as keeping at least one milk cow, keeping at least thirty laying hens, planting a vegetable garden and canning any excess produce, rebuilding land fertility and planting legumes and pastures in former corn and cotton fields, and reducing acreage.[40] To this end the Red Cross distributed over 100,000 four-pound boxes of garden seed to relief recipients.[41]

There were many Arkansans who definitely resented the fact that "outsiders" were sending them aid. This attitude is reflected by this editorial from the *Southwest American* of Fort Smith:

> Damaged by drought and financial difficulty, misrepresented by exaggerated stories of starvation and misery, Arkansas can only stand its ground, keep a stiff upper lip, plant and tend its crops, keep the wheels of industry turning, and pray, "Lord, save us from our friends!"[42]

Similarly, an article in *The Arkansas Farmer* decries the situation by stating: "Heaven forbid that the time will ever come again when Arkansas farmers will have to ask the peoples of the nation to contribute so that they may be fed."[43]

This "Arkansas pride" was also reflected by many families on relief. Some kept track of all the "gifts" they had received from the Red Cross with the intention of "paying it back" when they were able.[44] Many turned to the Red Cross only when they desperately needed food or clothing for their children and would accept nothing for themselves.

[40]Wilson, "Famine in Arkansas," *Outlook*, LVII, 596.
[41]"Good News from Arkansas," *Literary Digest*, CVIII (Mar. 28, 1931), 12.
[42]C. F. Byrns as quoted in *ibid.*, p. 12.
[43]Stanley Andrews as quoted in *ibid.*, p. 12.
[44]"Drought: Field Reports," *New Republic*, LXVI, 40.

In some cases, especially in the back country of the Ozarks, even when Red Cross volunteers actively sought out needy persons, their aid was rejected.[45] Naturally there were those who gladly accepted whatever assistance was offered. Probably, too, there were a few who preferred direct relief to work. It can be safely stated, however, that the vast majority wanted work and a chance to get "back on their feet."

By the spring of 1931 there seemed to be a general feeling of optimism There was a hope that spring and summer crops would be good, and the farmers would soon have incomes again. As early as January, Governor Parnell had confidently proclaimed that "already there are signs of the dawn."[46] The Eureka Springs *Daily Times Echo* expressed the hope that the "spirit of pessimism" had vanished and that the state would soon be "bringing order out of chaos."[47] *Collier's* magazine stated that: "Arkansas has staged a magnificent comeback. They've met the Federal government more than halfway in the work of rehabilitation as might be expected of a people who don't aim to be 'beholden'."[48]

The outlook indeed seemed promising when the Red Cross announced that it would pull its workers out of Arkansas by April 1, several weeks ahead of the anticipated date.[49] Also, most of the banks which had failed during the autumn of 1930 were open again by the spring of 1931. The state had plans to spend $2,000,000 on new college buildings and an equal amount on county roads. Thirty-two million dollars had also been appropriated for new state highways.[50]

These measures all seemed to indicate that Arkansas was on the way to economic recovery. Indeed, two visiting Congressmen from Minnesota proclaimed that "nothing but better times are in store for the state and its people."[51] An

---

[45]*Gazette*, Feb. 1, 1931, p. 1.
[46]Parnell, *Inaugural Address* (Little Rock, 1931), 5.
[47]"Good News from Arkansas," *Literary Digest*, CVIII, 12.
[48]Walter Davenport, "The Drought and other Blessings," *Colliers*, LXXXVIII (July 11, 1931), 10.
[49]Burton F. Vaughan, "Arkansas Makes a Brilliant Recovery," *Review of Reviews*, LXXXIII (June, 1931), 90.
[50]Wilson, "Famine in Arkansas," *Outlook*, LVII, 597.
[51]*Gazette*, Mar. 18, 1931, p. 1.

editorial in the *New York Times* even went so far as to suggest that in some ways the drought had been a "good thing" for Arkansas, beecause it had caused farmers to improve their methods.[52]

Unfortunately, most of this optimism was unwarranted and premature. An increase in the country's farming population of 206,000 between 1930 and 1931 meant that the market for agricultural products would be even more crowded than before.[53] In addition, money became tighter, unemployment increased, and business slowed considerably. What Arkansans didn't comprehend at the time was that they were only a part of a much larger picture.

GOVERNMENTAL ACTION, 1931

In the face of these many problems, much constructive state legislation was needed. Unfortunately, neither the legislature of 1931 nor Governor Parnell saw fit to provide it. The governor set the tone for the rather unproductive session in his inaugural address on January 14, 1931.

> The cause of our distress is economic, not legislative. The remedy for it lies in economic recovery, not in legislative palliative . . . . This is no time to try out fantastic schemes nor visionary panaceas for all our ills.[54]

Both the governor and the legislature agreed that this was no time to call for increased taxation. Yet educators insisted on additional funds to compensate for the lack of tax monies available to the schools in 1931. In a compromise move the legislature appropriated $5,000,000 from the state equalizing fund to pay teachers' salaries and authorized the state debt board to sell notes in order to raise funds for teacher relief. Individual school districts also began to sell notes. [55]

---

[52]"Better Days for Arkansas," *New York Times*, Feb. 15, 1931, Pt. III, p. 1.
[53]John D. Hicks, *Republican Ascendancy, 1921-1933* (New York, 1960), 264.
[54]Parnell, *Inaugural Address*, 5-6.
[55]*Arkansas Acts* (1931), 665-68, 673-77.

The only other constructive legislation of that session dealt with improving the regulation of banking policies and freeing county funds from closed banks. The remaining legislative business was quite routine. No state action was taken to join with the Red Cross in relief work, to seek ways to stem the growing unemployment in the cities, or to revise the state's outdated budget.

The picture was not much brighter on the national scene. As early as October 1930 President Hoover had set up a national Committee for Unemployment Relief under Arthur Woods. With its limited funds, however, about all it could do was offer encouragement to local efforts at voluntary relief or work-relief programs.[56] When the President's representative from this committee toured Arkansas in January 1931, he claimed that of the six southwestern states he had visited, Arkansas was doing the *least* to cut back unemployment, and he blamed this on the lack of responsible state leadership.[57] The criticism was undoubtedly deserved, for as yet the state had done very little in the area of state-sponsored relief except discuss it.

The Woods committee was replaced in August 1931 by a larger agency, the President's Unemployment Relief Organization, headed by Walter S. Gifford. It was still, however, aimed at promoting *local* work-relief projects and *voluntary* contributions to private welfare.[58] Its counterpart in Arkansas was an "Unemployment Commission" headed by Harvey C. Couch. This agency tried to organize committees in each county and petitioned the federal government for funds to employ men in the construction of roads and levees. They devised "A Day's Pay" campaigns, under which employed workers agreed to contribute one day's salary a month to help those less fortunate than themselves. Hot Springs for example, used such funds to sponsor local work-relief projects.[59] Most of their efforts, however, were directed to-

[56] Hicks, *Republican Ascendancy*, 270.
[57] *Gazette*, Jan. 14, 1931, p. 1.
[58] Arthur M. Schlesinger, Jr., *The Age of Roosevelt*, Vol. I: *The Crisis of the Old Order, 1919-1933* (Boston, 1957), 173.
[59] Sharp, *Traveling Recovery Road*, 17.

ward urging private industry to keep employment and wages up and making public appeals for Red Cross contributions. Some critics felt these programs were barely better than nothing. Broadus Mitchell, for example, has said: "Nothing more haphazard and trivial, in view of the magnitude of the need of the unemployed, could have been proposed."[60]

As the winter of 1931-32 appproached it became increasingly apparent that the outlook was not as sunny as had been forecast. The state was not receiving the anticipated income, and of the tax money received, almost 75 per cent was allocated to the highway department. As a result, education was facing a real crisis. Regular tax revenue was expected to be about $3,750,000 less than the previous year.[61] Superintendents reported that they received about one-half the amount of money they needed, and one teacher stated that he was paid only $25 in eight-months of work.[62]

This financial crisis in education, along with a growing general unrest in Arkansas, forced Governor Parnell to call a special session of the legislature in October 1931. These deliberations represented political bickering at its worst, for they reflected few of the real needs and concern for the sufferings of the citizens of the state. No action was taken on taxation or on reallocation of the existing tax funds. Most of the legislators' time was spent in long debates over an audit of the highway department and an investigation of some highway construction procedures. Once such an audit was approved both houses adjourned, leaving unconsidered a bill authorizing additional funds for school districts and counties.

DESPAIR OF 1932

The advent of 1932 brought little hope to the country. By the middle of that year the gross farm income had fallen

---

[60]Broadus Mitchell, *Depression Decade: From New Era Through New Deal, 1929-1941* (New York, 1947), 102.

[61]H. L. Lambert, "Our Educational Situation," *Journal of Arkansas Education*, X (Oct., 1931), 16.

[62]H. L. Lambert, "What May the Schools Expect," *Journal of Arkansas Education*, IX (Apr., 1931), 16.

to less than half of what it had been in 1929. The Department of Agriculture estimated that the average farmer's net annual income in 1932, after all production expenses were paid, was not over $230.[63] Contributing to the problems of overproduction and low agricultural prices was the fact that in all the southern states a great migration from urban to rural areas occurred during 1932.[64] The unemployed city worker hoped at least to be able to support his family on the farm. Usually he found conditions there no better than in the city. Transients were also becoming a major problem by 1932. It is variously estimated that from one and one-half to two million people were wandering the country in search of a few days' work. The cause of this phenomenon was, of course, unemployment. The nation's unemployment figures had risen from 4,000,000 in 1930 to 12,000,000 in 1932. Like the gross farm income, the gross national income had decreased by over half, dropping from 87.4 billions to 41.7 billions. The nation's net investment, based on 1929 prices was a *minus* $358,000,000.[65]

Undaunted by these facts, President Hoover was still able to proclaim in the summer of 1932 that he believed the United States "had reached the bottom of the depression pit" and move upwards from then on.[66]

Arkansans, however, had no such rosy-colored glasses. Their state's bureau of labor and statistics reported that "industry had about reached the bottom in Arkansas at the close of June, 1932."[67] For almost two years there had been steady drops in both payrolls and employment. Between 1930 and 1932 some 35,000 Arkansans out of a normal industrial work force of 283,000 lost their jobs. When the 1932 employment figures of Arkansas are compared to the 1929 figures only 62.7 percent of the laborers were still employed. In that same time period payrolls decreased

---

[63]*Gazette*, Oct. 22, 1931, p. 1.
[64]Ezell, *South Since 1865*, 429.
[65]Schlesinger, *Crisis of the Old Order*, 251, 248.
[66]Quoted in Hicks, *Republican Ascendancy*, 276.
[67]Arkansas Bureau of Labor and Statistics, *Eleventh Biennial Report, 1932-1934* (Little Rock, 1934), 23.

over 45 percent and private deposits in state banks fell from $137,000,000 to $62,000,000.[68] As of March 1 the state's outstanding bonded indebtedness was over $105,000,000, and five million had to be paid during 1932. About 85 per cent of this debt was from highway and bridge construction. The state had, in fact, sold so many bonds that its indebtedness equalled $2.92 per capita.[69] On March 15 post-dated vouchers for highway construction amounting to over $600,000 came due and there was no money in the treasury with which to meet the payments.[70] The assessed value of real and personal property decreased more than $124,000,000 between 1929 and 1932. As a result, the state's income dropped 25 per cent in that same time period.[71] In addition to the state indebtedness it was estimated that county and local governments were indebted for nearly $130,000,000 as of April, 1932.[72] Commissioner of Labor W. A. Rooksbery estimated that 60,000 Arkansans had depleted all their savings and assets and were unemployed as of October 1932. Since most welfare bureaus and private social agencies had exhausted their funds, he saw no hope for these thousands of unemployed.[73]

Still, President Hoover clung to his theory of relief: "It is not the function of the government," he said, "to relieve individuals of their responsibilities to their neighbors, or to relieve private institutions of their responsibilities to the public, or of local government to the states, or of state government to the Federal government."[74]

In his attempt to deal with the financial chaos of Arkansas Governor Parnell called a second special session of the legislature in March of 1932. For most of the session the debates in both houses revolved around the highway department and its budget. Should the huge highway debt

---

[68]Sharp, *Traveling Recovery Road*, 17-18.
[69]*Gazette*, Mar. 8, 1932, p. 1.
[70]*Ibid.*, Mar. 12, 1932, p. 1.
[71]*Ibid.*, Dec. 2, 1932, p. 1.
[72]*Ibid.*, May 1, 1932, p. 1.
[73]*Ibid.*, Oct. 14, 1932, p. 10.
[74]Quoted in Mitchell, *Depression Decade*, 87.

be refunded? Should taxes be raised to secure the money needed to pay off the state's debt? Should the highway audit commission be allowed to continue its work?

The resolution of the highway question did not come about until April 11, more than a month after the session began. (The daily salary of the legislators was an expense that the state could ill afford.) A compromise bill on refunding the highway debt was finally accepted by both houses.[75] The district highway bonds were refunded at 4.5 per cent interest and took priority over the maintenance and operation expenditures of the highway commission.[76]

The legislature defeated a bill to make the highway commission an elective body. It refused to take up the question of needed additional funds for the public school system, and it ignored the Governor's proposals on governmental reorganization. Because the house and senate could not even agree on a dismissal time, Governor Parnell was forced to dismiss the session himself.[77] Considering the enormity of the state's problems, this was not a fruitful session.

Even as the special session disbanded there was already growing sentiment among many Arkansans for another called legislative session in hopes of adopting more relief measures. An organization, composed of farmers and laborers, was formed to work toward this end. However, Governor Parnell refused to call the legislature back.[78]

RELIEF PROGRAMS, 1932

Probably the most outstanding relief program initiated by Congress during 1932 was the Reconstruction Finance Corporation (RFC) established as part of the emergency Relief and Construction Act signed by President Hoover

---

[75] *Gazette*, Apr. 12, 1932, p. 1.
[76] Reaves, "Highway Bond Refunding in Arkansas," *Arkansas Historical Quarterly*, II, 318-19. In less than a year, however, it was obvious the plan was a failure, for only $15,000,000 out of a total $47,000,000 outstanding bonds were exchanged.
[77] *Gazette*, Apr. 13, 1932, p. 1.
[78] *Ibid.*, Apr. 19, 1932, p. 1.

on January 22. Under the leadership of Charles Dawes, the RFC was given $500,000,000 from the federal treasury to begin operations. Its purpose was to make loans to banks and trust companies and other large organizations, with the belief that this would stimulate the economy in several areas and lead to increased employment. Only a minimal amount of its appropriation, however, was extended to agricultural credit, and the relief payments to the states did not equal the expectations.[79] Arkansas made greater use of farm loans than any other state, taking out 46,835 loans amounting to over $4,000,000 between February and August of 1932.[80] J. W. Jarrett, chief examiner for the state banking department, was appointed manager of the RFC in Arkansas.[81]

Governor Parnell appointed a state emergency relief commission headed by W. A. Rooksbery to work with Jarrett in obtaining RFC funds for the state.[82] The application stated that:

> the state government is wholly unable to care for the present relief needs in our industrial communities, and careful investigation convinces me that the communities themselves are making heroic struggles to meet the emergency but with only partial success, and are now in urgent need of financial assistance.[83]

From this application Arkansas received its first RFC grant for public work relief amounting to $502,500. Another loan in November brought the total received by the state to $1,319,168.[84] The state emergency relief commission set about establishing local commissions in every county. This task was accomplished by January 1933 and relief funds were obtained for all counties. The state commission

---

[79] Hicks, *Republican Ascendancy*, 271-72.
[80] *Gazette*, Oct. 5, 1932, p. 5.
[81] *Arkansas Gazette, Centennial Edition*, 174.
[82] Arkansas State Planning Board, *Progress Report* (Little Rock, 1936), 129.
[83] Governor Harvey Parnell, as quoted in Sharp, *Traveling Recovery Road*, 18.
[84] *Gazette*, Nov. 19, 1932, p. 1.

functioned for eight months, but was forced to disband when all RFC funds had been exhausted.[85]

Naturally, private welfare work did not come to a sudden halt just because the national RFC was established. Voluntary funds were difficult to secure, however, The Community Chest goal in Little Rock in 1932 was $190,000, but when the drive ended only $132,758 had been pledged.[86]

Most domestic workers lost their jobs early in the Depresson. This included many women who were the sole support of their families. In an effort to provide them with work relief, some privately financed programs were established. One of these was a canning project in North Lttle Rock in which one-hundred acres were turned into gardens. This provided labor for dozens of men. The produce grown in these gardens was then canned by the women and the canned foods distributed to those on relief. Sewing projects were also set up, using donated space, borrowed machines, and scrap material. Both skilled and unskilled seamstresses worked to turn out clothing that was given to the needy. In both projects the women were paid with public funds at the rate of $1.20 a day but could not work over three days a week. The equipment and materials all had to come from private donatons, however.[87]

Two or three additional work-relief programs were needed for every one that was begun during 1932 in Arkansas. No agency had enough funds to meet the demands made upon it. Work-relief provided jobs for only about one-third of the unemployed. As the winter of 1932-1933 approached, the mood of the state could be described as almost desperate.

The history of Arkansas between the years 1929 and 1933 is a study in discouragement, confusion, suffering, and political ineptness. Legislators met for weeks at a time in special session, paid daily by the state, engaged in trivial bickering, while other state employes, such as public

---
[85]Arkansas State Planning Board, *Progress Report*, 129.
[86]*Gazette*, Oct. 25, 1932, p. 1.
[87]Sharp, *Traveling Recovery Road*, 19, 93.

school teachers, went months without salaries. Governor Parnell belittled plans for state sponsored relief programs and insisted that hard work would end anyone's suffering, while hundreds were losing their jobs daily and roads were crowded with families leaving the cities for the farm in hopes of eking out a better existence than they had known. Schools closed for lack of funds and children fell a year or two behind in their eduction while state money was pocketed by contractors friendly to highway commission members. The state was criticized by a federal supervisor for not doing enough for the unemployed, but instead of taking positive steps to correct the problem, state officials spent their time making defensive statements.

Senators Caraway and Robinson fought and pleaded until Congress appropriated funds for drought relief, only to have many Arkansas leaders stoutly declare that the state could handle her own problems. The legislature refused to consider plans for governmental reorganization that would have resulted in more efficient, less expensive state government. One of the main arguments against the Parnell plan presented in 1932 was that it had been drawn up by "outsiders" from New York.[88] An orderly group of drought-stricken farmers who peaceably came to town to ask for food for their hungry families were labeled "troublemakers" and Communists by more prosperous citizens.

To be sure Arkansas was not the only state that made a painfully slow response to the suffering of its citizens during the early years of the Great Depression. There were, after all, very few Governor Roosevelts or Huey Longs in the 1930's, and many other states lacked capable leadership. Even a far-sighted governor could have relied on no large tax payments from industry in a poor agricultural state like Arkansas. There were many Arkansans who worked long and hard to bring private relief to the destitute, to enlist federal aid at every opportunity, and to fight inefficiency and corruption wherever it was discovered. Arkansas truly

---

[88]*Gazette*, May 27, 1932, p. 1.

had few resources upon which to draw in making a speedy recovery.

No excuse, however, can be made for the insensitivity and ineptness of many state officials or the disorganizaton and inefficiency of many committees. The confusion of the times does not provide license for dishonesty or mismanagement. Arkansans deserved far more from their state government than they received in those dismal days between 1930 and 1933.

The Anglo-American Trade Agreement and Cordell Hull's Search for Peace 1936-1938

ARTHUR W. SCHATZ

By the middle of the 1930s, policymakers in the United States had become deeply concerned about the possibility of war in Europe and Asia. President Franklin D. Roosevelt eagerly began to seek ways to prevent a major conflict or, at least, to insulate the Western Hemisphere against its direct effects. In 1936, he called the Buenos Aires Conference in an attempt to create a system of "collective neutrality." In 1937, he toyed with the idea of quarantining aggressor nations and, until 1938, mulled over the suggestion that he sponsor an international conference. While Roosevelt fitfully explored these avenues without notable success, Secretary of State Cordell Hull also sought a foreign policy initiative that would help to pull the world back from the "Niagara of war." Hull had long insisted as a general proposition that economic rivalry was a basic cause of war and that gradual international adherence to the principles of liberalized trade embodied in the Reciprocal Trade Agreements Program of 1934 would ultimately produce prosperity and peace.[1] By 1936, however, Hull began to suspect that long-range remedies were inadequate; and he launched a vigorous effort to forge an Anglo-American commercial alliance that would accelerate the movement toward freer trade and thus slow the drift toward war. This endeavor culminated in the Anglo-American trade agreement of 1938, and for Hull it seemed to provide a workable plan by which the United States could effectively exert a pacific influence on the course of world events.

Hull conceived of the Anglo-American agreement as the beginning of a worldwide campaign to restore political stability. Mutual adoption of liber-

Mr. Schatz is associate professor of history in San Diego State College.

[1] Julius W. Pratt, *Cordell Hull* (2 vols., New York, 1964), I, 29-30; Donald F. Drummond, "Cordell Hull, 1933-1944," Norman A. Graebner, ed., *An Uncertain Tradition: American Secretaries of State in the Twentieth Century* (New York, 1961), 185, 187; William R. Allen, "The International Trade Philosophy of Cordell Hull, 1907-1933," *American Economic Review*, XLIII (March 1953), 105-06; William L. Langer and S. Everett Gleason, *The Challenge to Isolation, 1937-1940* (New York, 1952), 7, 17.

alized trade practices by the two major western democracies, he argued, would provide the "central point or spearhead" for the creation of what he called a "world program" that "would emanate from many centers." Once the thirty to thirty-five peace-loving nations, led by the United States and Great Britain, agreed to a policy of freer trade and pacific conduct, they could "present it to Germany, Italy and other countries which may still be pursuing a policy of closed economy or autarchy." When these latter nations had rejoined the market economy, the economic causes for war would disappear; and the time would be appropriate, Hull believed, for "settlement among other nations abroad of any political questions or difficulties which may exist."[2] If the "jingoist nations" refused to seize this opportunity, the fact of Anglo-American cooperation might still deter war or, failing that, pave the way for wartime collaboration.

Reliance on economic means to accomplish political ends grew naturally out of Hull's concern with problems of international trade and tariff policy. As early as 1916, he had become convinced that "unhampered trade dovetailed with peace" and, since that time, had crusaded for reduced barriers to world trade.[3] He combined this economic orientation with firm adherence to Wilsonian moral principles, and by the 1930s he had developed a clearly defined set of injunctions to govern international conduct. All nations, he insisted, must abjure the use of force, adhere faithfully to agreements, observe international law, reduce arms, refrain from interference in the internal affairs of independent states, and establish fair-trade practices by adopting the policy of equality of treatment. Liberalized trade was the catalyst for achieving this reorientation, and Hull frequently referred to the trade agreements policy as "the most important part" of a general American program for world peace.

Although the Trade Agreements Act was passed as an emergency measure to aid domestic recovery, Hull insisted from the outset that the program was an important diplomatic tool. He defined the goal of American diplomacy as the "maintenance and promotion of peace throughout the world, both political and economical." Any political settlement, however, had to be predicated upon economic pacification. "The truth is universally recognized," he observed, "that trade between nations is the greatest peacemaker and civilizer within human experience"; and an administration program "to secure trade agreements with the principal nations is the first step in a broad movement to increase international trade. Upon this program,"

---

[2] Memorandum by Cordell Hull, undated, Folder 386, Cordell Hull Papers (Manuscript Division, Library of Congress). See also Cordell Hull, *The Memoirs of Cordell Hull* (2 vols., New York, 1948), I, 518-19; Pratt, *Cordell Hull*, I, 132-33.

[3] Hull, *Memoirs*, I, 81.

he concluded, "rests largely my hope of insured peace and the fullest measure of prosperity."[4]

Hull's insistence on the salutary effects of liberalized trade amazed and sometimes irked some of his New Deal colleagues,[5] but during the depression his views won increasing support among policymakers who were equally convinced that world problems were simple reflections of domestic problems. These men "found it difficult to believe that the states of Europe could be discontented for other than economic reasons."[6] Until the end of 1935, however, the international political ramifications of liberalized trade were discussed in only the most generalized and long-range terms. Generally, Hull and his supporters assumed that the major contribution which the United States could make to peace would be to generate a domestic prosperity that would spread to the rest of the world and create the proper climate for political pacification.[7]

The efficacy of this approach came under serious question during 1936. That year the Ethiopian crisis dominated the news, but details of German remilitarization of the Rhineland, civil war in Spain, and Japanese activities in North China also punctuated the headlines. By the end of the year, Lloyd's of London announced that it would no longer auction insurance on property against the hazards of war.[8] While leaders in Rome, Berlin, and Tokyo more clearly shaped the course of events, the major democratic states seemed diplomatically bankrupt. In Great Britain, Neville Chamberlain laconically noted: "we have no policy."[9] Again preoccupied with a serious domestic crisis, France was in no mood to accept any foreign challenge. In the United States, the isolationists insisted on noninvolvement and adopted strict neutrality laws. Action by the League of Nations was supported only halfheartedly by European countries and only morally by the United States.

Hull was convinced that the international peace-keeping machinery had

---

[4] Memorandum by Hull [1934], Folder 384, Hull Papers.

[5] Harold Ickes describes Hull as possessing "a one-track mind and the only station on his one-track mind is 'trade agreements.'" Harold L. Ickes, *The Secret Diary of Harold Ickes* (3 vols., New York, 1953-1954), II, 555. The journalist Ernest K. Lindley was relieved that Hull had been raised on Adam Smith instead of Karl Marx or "he might have been one of our foremost revolutionaries." Ernest K. Lindley, *Half Way with Roosevelt* (New York, 1936), 345.

[6] William W. Kaufmann, "Two American Ambassadors: Bullitt and Kennedy," Gordon A. Craig and Felix Gilbert, eds., *The Diplomats: 1919-1939* (Princeton, 1953), 654.

[7] For the campaign to promote this idea, see New York *Times*, Nov. 2, 1934, April 7, May 24, 1935; Department of State, *Press Release*, March 2, 1935, pp. 154, 158; *ibid.*, April 6, 1935, pp. 223-24; *ibid.*, Dec. 7, 1935, pp. 473-77; Willard L. Thorp, "International Economic Policy," Clair Wilcox and others, eds., *America's Recovery Program* (London, 1934), 238-39.

[8] Whitney H. Shepardson and William O. Scroggs, *The United States in World Affairs: An Account of American Foreign Relations, 1936* (New York, 1937), 3.

[9] Quoted in Keith Feiling, *The Life of Neville Chamberlain* (London, 1946), 295.

broken down and that Germany, Italy, and Japan were "definitely revealing intentions of aggression." The collapse of the London Naval Conference coupled with the accelerated rearmament programs of Germany and Italy made impossible a peace based entirely on disarmament. In April, Hull began to advocate the immediate rearmament of the United States and the adoption of common attitudes and policies among the "law-abiding nations."[10] As a practical matter, the United States tried to support the League by imposing a moral embargo against Italy and by helping to isolate the Spanish Civil War. During the debate on the neutrality bill of 1936, the Department of State urged Congress to permit greater presidential discretion in applying the law. To Hull, however, such actions were essentially passive responses, and he was convinced that there had to be a more effective means of easing the discontent that was provoking aggression.

Hull had hoped that the adoption of common attitudes by Great Britain, France, and the United States would "bring the three jingoist nations to their senses." In the early 1930s, however, there was little cordiality and less cooperation among the democracies. Friction resulted from the events at the London Conference of 1933, the debt question, and the American neutrality laws. Furthermore, Hull believed that any overt political cooperation with Britain and France was barred "for the patent reason that public opinion . . . was, in majority, militantly and almost violently against our entering into any such joint undertakings."[11] He decried the stringent requirements of the neutrality acts and chided the Senate for rejecting membership in the World Court. These actions, he complained, were the result of "blind and extreme nationalistic sentiment" that made it "exceedingly difficult for our government to function in a constructive, sane and practical way at all to the extent desirable."[12] But while Hull talked at length about the wisdom of more political cooperation and frequently criticized the isolationists, he usually advised caution and delay on specific questions of policy. The isolationists, in fact, seem to have provided, more often than not, a convenient excuse for inaction and negativism.[13]

Because of obvious political difficulties at home, Hull's efforts to promote common foreign policy attitudes remained firmly fixed within the more comfortable framework of his moral and economic preference. When the Norwegian minister inquired whether the administration contemplated any policy changes in view of recent European events, Hull replied that the

[10] Hull, *Memoirs*, I, 455; Pratt, *Cordell Hull*, I, 104-05.
[11] Hull, *Memoirs*, I, 455.
[12] Hull to Henry L. Stimson, Aug. 22, 1935, Folder 85, Hull Papers.
[13] Drummond, "Cordell Hull," 199-200; William E. Leuchtenburg, *Franklin D. Roosevelt and the New Deal 1932-1940* (New York, 1963), 203.

United States intended to continue its efforts to restore the practice of "international law, morals, [and] the sanctity of contracts and agreements . . . with the primary object of developing the most solid foundation for a permanent peace structure. . . ." To achieve these ends, the American government would also continue to work for a "comprehensive and basic economic program for world economic rehabilitation. . . ."[14]

Although this seemed little different from his earlier position, Hull was already in the process of redefining the means by which liberalized trade might be made immediately applicable to the German and Italian problem. For Hull, as for many Americans in the early 1930s, there seemed to be some validity in the claim that Germany and Italy were "have not" nations and that they were prevented from achieving their rightful international status by the restrictive trade policies of the western nations. Fascism and nazism were not simple symptoms of political nationalism, but the "characteristic expressions of great people in revolt against the limitations placed upon their national prosperity by their poverty in natural resources." The democratic nations could not expect a change in German and Italian political attitudes so long as the system of peace was based "upon the contemporary status quo of material inequality." Freer trade provided a solution. When the policies of exclusion and discrimination disappeared, all peoples would be able to obtain their economic needs without resorting to force.[15]

Hull openly supported this analysis by early 1936 and began to argue that as living standards rose "discontent will fade and dictators will not have to brandish the sword and appeal to patriotism to stay in power." He uncritically accepted at face value the colonial demands of Hitler and Mussolini: they did not want colonies for settlement, "but because they must find some path to the raw materials they need to sustain growing nations." They needed raw materials and markets, and the United States and Great Britain had more than enough. For Hull, the acceptance or rejection of liberalized trade was a matter of making a choice between "business or bullets."[16]

[14] Memorandum by Hull, May 20, 1936, Department of State, *Foreign Relations of the United States, Diplomatic Papers: Europe, 1936* (Washington, 1953-1954), 397-98.

[15] Frank H. Simonds and Brooks Emeny, *The Price of Peace: The Challenge of Economic Nationalism* (New York, 1935), xi, 328-29; Hull, *Memoirs*, I, 235. See also Samuel Lubell, "World Trade or Smash," *North American Review*, CCXXXIX (Jan. 1935), 71; John D. Condliffe, *War and Depression* (Boston, 1935), 5-6; Marc A. Rose, *Economics and Peace: A Primer and a Program* (New York, 1937).

[16] New York *Times*, March 7, 1936. This argument was rather widely supported by other government officials. See, for example, Henry L. Deimel, Jr., "Commercial Policy Under the Trade Agreements Program," *Annals of the American Academy of Political and Social Science*, 186 (July 1936), 16-23; Sumner Welles, "Present Aspects of World Peace," *World Affairs*, 100 (Sept. 1937), 166-67; Francis Bowes Sayre, *The Way Forward: The American Trade Agreements Program* (New York, 1939), 39-40.

Although such considerations became increasingly popular, Anglo-American trade cooperation also promised important domestic benefits. By the middle of the year the state department had concluded fourteen trade agreements with nations that exported primarily raw materials and foodstuffs, but the American negotiators had been unable to conclude any pact of major benefit to the domestic producer of farm products. As a result agrarian representatives in Congress openly asserted that the program was designed to "sell-out" the farmer for the advantage of the manufacturer by granting tariff reductions on agricultural imports in exchange for improved treatment of industrial exports. Unless the administration could negotiate an agreement with a major industrial nation that would offer significant concessions on agricultural exports, the farm lobbies threatened to put an end to the program.[17] Given his assumptions about the relationship between trade and peace and the fact of domestic discontent with the reciprocity program, Hull began to increase pressure on Great Britain to conclude a trade agreement with the United States.

In August 1935, commodity study groups in the United States and Great Britain had begun preliminary investigations to seek a basis for possible agreement, but serious economic and political problems precluded early action. During the depression, the British had ignored American foreign economic policy and had signed a series of preferential payments and clearing agreements that, by 1936, effectively blocked American negotiations with Argentina, Spain, Denmark, and the Baltic States.[18] The British were also reluctant to discuss any modification in the system of imperial preferences, which they had established at Ottawa in 1932. They would not in any case consider changes in this system without the consent of the Commonwealth nations, and such consent had to await the decision of the Imperial Conference scheduled for May 1937. In Washington, Hull and Roosevelt wanted to postpone formal negotiations until after the election of 1936 and the renewal of the trade program in early 1937; but Hull believed that the need

[17] New York *Times*, Nov. 24, 1935, Jan. 8, 1936; Gilbert C. Fite, *George N. Peek and the Fight for Farm Parity* (Norman, 1954), 286-88; William E. Borah to O. L. Roberts, Sept. 16, 1936, William E. Borah Papers (Manuscript Division, Library of Congress); Walter M. Pierce to Hull, July 18, 1936, Issues File, 2.99, Walter M. Pierce Papers (University of Oregon); memorandum by Eugene H. Dooman, June 20, 1936, Department of State, *Foreign Relations of the United States, Diplomatic Papers: The Far East, 1936* (Washington, 1954), 908.

[18] For a general description of British commercial policy during the depression, see Carl Kreider, *The Anglo-American Trade Agreement: A Study of British and American Commercial Policies 1934-1939* (Princeton, 1943), 19-25; Frederic Benham, *Great Britain Under Protection* (New York, 1941), 90-102; Henry J. Tasca, *World Trading Systems: A Study of American and British Commercial Policies* (Paris, 1939), 145-47, 151-52.

for action was so important that he attempted to persuade the British government to revise its trade policies unilaterally or, at least, to give verbal support to the doctrine of liberalized trade as a means of creating an economic foundation for peace.

Late in January 1936, Hull gave the British ambassador, Sir Ronald Lindsay, the first of a series of lectures on the ill-conceived course of British trade policy. By refusing to cooperate with the United States, he asserted, Great Britain was frustrating its desire for peace and prosperity—Italian aggression in Ethiopia made this sufficiently clear. Lindsay must certainly agree, Hull declared, "that if Italy had had even near her pre-panic quantity of exports there was a real possibility that her armies would not be on the march today. . . ." German military forces, moreover, probably "would be on the march before this leisurely policy of restoring trade and employment had been taken sufficiently in hand and dealt with. . . ." In view of the unsettled political situation, Anglo-American trade cooperation "would probably mark the difference between war and peace in Europe in the not distant future. . . ."[19]

British reaction to this analysis was polite but noncommittal, and Hull sought ways to provoke some change in British policy. During the early years of the depression, British policymakers had tended to subordinate current trade to debt collection. They had been interested primarily in concluding preferential agreements to liquidate blocked accounts and to facilitate the transfer of sterling payments to British creditors.[20] In February 1936, the Department of State Division of Trade Agreements became concerned about the possibility that the United Kingdom intended to reverse these priorities and utilize its generally unfavorable balance of visible trade to gain preferential treatment for British exports. If it did so, the possibility of expanding the American trade program "would be either excluded or rendered extremely difficult." The Division suggested, therefore, that the United States propose a joint statement announcing that the perpetuation and expansion of discriminatory policies were inimical to the "proper functioning of international trade relations." Such a statement would permit Britain to retain existing practices that it deemed necessary to meet the current emergency, but would also demonstrate disapproval of preferential agreements in general and emphasize Britain's willingness to join the American drive for liberalized trade. The United States would gain most

---

[19] Memorandum by Hull, Jan. 22, 1936, Department of State, *Foreign Relations of the United States, Diplomatic Papers: General, The British Commonwealth, 1936* (Washington, 1953-1954), 629-32. For other statements on the same theme, see memorandum by Hull, Feb. 5, 1936, *ibid.*, 633-34; Hull to Ray Atherton, Feb. 13, 1936, *ibid.*, 635-43.
[20] Tasca, *World Trading Systems*, 94-96.

from a change in British policy, but the Division believed that Great Britain was anxious "to emphasize, in some form or other, the cordiality of its relations with the Government of the United States." Political considerations, therefore, might prompt the British to agree to such a statement and to begin a reorientation of their commercial policy.[21]

Pending the start of negotiations for an Anglo-American trade agreement, the idea of a joint statement of principle appealed strongly to the secretary of state. He not only importuned the British for such a declaration but also attempted to expand its application. On March 11, he directed American embassies in Latin America and in most non-fascist European capitals to urge upon these governments economic cooperation with the United States. The American government, Hull explained, desired to place international economic relationships on a "higher plane than the narrow national and international political considerations which have been brought forward by certain nations to justify the extreme steps taken in the economic and financial field." Only through universal acceptance of the United States' commercial policy could the community of nations restore prosperity and construct "a sound foundation for building a permanent structure of world peace." The alternative was continued depression, increased arms production, and ultimate political upheaval. Hull did not intend this declaration as an open invitation to undertake trade negotiations with the United States; he simply wanted all nations to realign their policies unilaterally or to announce support for the policy of equality of trade treatment.[22]

Hull's attempt to provoke an international stampede toward liberalized trade demanded British participation. He urged the British to agree to a joint policy statement and pointed out that, once the major democratic states achieved unanimity, similar announcements by other nations would follow, "with the result that the moral effect of a pronouncement so urgent and so sound would inspire confidence in the industrial and business world as it would also quiet much of the high tension in both the economic and political situations." Hull was careful to explain that the British government would not be expected to surrender immediately any practices deemed necessary to meet short-run emergencies.[23]

British Foreign Secretary Anthony Eden believed that political settlement in Europe must precede economic arrangements and suspected that Hull's

---

[21] Division of Trade Agreements, "Memo on Possibilities of Concerted Action with the British Government in the Sphere of Commercial Policy," Feb. 27, 1936, File No. 641.0031/69, General Records of the Department of State, RG 59 (National Archives).

[22] See Hull to Jesse I. Straus, March 11, 1936, *Foreign Relations . . . British Commonwealth, 1936*, pp. 486-88.

[23] Hull to Ronald Lindsay, March 28, 1936, File No. 611.4131/155B, General Records of the Department of State.

suggestion was probably pointless. But Eden also believed that it was important to promote Anglo-American cooperation in every possible way, and he agreed to make the desired statement if "this would be regarded by the United States Government as a useful contribution to the attainment of the objects which Mr. Hull has in mind." The United States should expect no immediate change in practice.[24]

Although this answer was far from enthusiastic, Hull was satisfied. He assumed that earlier Anglo-American cooperation would have eased the economic pressure that drove Italy to war against Ethiopia. The democracies could not now sit idly by and await a similar explosion by the Germans. The British statement, however guarded, was important.[25] Roosevelt also thought the statement would be a "very useful beginning" so long as it was couched in general terms. He reminded Hull that "we do not want to involve ourselves [before November] in controversy over specific details which would narrow the field to specific commodities."[26]

During the depression, ministries of economic affairs and foreign ministries frequently pursued contradictory policies; and the Foreign Office decision fell afoul of opposition from Neville Chamberlain, chancellor of the exchequer, and Walter Runciman, president of the Board of Trade. Neville Chamberlain rejected flatly any compromise of Commonwealth trade ties, whether real or implied. Runciman insisted that Great Britain must use its economic power to increase its foreign trade and promised that he would achieve this objective by whatever means necessary. He agreed that Hull's ideas were theoretically sound, but he argued that no nation could follow them under the circumstances, especially one with a chronically unfavorable balance of trade. Moreover, Runciman did not think that the United States intended to reduce its tariff rates substantially. If the Americans really wanted economic cooperation, they should begin with a settlement of the war debt question and a change in financial policy that would release some of the gold held in American vaults.[27] In June, the objections of Neville Chamberlain and Runciman prompted the cabinet to overrule the Foreign Office.

In explaining this decision, Runciman told the counselor of the American

[24] British Foreign Office to Department of State, May 26, 1936, File No. 641.0031/61, *ibid*. For the American analysis of Anthony Eden's reasoning, see James C. Dunn to Hull, Dec. 4, 1936, File No. 611.4131/208A, *ibid*.; Atherton to Hull, Feb. 26, 1936, *Foreign Relations . . . British Commonwealth, 1936*, p. 644.

[25] Memorandum by Hull, June 20, 1936, *Foreign Relations . . . Europe, 1936*, pp. 398-99.

[26] Franklin D. Roosevelt to Hull, May 28, 1936, OF 20, Franklin D. Roosevelt Papers (Franklin D. Roosevelt Library, Hyde Park).

[27] Atherton to Hull, Feb. 28, 1936, *Foreign Relations . . . British Commonwealth, 1936*, pp. 644-45; Robert W. Bingham to Hull, July 16, 1936, File No. 641.0031/67, General Records of the Department of State.

embassy that the proposed declaration should be postponed until there was some "practical achievement between the two countries." Neville Chamberlain offered a more immediate reason: the political situation in Europe was deteriorating rapidly and the German economic offensive then under way was threatening some British markets. Before the end of 1936, the government intended to renew the payment agreement with Argentina and the clearing agreements with the Scandinavian countries. The cabinet was afraid that Great Britain would be charged with inconsistency and bad faith if it made "the contemplated public policy declaration while these arrangements were in process of negotiation."[28]

Disappointed but undismayed by this reversal, Hull continued throughout the summer and fall to warn London of a military explosion in Central Europe unless Great Britain changed its policy of diplomatic drift. Repeated refusal of commercial cooperation might be disastrous because, he cautioned, it could drive the United States into "isolation in virtually every way in time of war, including such vital phases as international credit and trade policies...."[29] The statement of principle, however, had always been a stopgap measure preliminary to the conclusion of a formal trade pact. With the presidential election campaign successfully concluded, the state department decided in November to submit to Great Britain a list of desired commercial concessions to determine whether a basis existed for formal negotiations.

The United States asked Great Britain for extensive duty reductions on such essential exports as hog products, grains, fruits, tobacco, and soft lumber and offered in return reduced tariffs on a number of British manufactures, primarily textile specialties.[30] Ambassador Lindsay was shocked at the American proposal and complained that it hit the imperial preferences squarely in the "solar plexus." But the London government did not reject further discussions. The ominous events of 1936 undoubtedly influenced the British as "Europe moved from belief in the possibility of peace to a dull acceptance of the certainty of war."[31] It was painfully apparent, too, that no strong leadership had yet emerged to avert catastrophe. Under the

---

[28] Atherton to Hull, June 25, 1936, File No. 641.0031/64, General Records of the Department of State; June 26, 1936, File No. 641.0031/65, *ibid.*

[29] Memorandum by Hull, July 20, 1936, *Foreign Relations . . . British Commonwealth, 1936*, pp. 677-78. See also Hull to Bingham, July 3, 1936, File No. 641.0031/64, General Records of the Department of State; Hull to Bingham, Sept. 3, 1936, File No. 611.4131/182A, *ibid.*; Memorandum by Hull, Oct. 22, 1936, *Foreign Relations . . . British Commonwealth, 1936*, pp. 688-89.

[30] "List of Products Presented to Mr. [H. O.] Chalkley," Nov. 16, 1936, File No. 611.4131/205, General Records of the Department of State.

[31] Charles Loch Mowat, *Britain Between the Wars 1918-1940* (London, 1955), 532.

circumstances, Hull's offer of cooperation, however unpalatable in form, probably seemed to contain important possibilities.

During 1936, the British had moved from a position of opposition to trade negotiations to a position of doubt, and by the end of the year they were ready to admit the possibility of an agreement. As early as May, state department economic adviser Herbert Feis had reported after a visit to London that British officials were indifferent to economic arguments for a trade agreement. Interest promptly sharpened, however, at the suggestion that such a pact might throw the weight of the United States into the political balance of Europe.[32] By the beginning of 1937 this attitude seemed firmly established. Robert W. Bingham, American ambassador to Great Britain, reported that Eden had always favored an agreement, that Prime Minister Stanley Baldwin had agreed in principle, and that Runciman now had committed himself to the possibility of negotiations. Bingham explained this movement as part of a British drive "to persuade the United States . . . that the frontier of democracy lies somewhere in the North Sea; that England, the outpost of democracy in Europe, is a small island . . . without raw materials and dependent upon the United States and the British Dominions for war materials and foodstuffs." British officials, Bingham told Roosevelt, were beginning to exhibit "a progressive and almost bewildering friendliness."[33]

Although cooperation with the United States seemed more desirable, it also created some problems for the British. For example, concessions on agricultural imports would surely raise strong opposition from both English and Commonwealth farmers. Despite Bingham's optimistic predictions, Runciman still disliked the American conditions. He was willing to proceed with preliminary discussions, but he continued to insist that every agreement ought to narrow significantly the balance of visible trade between the nations involved. Since the United States sold considerably more to Great Britain than it purchased, Runciman argued that the United States should make the greatest concessions. His position insured a good deal of hard bargaining and could even prevent formal negotiations.[34] Moreover, there was the possibility that the American desire for modification, if not abolition, of imperial preferences might destroy any possible political advantages. H. O. Chalkley, counselor of the British embassy in Washington, emphasized the

[32] Herbert Feis to Hull, May 20, 1936, Folder 90, Hull Papers.
[33] Bingham to Hull, Jan. 4, 1937, Folder 97, *ibid.*, Bingham to Roosevelt, Jan. 5, 1937, *ibid.*
[34] Memorandum by Bingham, Dec. 18, 1936, *Foreign Relations . . . British Commonwealth, 1936*, pp. 702-03. See also Feis to Hull, Oct. 31, 1936, File No. 611.4131/197, General Records of the Department of State.

importance of maintaining firm relations with the Commonwealth nations "at this time when the political situation in Europe is so uncertain." In the event of war, Chalkley explained, the neutrality laws might close American markets to the United Kingdom. British diplomats, therefore, had to assume "that the United Kingdom itself and the Dominions represent the only assured sources of supply in the time of war. . . ." Great Britain could not abolish the preferences and would not modify them without the consent of the Dominions.[35]

Runciman's visit to Washington in January allowed extensive discussion of these problems. The neutrality laws especially caused the British government "considerable anxiety," and, before a meeting with Roosevelt, Runciman publicly asserted that this legislation would be a major subject of conversation. The press promptly concluded that the British would not discuss a trade agreement unless they were assured that continued trade would not be endangered by war. Runciman undoubtedly impressed on Roosevelt the desirability of cash-and-carry from the British point of view.[36]

Presumably, this subject was also a topic of the talks between Runciman and Hull, but the secretary of state preferred to concentrate on the more general ramifications of commercial cooperation. Runciman made a tactical error at the outset when he suggested that the British were waiting to see what Germany would do. Hull promptly seized this opening to restate his conviction that the current United States effort to liberalize trade at least offered a positive alternative to this passive attitude. If the United Kingdom would lead Europe "in proclaiming a program of liberal economic relations, on a basis of world order under law," as the United States had done in the Western Hemisphere at the Buenos Aires Conference, it would get the cooperation of the Scandinavian countries, all the nations from Holland to Switzerland, "some of the Balkans, possibly Poland, and certainly the twenty-two American nations." Nearly forty nations then "would be marching across the Western World proclaiming a broad, concrete, basic program to restore international order and promote and preserve peace and the economic well-being of people everywhere." Hull further predicted that Germany and Italy would undoubtedly recognize the benefits of such a program and join out of self-interest. Once the Axis nations began to cooperate commercially, "the gate would be wide open for a discussion of political problems."[37] Certainly a trade agreement was vital in this respect. It was also desirable "from the point of view both of symbolizing community of basic

[35] Memorandum by Henry C. Hawkins, Dec. 26, 1936, *Foreign Relations . . . British Commonwealth, 1936*, p. 705.
[36] Robert A. Divine, *The Illusion of Neutrality* (Chicago, 1962), 174-75.
[37] Hull, *Memoirs*, I, 524-25.

views and policies as between the two countries and of improving Anglo-American trade relations." Finally, Hull told Runciman that the United States would not demand the abolition of imperial preferences but only modification to insure that they would not be used to divert international trade into unnatural channels.[38]

Still unconvinced, Runciman returned to England. He continued to argue that any trade agreement must be considered solely on its economic merits and that any agreement with the United States should bring visible trade more nearly into balance. At the very least, the United Kingdom could not continue to permit the United States to enjoy a large favorable balance of trade unless the Americans gave some guarantee that in case of war any new sources of supply would not be abruptly closed.[39] The cabinet was apparently more impressed with the merit of the political argument. By April only Runciman remained skeptical about further negotiations, and Baldwin was afraid that if he tried to force him to accede to American terms, Runciman might resign and provoke a cabinet crisis.[40]

This deadlock was broken in May: Congress adopted a permanent neutrality law that incorporated cash-and-carry provisions; Neville Chamberlain replaced Baldwin as prime minister; and, in the cabinet shakeup that followed, Oliver Stanley, who openly favored an agreement with the United States, succeeded Runciman as president of the Board of Trade. In addition, when the London Imperial Conference convened in May, the Dominion governments agreed to support Neville Chamberlain's appeasement policy and consented in principle to modification of some imperial preferences in order to facilitate a trade agreement with the United States.[41] To the surprise of many at the Conference, Neville Chamberlain emerged as the most ardent champion of commercial cooperation with the United States. The South African prime minister was amazed at the abrupt change in the British prime minister's attitude, and he told Bingham that he saw no reason to question Neville Chamberlain's sincerity.[42] Another official re-

[38] Hull to Atherton, Feb. 12, 1937, Department of State, *Foreign Relations of the United States, Diplomatic Papers: The British Commonwealth, Europe, Near East and Africa, 1937* (Washington, 1954), 12; memorandum by Hull, March 5, 1937, Department of State, *Foreign Relations of the United States, Diplomatic Papers: General, 1937*, pp. 644-45.

[39] Pencil notes, Norman H. Davis, undated, Norman H. Davis Papers (Manuscript Division, Library of Congress).

[40] Davis to Hull, April 5, 1937, Folder 98A, Hull Papers; Davis to Roosevelt, April 13, 1937, PSF: State Department, 1933-37, Roosevelt Papers.

[41] Nicholas Mansergh, *Survey of British Commonwealth Affairs: Problems of External Policy 1931-1939* (2 vols., London, 1952), I, 88-92; Arnold J. Toynbee, *Survey of International Affairs: 1937* (2 vols., London, 1938), I, 63.

[42] Bingham to Hull, June 16, 1937, Folder 99A, Hull Papers.

marked that for the son of Joseph Chamberlain to change his mind on imperial preferences was "nothing short of a miracle."[43]

Although the prime ministers of the Dominions generally sympathized with the plan to negotiate with the United States, some of them were unwilling to release Great Britain immediately from its imperial commitments. This hesitation arose as much from political as from economic causes however. Canada's Mackenzie King pointed out that his government could not consent to reduced preferences in the English market without some *quid pro quo* from the United States. An impending election in Australia prompted that government to withhold any specific commitment at London. But by the end of the year the Australian elections were successfully completed, the United States had opened negotiations with Canada to revise the Canadian-American agreement of 1935, and the British government had dealt separately with the other Dominions to secure agreement for specific reductions in the preferences.[44]

During these months, Hull continued to chafe at the delay and repeatedly predicted disaster unless swift action were taken. In August, he told a British embassy official that continued procrastination might result in "more troubles similar to that in the Spanish Mediterranean and between China and Japan with absolutely no remedies for them except to the extent that rearmament might serve as a restraining and restricting factor." It was no longer possible, he asserted, for the British navy "to prevent 70 million hungry Germans from going on the march when they become sufficiently destitute; nor would it be possible for a rearmed Great Britain to prevent an economic collapse and cave-in, beginning in the German area, within another two years."[45] In September, he urged that his commercial program be adopted to support the appeasement policy because an agreement would immediately mobilize "some forty nations behind a definite policy of economic appeasement, which in turn would facilitate political appeasement."[46]

With the preliminary arrangements finally completed, Hull announced on November 18, 1937, that the United States and Great Britain intended to open formal negotiations. Before making this announcement, he and Ambassador Lindsay considered the wisdom of either Neville Chamberlain

---

[43] Memorandum by Hawkins, Aug. 6, 1937, File No. 611.4231/2031, Records of the Department of State.

[44] Henry F. Grady to Prentiss Gilbert, June 15, 1937, File No. 611.4131/312, *ibid.*; Bingham to Hull, June 4, 1937, *Foreign Relations . . . Africa, 1937,* pp. 38-39; Kreider, *Anglo-American Trade Agreement,* 39.

[45] Memorandum by Hull, Aug. 9, 1937, *Foreign Relations . . . Africa, 1937,* p. 65.

[46] Memorandum by William B. Butterworth, Sept. 22, 1937, *ibid.,* 66.

or Hull making a public comment on the political significance of the agreement. After some discussion, they decided to forego such action because, as Hull remarked, "it might be more dignified and effective merely to announce the bare fact without comment."[47]

But more than dignity was involved. Influential economic groups on both sides of the Atlantic were certain to object strenuously to the agreement and to insist that their interests be fully protected. In the United States several eastern industrialists had already protested against the pact, and British producers were equally upset about the prospects of increased competition.[48] There was, moreover, some sensitivity to the political implications involved. In November, Massachusetts Senator Henry Cabot Lodge introduced a Senate resolution to suspend negotiations for a year. "With the mad race in armaments and threats of war looming large," he warned, "it is vital that the United States should not increase her entanglements with other nations."[49] Some English producers complained that their government intended to sacrifice them for political advantage. A crescendo of protest eventually forced Stanley to deny that the trade negotiations had any political motives. However much the statesmen on both sides might believe in the wisdom or rhetoric of liberalized trade, they could neither ride roughshod over domestic sensitivities nor ignore international political realities. For these reasons, the agreement had to be defended on its economic merits, irrespective of any broader considerations.

Although both nations agreed, as Hull later remarked, "to minimize the political nuances of the prospective agreement," political factors increasingly dominated the thinking of both Neville Chamberlain and Hull. The prime minister had never fully accepted Hull's assumption that economic improvement was a prerequisite to political settlement; indeed, he tended to reverse the priorities. Nonetheless, the idea of economic appeasement was never an American monopoly. There were two forms such appeasement could take: collaboration might be used either to conciliate and mollify the dictators, or it could aim at unifying the non-authoritarian states. Until late

[47] Memorandum by Francis B. Sayre, Nov. 16, 1937, *ibid.*, 84. For some contemporary comment on the political significance of the negotiations, see Percy W. Bidwell, "Prospects of a Trade Agreement with England," *Foreign Affairs*, 16 (Oct. 1937), 103-14; "World Politics and World Recovery," *Commercial and Financial Chronicle*, 145 (Dec. 11, 1937), 3715-16.

[48] Herschel V. Johnson to Hull, Dec. 31, 1937, *Foreign Relations . . . General, 1937*, p. 694; Memorandum by Sayre, March 11, 1938, Department of State, *Foreign Relations of the United States, Diplomatic Papers: The British Commonwealth, Europe, Near East and Africa, 1938* (Washington, 1955-1956), 141; New York *Times*, March 16, 1938; Toynbee, *Survey of International Affairs: 1937*, I, 104-07.

[49] Quoted in Toynbee, *Survey of International Affairs: 1937*, I, 106; *Cong. Record*, 75 Cong., 2 Sess., 205 (Nov. 22, 1937).

1936, the British generally had taken the first position. By 1937, it became increasingly apparent that economic appeasement of the dictators was of little value without a political settlement; and it might even contribute to German rearmament.[50] When formal negotiations began, Neville Chamberlain admitted privately that "the reason why I have been prepared . . . to go a long way to get this treaty, is precisely because I reckoned it would help to educate American opinion to act more and more with us, and because I felt sure it would frighten the totalitarians. Coming at this moment, it looks just like an answer to the Berlin-Rome-Tokyo axis."[51] For the British prime minister, the fact of an agreement was as important as the substance.

Hull agreed that cooperation was desirable in itself, but he continued to cling to the proposition that economic appeasement could not work unless accompanied by economic improvement. For this reason, an Anglo-American agreement had to cut deeply enough to increase world trade substantially. Hull admitted that the United States was highly interested in monetary results, and, unless the negotiators significantly reduced trade barriers, the Axis powers might conclude that cooperation was merely a sham. "I have not the slightest doubt," he told the British in July 1938, "that these negotiations are being watched by those countries to see whether we are capable of working out an agreement that is really worth while, and I think that it would greatly harm not only our two countries but also the whole outlook for peace and economic improvement if we, after months of haggling, should turn out a little, narrow, picayunish trade agreement."[52]

Several times, however, dollars and cents considerations brought the negotiations close to collapse. The difficulty lay in the conflict between the American insistence on sweeping agricultural concessions and the British desire to protect their own farmers and to limit modification of the imperial preferences. By October, the president of the Board of Trade advised rejecting the American demands. Lord Halifax, the new foreign secretary, agreed that the proposals contained serious economic disadvantages, but he was uncertain that these offset the political advantages.[53] Ultimately, Neville Chamberlain had to make the final decision.

He was unwilling to reject an agreement at the eleventh hour despite the skepticism of several of his cabinet members. He apparently continued to

---

[50] Toynbee, *Survey of International Affairs: 1937*, I, 63-64; R. W. Seton-Watson, *Britain and the Dictators: A Survey of Post-War British Policy* (New York, 1938), 264-65.
[51] Quoted in Feiling, *Chamberlain*, 308.
[52] Hull to Joseph P. Kennedy, July 25, 1938, *Foreign Relations . . . Africa, 1938*, p. 41.
[53] Kennedy to Hull, Oct. 7, 1938, *ibid.*, 59-60; Kennedy to Hull, Oct. 12, 1938, File No. 611.4131/1812, Records of the Department of State.

believe that the agreement would demonstrate "the possibility of these two great countries working together on a subject which . . . may prove a forerunner of a policy of wider application."⁵⁴ Although Hull was not entirely satisfied with the terms, he made Neville Chamberlain's decision easier by agreeing, in early November 1938, to reduce the American demands and sign the pact. Hull was impressed by reports that there was "a growing doubt in British official circles as to the desirability of this agreement." At the White House on November 17, in a setting carefully designed to emphasize the political significance that the statesmen attached to the agreement, representatives of the two governments signed the treaty.⁵⁵

The hopes that Hull had harbored in the two years before the agreement was signed were obviously beyond fulfillment by November 1938. In the months after Munich, Roosevelt began more surely to take the policymaking initiative. By October, he was smoothing the way for purchase of munitions by the French and British; and in his annual message to Congress in January 1939, he warned that aggression had to be opposed by "methods short of war, but stronger and more effective than mere words. . . ." After this, the administration began to grope steadily toward a program of active support for the European democracies.⁵⁶ Within this program there existed little place for Hull's commercial emphasis, because, as Roosevelt remarked to his cabinet in the summer of 1939, it bordered on the ridiculous to assume that the United States could accomplish any immediate good by selling "a few barrels of apples here and a couple of automobiles there."⁵⁷ But the more important problem, he later told Henry Morgenthau, Jr., was that the "trade treaties are just too goddamned slow. The world is marching too fast."⁵⁸

The President does not seem to have challenged Hull's course of action until late 1938, probably because in part he agreed with the basic assumptions and in part because he had no clear alternative to offer. Roosevelt had been searching since 1936, as he said in his "Quarantine Speech," for

⁵⁴ New York *Times*, July 27, 1938. See also Kennedy to Hull, Oct. 12, 1938, File No. 611.4131/1828, Records of the Department of State.

⁵⁵ Hull to Kennedy, Nov. 3, 1938, *Foreign Relations . . . Africa, 1938*, pp. 70-71; New York *Times*, Nov. 18, 1938. The *Times* duly noted that the signing of the trade pact marked the first time since the exchange of ratifications of the Kellogg-Briand Pact in 1928 that the East Room of the White House had been used for a diplomatic ceremony.

⁵⁶ Langer and Gleason, *Challenge to Isolation*, 45-51. Apparently, Hull reluctantly conceded that economic conditions had become largely irrelevant. At a cabinet meeting in May 1939, he predicted the total economic collapse of Germany within eighteen months. Ickes, *Secret Diary*, II, 636.

⁵⁷ Ickes, *Secret Diary*, II, 568.

⁵⁸ John Morton Blum, *From the Morgenthau Diaries* (3 vols., Boston, 1959), I, 524.

"some positive endeavors to preserve peace," but his efforts had produced few positive results.[59] Perhaps as a consequence of this, he frequently repaired to the Hull formula. In considering the suggestion that he sponsor an international conference in late 1937, for example, Roosevelt gave considerable prominence to economic questions; and, according to Sumner Welles, he intended to discuss the possibility of "an agreement upon a gradual transition period toward an ultimate international economy based upon reduced armaments, a greater common use of world resources, and the improvement and simplification of economic relationships between all peoples."[60] And as late as October 1938, Roosevelt told King "that unless very soon Europe as a whole takes up important changes in two companion directions—reduction of armaments and lowering of trade barriers—a new crisis will come."[61]

Despite this acquiescence, Roosevelt always doubted that the trade agreements program was sufficient by itself. He had long believed that disarmament and economic settlement went hand-in-hand, and he tended at times to stress the former. By 1937, he was concerned that the restoration of international trade along the lines Hull suggested might make rearmament easier for Germany and Italy. Furthermore, a continuing arms race might prevent any economic settlement. In the fall of 1937, for example, he noted that "an economic approach is a pretty weak reed for Europe to lean on." How, he asked, could there by any progress "if England and France say we cannot help Germany and Italy to achieve economic security if they continue to arm and threaten, while simultaneously Germany and Italy say we must continue to arm and threaten because they will not give us economic security."[62] But publicly Roosevelt expressed no such doubts and generally gave the impression, as Anne O'Hare McCormick wrote, that the main line of American foreign policy was adherence to the trade agreements in the hope of removing "the causes of war by reopening the dammed-up channels of international commerce."[63]

Hull's efforts between 1936 and 1938 provided in a period of uncertainty and withdrawal at least the semblance of a positive and consistent foreign policy for the United States. In retrospect, however, this attempt to

---

[59] See Dorothy Borg, "Notes on Roosevelt's 'Quarantine' Speech," *Political Science Quarterly*, LXXII (Sept. 1957), 405-33.

[60] Sumner Welles, *The Time for Decision* (New York, 1944), 65.

[61] Elliott Roosevelt and Joseph P. Lash, eds., *F.D.R.: His Personal Letters 1928-1945* (2 vols., New York, 1950), II, 816.

[62] *Ibid.*, I, 680.

[63] New York *Times*, Aug. 9, 1937. See also S. Shepard Jones and Denys P. Myers, eds., *Documents on American Foreign Relations: January 1938-June 1939* (Boston, 1939-), 276.

achieve economic solutions for highly charged political problems seems to reflect a fundamental misreading of European conditions. He failed to recognize that centralized control of foreign trade might stem from political as well as economic motives. His insistence on gaining the greatest possible opportunities for American producers while holding concessions to a minimum at times made his political arguments seem little more than rationalizations. But he clearly believed that he was doing something of immense importance, and he was convinced that in the long run what was good for the United States was good for the world. Commercial diplomacy, moreover, appeared to offer the only acceptable course of action because of the prevailing domestic isolationism and the inability of the democracies to agree on a common policy.

Despite the immediate failure of his campaign, Hull never wavered in his assumptions about the relationship of freer trade to world peace. Certainly, the Anglo-American agreement had no discernible effect on Hitler's actions, nor in fact did it change substantially British commercial policy. Hull never repudiated his assumptions, and he attributed the failure of his policy to circumstances which prevented it from having genuine results.[64] When Congress considered the third renewal of the trade act in 1940, Hull argued vigorously that the program was essential to lasting peace in the postwar world, and during the war he insisted at every turn that the United States use its diplomatic leverage to insure the establishment of liberalized trade once peace was restored.

[64] Hull later asserted that, if the British agreement had been among the first signed instead of among the last, the course of events in Europe might have been far different. Hull, *Memoirs*, I, 530.

Roosevelt, Willkie, and the TVA

By James D. Bennett

The passage of a public power measure during the "Hundred Days" of the 73rd Congress dictated a clash between Franklin D. Roosevelt and Wendell Willkie, a spokesman for the private power industry, over the New Deal's attempt to create a national power program. There was no immediate indication, however, that the conflict between these two men would become what one of Willkie's biographers has called "a sort of focus for the opposition of American business to certain aspects of the New Deal."[1] When the battle got fully underway, it was one of intense interest for Tennesseans.

In calling for the creation of the Tennessee Valley Authority, in April of 1933, President Roosevelt said: "The continued idleness of a great national investment in the Tennessee Valley leads me to ask the Congress for legislation necessary to enlist the project in the service of the people." The President suggested a corporation "clothed with the power of government but possessed of the flexibility and initiative of a private enterprise . . . charted with the broadest duty of planning for the proper use, conservation, and development of the natural resources of the Tennessee River drainage basin."[2]

This message envisioned a program more comprehensive than any previous measure for the development of the government's installations at Muscle Shoals. The idea itself, of course, was not new. Throughout the 1920's Senator George W. Norris of Nebraska had waged a losing battle to salvage government investments which included Wilson Dam and nitrate plants constructed as wartime measures under the National Defense Act of 1916. At first with the support of various liberals, and later almost alone, Norris had introduced measures, only to see them fail of passage or be vetoed by various Presidents. The Senator took heart upon the election of Roosevelt, for the latter's position on public power was already well known, and Roosevelt and Norris had discussed the develop-

---

[1] Ellsworth Barnard, *Wendell Willkie, Fighter for Freedom* (Marquette, Michigan, 1966), 84.
[2] U.S. Congress, House Document No. 15, 73rd Cong., 1st Sess. (1933); *Congressional Record*, 73rd Congress, 1st Sess., 1423 ff.

ment of the Tennessee River Valley when they visited the area prior to the inauguration.[3]

Senator Norris introduced a new measure on the day following the President's message. Two sections were particularly pertinent to the future controversy over power development: first, the objectives of the proposed law were to be flood control, national defense, promotion of agricultural and industrial development, improvement of navigation, development of hydroelectric power, reforestation and proper use of marginal lands; second, the Authority should have power to produce and sell electrical energy, and in the sale of such energy, preference should be given to publicly owned organizations. To facilitate these sales, the Authority might construct transmission lines.[4]

Hearings on this and similar Muscle Shoals bills were conducted by the House Committee on Military Affairs in April, 1933.[5] Wendell Willkie, President of Commonwealth and Southern Company, the largest electric utilities holding company in the Southeast, was the chief witness to testify before the Committee. In these hearings, Willkie favored the proposed measure: "I want to say, Mr. Chairman, that no one has read or referred with more gratification than we have of this magnificent proposed development of the Tennessee Valley... We view with a great deal of anticipation the proposed program of the President... with reference to that valley. And we do not come here either as opposition or protestant witnesses against that proposition."[6]

Optimists saw in this endorsement the makings of a peaceful settlement of the Muscle Shoals problems. Willkie's affirmative, however, was coupled with a recommendation that the authority's power activities be limited to sale at the switchboard of electric energy produced as a result of a comprehensive water program. Thus, passage of the Tennessee Valley Authority Act brought the ancient conflict between public and private power operation to a climax. There already existed a number of privately-owned utilities

---

[3] Carlton Jackson, *Presidential Vetoes, 1792-1945* (Athens, Georgia, 1967), 196-99; Preston J. Hubbard, *Origins of the TVA, The Muscle Shoals Controversy, 1920-1932* (Nashville, 1961), 314.

[4] U.S. Congress, Senate, S.B. 1272, 73rd Cong., 1st Sess.

[5] The only other measure to receive serious attention was H. R. 5081, introduced by Representative Lister Hill of Alabama.

[6] U.S. Congress, House, *Muscle Shoals*, Hearings before Committee on Military Affairs, 73rd Cong., 1st Sess., (1933), 107.

in the Tennessee Valley, and when the TVA began to seek markets for its power, the one would almost certainly collide with the other.

Nevertheless, following the signing of the TVA Act on May 18, Commonwealth and Southern and the TVA sought to arrange a working agreement. The negotiations, while never pleasant, resulted in a contract on January 4, 1934.[7] This agreement involved a division of territory whereby Willkie agreed to sell to the Authority a transmission line near the Norris Dam site, together with certain properties in Alabama, Georgia, Mississippi, and Tennessee; the Authority agreed that it would not sell power outside the ceded territory to any customer already being supplied by Commonwealth and Southern. This arrangement gave the Authority an area in which it could control distribution as well as generation of power, thus providing an opportunity to regulate power rates to the individual consumer. Such regulation was vital to the "yardstick" theory, a formula for ascertaining the precise cost of producing and distributing a unit of electric energy.

Except for a small area of northern Mississippi, however, the property transfers were never made. The "truce" of January 4, 1934, was short-lived. Willkie was totally unable to accept the idea of the "yardstick," to which he referred as an "unrealistic proposal."[8] Litigation hampering the Authority's power program was begun in September, and by the end of 1934 the public *vs.* private power issue was clearly joined on all fronts between the TVA and C & S. The battle degenerated from legal attacks to the building of "spite lines" by the private power companies in the Valley. Willkie emerged in late 1934 as the acknowledged leader of the private electric power industry.[9]

The change in Willkie's position may be traced through his public statements during the latter months of 1934. Speaking to the American Statistical Association in New York on September 26, he criticized those who had attacked New Deal personalities. "To abuse men in public office will not save us," he stated. His position had changed greatly by November 7, when he spoke to the Rotary Club in Birmingham. Here his references to David Lilienthal, TVA direc-

---

[7] This contract labeled "Exibit 1-4-34a" is in *Minutes,* Board of Directors, Tennessee Valley Authority, January 4, 1934, and is on file in the offices of the Authority.

[8] Joseph Barnes, *Willkie, The Events He was Part of, the Ideas He Fought For* (New York, 1952), 65.

[9] Arthur M. Schlesinger, Jr., *The Coming of the New Deal* (Boston, 1959), 308-10, 324; Barnes, *Willkie,* 74-75, 97.

tor in charge of power, were sarcastic and bitter, his appeal was to Southern loyalties, and his threat was a boycott of the Valley region by New York capital if the TVA experiment continued.[10] On November 20, he issued a public statement critical of President Roosevelt's Tupelo, Mississippi, speech of November 18. In this first city to distribute TVA power, the President applauded the remarkable increase in the sale of electric energy by Tupelo since it had been under the supervision of the Authority. He predicted that this distribution of cheap power would be "copied in every State of the Union before we get through," and he paid a glowing tribute to the "extraordinary sale of electrical appliances under the TVA plan." Willkie commented that the President was "obviously misinformed" in ascribing power sales gains to TVA operation and held that had the private utility companies received as much government subsidy as the Authority, they could charge lower rates than those of TVA. His corporation appreciated very much the President's tribute to the increased sale of electrical appliances because, Willkie said, "90 per cent of the appliances sold under that plan were sold by the operating units of the Commonwealth and Southern Corporation."[11] This episode grew out of the two men's principles: Roosevelt felt that a cooperative effort was necessary for utility regulation and for the best use of resources, and Willkie was sincerely alarmed at what he considered a major threat to private business activity.

The growing split between Roosevelt's TVA and Willkie's operating companies entered a legal phase in September, 1934. Under George Ashwander's leadership, some preferred stockholders of Alabama Power Company, a C & S affiliate, questioned the company's right to carry out the terms of the January 4 contract to which it was a party.[12] The suit, as a result of the sweeping injunctions secured in connection with it, prevented the implementation of the Authority's power policy until February 1936, when the Supreme Court sustained its right to sell Wilson Dam power. Three months after the decision in the *Ashwander* case another suit, known as the Nineteen Companies, or the Eighteen Companies case, attacked the broad development of the Authority.[13] Several

---

[10] Barnes, *Willkie*, 75.
[11] New York *Times*, November 20, 1934.
[12] *Ashwander* v. *Tennessee Valley Authority*, 8 F. Supp. 965 (N. D. Ala. 1935).
[13] Georgia Power Company, one of the plaintiffs, was prohibited by the court from participating in the suit.

C & S operating companies were included among the list of plaintiffs and the suit was pushed vigorously by the Edison Electric Institute, the electrical industry's organization, of which Willkie was a board member. The case was finally decided in favor of the Authority in 1939, but until that time the threat of the Authority's being declared unconstitutional hovered over the New Deal. In the meantime, Roosevelt attempted a conciliation, the utilities resorted to building spite lines, and Willkie and Lilienthal continued negotiations. But at the same time Lilienthal urged municipalities and rural cooperatives to undertake the distribution of TVA power, and Willkie urged anti-TVA propaganda upon his operating companies.

Roosevelt's attempt at conciliation produced a conference between himself and Willkie in December, 1934. Little was a accomplished by this meeting. The President told a press conference that the proceedings were entirely "amicable," and optimistic reports were made about the possibilities of cooperation on the sale of appliances, but reporters found little of significance in the subsequent press conference.[14] Willkie did send what was to become a famous telegram to his wife, a strong opponent of the New Deal: "CHARM EXAGGERATED STOP I DIDN'T TELL HIM WHAT YOU THINK OF HIM."[15]

While the Eighteen Companies case made its slow way through the courts, Tennessee's attention was attracted to the flurry of building by Commonwealth and Southern companies in the Tennessee Valley. This was, of course, the spite lines.

Under the terms of the TVA Act and the January 4, 1934, contract, the Authority could encourage the development of rural cooperatives for distributing power, but could not serve areas where private lines were already located. This stipulation encouraged private companies in the Valley to erect poles, hastily strung with "dead" wires in places where cooperatives were planned. In some instances rival crews strung parallel lines through rural sections previously neglected by the private companies. The existence of these lines thwarted Authority development, and sometimes made it impossible for farmer cooperatives to qualify for construction funds from the Rural Electrification Administration.[16] A variation of the

---

[14] Barnes, *Willkie*, 77-78; New York *Times*, December 15, 1934.

[15] Barnes, *Willkie*, 77.

[16] J. Charles Poe, "The Morgan-Lilienthal Feud," in *Nation*, CXLIII (October 3, 1936), 385; Rebecca R. Wise (ed.), *Rural Electric Fact Book* (Washington, 1965), 16-17; John Dean Minton, The New Deal in Tennessee, 1932-1938, (Ph.D. dissertation, Vanderbilt University 1959), 289-91.

spite line was "cream skimming," by which a private power company hastily installed temporary facilities to serve only those sections of a rural area which were profitable, leaving outlying sections unserved and ineligible for construction loans.[17]

While spite lines and cream skimming continued, private utility representatives toured the countryside undermining the cooperative movement. Farmers were told that they would lose all they owned if they enrolled in the cooperatives; that their electricity would not be as strong as that from private utilities. One farmer was told by a utility representative that "the power will be so weak out there that your light bulb will only glow red."[18]

The completion of Norris Dam in August, 1936, meant that the territorial agreement the Authority and C & S would terminate, under terms of the January 4 contract, in ninety days. Negotiations were intensified to produce a new basis of cooperation before the contract expired. Willkie proposed a new division to follow the watershed of the Valley. Contending that this would prevent the Authority from fulfilling its legal obligation to give preference to public agencies, the TVA Board of Directors abandoned a Southeastern "power pool" with a unified distribution system and uniform rates, with municipalities permitted to choose between the TVA and C & S.[19] This proposal was unacceptable to Willkie, while Roosevelt favored it. With contract negotiations stalemated, and with the fight taking on what the *Nation* called the "colorful aspects of the old railway right-of-way battles," Roosevelt again attempted compromise.[20]

On September 17, 1936, the President invited Willkie, the TVA Directors, and other power and financial leaders to consider a power pool or grid system as a basis of cooperative operation in the Valley.[21] The conferees met on September 30. They included Thomas W. Lamont of J. P. Morgan and Company, Owen D. Young of General Electric, Alexander Sachs of the Lehman Corporation, and top government officials concerned with power. Yet Roosevelt

---

[17] Wise (ed.), *Rural Electric Fact Book*, 17.

[18] *Ibid.*, 21-24; Marquis Childs, *The Farmer Takes a Hand: The Electric Power Revolution in Rural America* (New York, 1952), 77.

[19] Board of Directors, Tennessee Valley Authority, *Minutes*, December 4, 1935; August 4, 1936; February 2, 1937.

[20] *Nation*, CXLIII (October 3, 1936), 383.

[21] New York *Times*, September 20, 1936.

and Willkie dominated the conference."² The President favored consolidating the energy produced by the Authority and operating companies, selling it at a uniform rate. Roosevelt was reluctant to have the Authority construct duplicate transmission lines except as a last resort. Hence, he figuratively offered an olive branch, hoping that Willkie would withdraw his suit against the Authority's constitutionality. There was some reason for Willkie to accept such a negotiated settlement. The TVA had supplied large quantities of cheap power to his operating companies in the Valley, and withdrawal of this wholesale source would force C & S to spend over $10,000,000 to construct steam generating plants to supply its immediate needs."³ The first reports on the conference were mildly optimistic. A statement issued jointly by Willkie and Frank R. McNinch, chairman of the Federal Power Commission, indicated that the conferees recognized that savings might be made under a pooling arrangement, and that further discussion would occur.²⁴

The possibility of successful negotiations, however, was clouded by the impending presidential election. The existing agreement, ironically, would terminate on election day. It was not the most propitious time for discussions, although C & S needed the power and the Authority needed the revenue from its sale. Therefore, as a stopgap measure, a truce agreement extending the January 4, 1934, contract for an additional three months, was signed on October 7, 1936.²⁵

These negotiations ended abruptly on January 25, 1937, because of an injunction secured by C & S on December 22, 1936, from Federal District Judge John Gore at Nashville. Roosevelt wrote Lilienthal that "the securing of an injunction of this broad character, under the circumstances, precluded a joint transmission facility arrangement, and makes it advisable to discontinue these conferences."²⁶

With the failure of the power-pooling proposal, the Authority turned to active marketing operations. These were limited until the

---

²² Barnes, *Willkie*, 106
²³ *Ibid.*, 109.
²⁴ New York *Times*, October 1, 2, 11, 18, 1936. Pertinent material on this conference is in *Franklin D. Roosevelt Papers*, Office File 42, Franklin D. Roosevelt Library, Hyde Park, New York.
²⁵ Board of Directors, Tennessee Valley Authority, *Minutes*, February 2, 1937; U. S. Congress, Senate Doc. 56, *Report of the Joint Committee on the Investigation of the Tennessee Valley Authority*, 76th Cong., 1st Sess., (1939), 204-205.
²⁶ *Investigation of the Tennessee Valley Authority*, 204.

spring of 1938, when the injunction was lifted by the appeals court, and validity of the Public Works Administration's loans to Alabama municipalities was upheld by the Supreme Court.[27] Following these decisions the Authority resumed negotiations with the private power companies and several properties were acquired before the end of the year.

In January, 1939, the Supreme Court adjudicated the power companies' suit against the TVA, by saying that there was no basis for a suit. The sale of the Tennessee Electric Power Company to the Tennessee Valley Authority was announced on February 4. This was the largest single acquisition made by the Authority and, together with a few additional properties, provided the Authority with lines covering substantially all of Tennessee, large portions of Mississippi and Alabama, and smaller areas of operation in other Valley states.[28]

With this acquisition, the Authority's operations could be stabilized. The 1939 Supreme Court decision removed the basis for most controversy over electric power generation and distribution and the Tennessee Valley Authority had acquired its basic shape by 1940. The Authority never fulfilled its dream of regional planning and no satisfactory power "yardstick" was developed but, as George B. Tindall has noted, it "wrought a transformation nonetheless with its dams, lakes, and transmission lines."[29] Already by 1940 it was a powerful tool for achieving "the economic development and social progress of the Valley and its people."[30]

But the Authority's influence was not limited to the Valley; it had stimulated rural development throughout the country. Within twenty years after the inception of the Tennessee Valley Authority, almost unbelievable strides were made in the electrification of the rural United States: over ninety-seven percent of the nation's farms were receiving central station electric service compared with only

---

[27] *Tennessee Electric Power Company* v. *Tennessee Valley Authority*, 90 F. (2nd) 885; *Alabama Power Company* v. *Ickes*, 302 U. S. 464.

[28] *Annual Report of the Tennessee Valley Authority for the Fiscal Year Ended June 30, 1939* (1940), 163-412; Twentieth Century Fund, *Electric Power and Government Policy: A Survey of the Relations Between the Government and the Electric Power Industry* (New York, 1948), 609.

[29] George B. Tindall, *The Emergence of the New South, 1913-1945*, Vol. X of *A History of the South*, edited by Wendell Homes Stephenson and E. Merton Coulter (10 vols; Baton Rouge, 1948-67), 454.

[30] George Fort Milton, "A Consumer's View of the TVA," in *Atlantic Monthly*, CLX (1937), 654.

eleven per cent in 1934. The rapidity with which electrification was achieved in the Valley of the Tennessee was often equaled, and sometimes surpassed, in other areas of the country. In less than one generation the American farmer had moved from dependence upon animal and hand power for energy and upon kerosene lanterns for light to extensive use of cheap electric power to lighten the work of the farm and eliminate the drudgery of the farm wife, to illuminate his home, and to furnish him entertainment and recreation.

As for Roosevelt and Willkie, they had fared well through the vicissitudes of the power controversy. Roosevelt's plan for providing cheap power for the electrification of rural America had been accepted and it had proved itself workable; Willkie had proved himself a shrewd and worthy guardian of the stockholders' capital and a capable spokesman for a giant industry. Moreover, he had become nationally known as a public figure of real ability, a man who, as a political candidate, might prove a match for his recent adversary. They would soon confront each other again on another battlefield.

SOME OBSERVATIONS ON RATIONING

CHARLES F. PHILLIPS

THE first formal rationing program for consumer goods which this country undertook was introduced by "freezing" stocks of new rubber tires and tubes on December 10, 1941. Actual rationing of these commodities began on January 5, 1942. At the time of this writing (July, 1944), therefore, we have had about thirty months of experience with rationing. A few observations based on this experience are indicated in the paragraphs which follow.[2]

RATIONING COSTS

In England commodities have been rationed when either of two conditions have existed: either there was a shortage of a commodity deemed essential to civilians, or it was decided that a deliberate shortage should be created in an effort to release man-power, raw materials, and equipment for use in other industries. These two conditions have been so widespread that rationing has been extended to a large number of goods. Without attempting a complete list of such rationed commodities, it may be pointed out that the list would include candy, chewing gum, clothing (even secondhand clothes above a certain price), domestic pottery, practically all foods (even including shell and dried eggs, jams and marmalades, salt, cereals, and syrups but excluding bread, flour, potatoes, fresh fish, poultry, and most fresh fruits and vegetables), footwear, furniture ("utility" type), gasoline, hollow ware and cutlery, musical instruments, rubber gloves, soap, sport goods and games, and tires.

In contrast with England, at least up to the present time, rationing in this country has been dictated solely by shortage considerations. In view of our resources, the result is a much more limited use of rationing than exists in England. To date, rationing is limited to automobiles, bicycles, firewood, some foods, fuel oil, gasoline, shoes (including rubber footwear), stoves, and tires.

While shortage conditions, plus perhaps the eventual need to create shortages to relieve man-power, raw materials, and so on, may gradually add other goods to the list of rationed commodities, it seems likely that rationing in this country will never proceed so far as it has abroad.[3] This, indeed, is fortunate, since one of the first lessons which experience has taught us about rationing is that it involves a substantial cost. On the one hand, it may be pointed out that during 1942 and 1943 the Office of Price Administration conducted all its rationing programs at a taxpayer's annual cost of approximately 50 cents per capita.[4] On the other hand, however, when this cost is translated into man-power, it rep-

---

[2] Some matters not covered in this article are included in the writer's article, "The Impact of Shortages on Marketing," *Harvard Business Review*, summer, 1943, pp. 432–42.

[3] Victor Abramson and Charles F. Phillips, "The Rationing of Consumer Goods," *Journal of Business*, January, 1942, p. 4.

[4] For the year ending June 30, 1944, the total budget for O.P.A. of $155,000,000 was equal to $1.14 per person.

resents a substantial real cost to the nation. In its Washington office the Rationing Department of the O.P.A. has had a staff of as many as 900 with an additional 3,000 in its field offices. The paid staff (mainly clerks, typists, and secretaries) in its local war price and rationing boards numbers about 34,000, and the majority of these employees are engaged largely with rationing rather than with price matters. In addition, there are about 53,000 individuals who devote a substantial part of their time to rationing work, serving on a volunteer basis as board members of the 5,500 rationing boards—"the neighbors who determine the needs of other neighbors." Finally, there are several times more than the foregoing put together who serve from time to time as volunteers, doing various clerical, filing, and registering jobs. For example, the O.P.A. estimates that more than 500,000 volunteers were engaged in the job of issuing War Ration Book II.

In addition to the dollar and man-power costs of those engaged directly in the program, we must consider the costs borne by manufacturers, wholesalers, and retailers. The introduction of point rationing of food, for example, reduced substantially the sales per employee (that is, increased the sales cost ratio) in the retail grocery store—this reduction being occasioned by additional time spent in educating the public to the point system, in putting up signs showing point values, in securing points from customers, and in counting and banking the points. Since the sale of merchandise from manufacturer to wholesaler and from wholesaler to retailer can be made only against valid ration evidence, additional account-keeping functions also fall to the wholesaler and the manufacturer.

Rationing also leads to the absorption of man-power by printers and bankers, who perform important services in modern rationing programs. The printing order for War Ration Book III was one of the largest printing orders on record, involved the printing of 150,000,000 books. The order required 100 cars of paper and absorbed considerable man-power for many days in 18 printing establishments, and this covers but a very small part of the printing required by our rationing programs to date. While the development of ration banking has relieved some of the burdens formerly placed upon manufacturers, wholesalers, and retailers, it in itself requires additional personnel in banks. Some indication of the ration banking job is the sum of $15,000,000 in the O.P.A. budget to cover this cost for a twelve-month period.

Without trying to exhaust the list of man-power costs of rationing, we must also emphasize the cost to the general public. If a total man-hours figure could be obtained covering the hours spent by the public, it would be colossal, even if it covered no more than the time involved in filling in ration forms, taking or mailing them to local boards, standing in line for rations, returning to boards for various adjustments in rations, and waiting in line in stores because clerks can serve fewer customers per hour. Just take one example. Every 3 months, at least 12,000,000 motorists make application for supplemental gasoline rations. Probably a figure of 15 minutes is the minimum in which all the steps ending in the receipt of the ration can be performed —a total of 3,000,000 man-hours.

Any kind of program which requires such a substantial amount of man-power in time of war as rationing requires must present a very strong reason for its existence. While it is widely agreed that some rationing is necessary, the costs of rationing in terms of man-power are so great that it should not be extended to additional commodities without a careful weighing of the costs as compared with the probable results. The public can undergo a substantial inconvenience in order to avoid the costs of rationing.

RATIONING DURABLE GOODS

Durable goods, particularly, should be examined with the utmost care before they are rationed. In this country some durable goods have been rationed, notably automobiles. Looking back over the experience of the last thirty months, it seems that a good argument can be made for not rationing such goods except when the supply to be rationed is a substantial part of the supply in use. This argument is especially strong when the commodity is one for which there is an active used-commodity market. To continue with automobiles as an example, at the time the rationing of automobiles was announced we had some 27,000,000 automobiles on the roads of this country. While they varied widely in their age and operating condition, most of these automobiles could be made to last several years. The stockpile of new cars available for rationing, however, was but some 500,000, or less than 2 per cent of the number on the road. When the supply to be rationed is such a small part of the total supply, rationing is hardly worth while. It probably would have been sufficient to have stopped the sale of new cars until such time as the military forces and the various units of government had purchased all the vehicles they felt they might need for their essential operations. Following such a period, the balance of the supply might have been made available to the public without rationing.

FORECASTING "SHORT" COMMODITIES

While the decision of what to ration can usually be made several months before the shortage reaches the point at which rationing must be introduced, we have learned from experience that not infrequently the shortage condition leading to rationing develops almost "overnight." The best illustration of this was, of course, our experience with tires. In the matter of a few hours on Sunday, December 7, 1941, the outlook for the supply of crude rubber changed so drastically that it was necessary to institute a rationing system without the benefit of a period for careful and detailed planning.

Other examples of inability to forecast a shortage are not lacking. When the processed-food program was being planned, the statistical position on dried beans was such that it was decided to leave them unrationed. Three days before the program was to be announced, however, the Russian purchasing authorities made arrangements for the purchase of a large quantity of beans—with the end result that last-minute changes had to be made to include dried beans in the rationing program.

In view of such rapid shifts in supply situations, advance planning for rationing is difficult. On the one hand, the rationing authority finds that frequently it has to ration a commodity concerning which little planning has been done, with the result that administrative difficulties are further increased. On the other hand, the authority will find that some commodities which it expects to ration and for which planning and staffing have progressed are subject to supply changes which make rationing unnecessary.

WHEN TO RATION

Once it is decided that a commodity should be rationed, it is desirable that the rationing program, or at least the "freeze" stage, be made effective as soon as possible. Otherwise, the rationing authority faces the possibility that information "leaks" concerning the forthcoming program will result in "runs." Such runs may prove disastrous to any program, but they are especially harmful in those cases in which it takes an appreciable amount of time to rebuild the stock. By way of illustration, in the tire-rationing program it was necessary to freeze stocks immediately after Pearl Harbor. Otherwise the tires on hand might have been purchased by the public in a short time. Had this happened, no tires would have been available for rationing until synthetic rubber plants were built, unless, of course, the government had undertaken a widespread requisitioning program.

There is still another reason for rapid movement into the operating phase of a rationing program as soon as it has been decided upon. The success of a rationing program depends heavily upon the degree of public acceptance which it secures. Public acceptance, in turn, is strengthened if, once a program is decided upon and announced, the rationing authority moves rapidly and with bold strides to get the rationing program into effect.

In this connection, it must be emphasized that, in order to make a rationing program effective in the time available, the rationing authority does not have time to achieve perfection. In other words, the authority must make broad decisions and draft regulations and forms to make these decisions effective. There is no time for all the careful detailed studies which one might like to make. Neither is there time for polishing regulations and forms. As a consequence, of course, the rationing authority will make mistakes. Furthermore, the authority will be criticized for having its forms and regulations either too complicated or too incomplete for use. Such criticism is unavoidable and must be accepted as a part of the price a nation pays for the benefits of moving rapidly into the rationing of limited supplies of essential commodities.

ANNOUNCING THE PROGRAM

Although the general rule should be to begin rationing as soon as possible after the decision to ration has been made, frequently the rationing authority finds that programs should be undertaken at a faster rate than is administratively feasible. In this country the press has quite consistently taken the position that even in such cases "sprung" rationing is preferable as compared with prior announcement of rationing programs. In other words, the press has argued that the authority should say nothing about rationing a commodity until the freeze can be made effective. The argument runs as follows: If the public is not aware that a particular program is to be made effective, hoarding will be decreased substantially. This means, in turn, that citizens will be on more equal terms as far as their supplies are concerned when the rationing program is made effective. Newspapers have taken particular delight in pointing out how announcements of forthcoming programs have caused heavy buying of one commodity after another.

It will be noted that the fundamental assumption of sprung rationing for programs which are delayed for administrative reasons is that it decreases prerationing hoarding. This assumption may be questioned. It is, of course, impossible to keep from the citizens of any country all statistics as to supplies on hand. Furthermore, it is impossible to take from citizens their knowledge of what goods are essential to their welfare. With a basis of knowledge of essential goods plus a knowledge of requirements and stocks on hand, it is not difficult for the general public to know far in advance of a critical situation that such a situation is developing. Moreover, this is the type of information which is widely circulated through our press. In other words, without any prior announcement of rationing on the part of the rationing authority, most of the factors which encourage hoarding are present.

In such a situation, prior announcement of a forthcoming rationing system may actually decrease hoarding, especially after several rationing programs are in effect and have convinced the public that rationing does guarantee that each will receive his share. Put in other words, as a critical situation develops and no announcement on rationing is forthcoming, each citizen, in an effort to protect himself, must practice hoarding. If, however, he is told that on a given date rationing will start and that from that time on his interest will be protected in so far as that is possible, part of his incentive to hoard is destroyed.

His incentive to hoard may be further minimized if the government makes it clear that all excess stocks in his hands on the day rationing begins must be declared and will be deducted from his ration. We have some specific evidence on this point for sugar, gathered by the consumer-panel technique and reported by Franklin R. Cawl.[5] He writes as follows:

After Pearl Harbor (January, 1942) approximately 24% of the families had purchased extra sugar to an extent which made the average of January, 1942, 14% above that of January, 1941. We were interested and pulled out the specific records to see how it was recorded.

[5] "Recent Changes in Farm Economic Levels," *Journal of Marketing*, April, 1943, pp. 362-63.

Frankly, these housewives made no bones at all about recording as high as six purchases from six stores in one day and telling us the amount of each purchase which usually ran between ten and twenty-five pounds. About the end of January a statement from Washington made it clear that when rationing actually came it would be necessary to declare all the sugar on hand. This statement was evidently taken very seriously for there was practically a complete stoppage by those women who did the multiple purchasing in January, a lessening of the number of purchases on the part of others, and a drop of over nine per cent in the per capita poundage in February over the February, 1941, figures. This illustration shows that while there was not a complete stoppage of sugar buying in February to offset the January overbuying, people will in general follow the directions given to them, especially if it is made clear to them that they may lose if the directions are not followed.

It may even be pointed out that on occasion the use of sprung rationing may actually encourage hoarding of some goods other than those covered by the immediate announcement. We also have evidence on this point. The sudden announcement of shoe rationing led to widespread belief that the day of clothing rationing was "just around the corner." The immediate result was a widespread buying movement which took thousands of dollars' worth of clothing off the retailers' shelves.

In view of the foregoing it cannot be glibly stated that it is always better to introduce a rationing program without giving the nation prior warning. As a matter of fact, which technique to use is one of those difficult decisions which must be made by the rationing authority, based on an analysis of the entire situation which confronts it at the time the decision must be made. If it is found that widespread runs are developing because the government is not moving fast enough to instal rationing programs to protect consumers, such runs may be minimized by the announcement that specific rationing programs will be introduced on specific future dates. On the other hand, if the authority finds that the people are failing to declare stocks on hand and that public temper is such that the sudden announcement of one program does not lead to a buying movement for some other commodity, then sprung rationing is acceptable.

IMMEDIATE IMPACT OF RATIONING PROGRAM ON DEMAND

In the preceding section it has been pointed out that the announcement of one rationing program—for example, shoes—may have the effect of substantially increasing the demand for some unrationed but related good—for example, clothing. The rationing of a commodity also has an immediate effect on the demand for the rationed product. Experience has shown that when a particular commodity is rationed, especially if it is rationed by a certificate scheme under which the individual must appear before a board made up of his neighbors, demand may decline rapidly.

Both the automobile and the bicycle programs are good illustrations of this decline in demand. During 1941 about 4,000,000 automobiles were sold in this country. Although under rationing the potential number of purchases was reduced substantially through a limited eligibility list, the list was left broad enough, it was thought, to move all the automobiles on hand in a few months. Actually, the rationing authority gave quotas to its local boards which during 1942 totaled 334,569 cars. However, purchase certificates for but 219,652 cars were received by people from the local boards. While this discrepancy was, in part, a result of a more strict interpretation of the regulations on the boards' part than was intended, in major part it was the result of the failure of eligible people to approach the board and ask for certificates. The mere fact that the commodity was rationed made the purchase of a car sound complicated. Many people who were actually eligible under the regulations never looked at the regulations to discover this fact. Finally, as indicated above, thousands of citizens hesitated to go before a group of their neighbors and attempt to prove that they actually needed a scarce commodity. In these respects our experience with bicycle rationing has been similar to our automobile-rationing experiences.

LOCAL BOARDS

From the very beginning of rationing in this country the rationing authority has relied for the major part of its work at the community level upon a local-board organization composed mainly of volunteers. In other words, instead of establishing a system of government employees for rationing work at the local level, it was decided that neighbors should ration neighbors.

Several considerations were involved in making this decision in favor of volunteer local boards. In the first place, the first rationing program—the tire program—was delegated to the O.P.A. at a time when that organization had no field staff. In order to put the program into effect with a minimum amount of time, a large local organization was necessary. It was felt that such an organization could be recruited most quickly by asking for volunteer aid. Second, a local-board organization based upon volunteers results in a substantial reduction in the cost of rationing. In view of O.P.A.'s persistent budget problem, this was no small consideration. Third, and most important, however, was the fact that by relying on volunteers, the caliber of the local organization could be held at a much higher level than if paid employees were involved. In other words, by making a request in each local community for volunteers it was possible to secure the leading citizens of each community to serve as board members. In practically all cases these individuals had other full-time occupations so that they would not have been interested in a paid job. However, for patriotic reasons they were quite willing to devote a substantial part (in many cases, all) of their time to the necessary job of rationing. The result was that the local war price and rationing board obtained a personnel which was far superior to what could have been built on a pay basis. Moreover, the volunteers frequently were community leaders, and their acceptance of rationing, as shown by their willingness to devote time to a laborious task, was an important factor in securing general community acceptance of the rationing program.

In spite of the obvious advantages of having neighbors ration neighbors, this system of local-board organization gave rise to certain problems. Perhaps the most important of these problems was that frequently the members of the board felt a greater responsibility to their immediate neighbors than they did to the national program. In practice, this meant that some boards found it very difficult to resist local pressure, especially in those programs in which the board exercised a tremendous amount of individual judgment. The gasoline program, since it involved the issuance of supplemental rations, is a good case in point. Many boards disagreed from time to time with certain of the decisions made by the national office; and, while in such cases the majority of the boards would still follow national-office policy, there were far too many exceptions to this general rule. Moreover, knowledge that local boards could not resist certain local pressure was one of the factors which the national office had to take into account in determining policy, with the end result that the "realistic" decision which was required provided, in some cases, a more lax and, in other cases, a more arbitrary program for rationing than was called for by the demand and supply situation.

The foregoing should not be construed as meaning that the local organization of the rationing authority should not possess an appreciable amount of discretion. The granting of discretion to a local board in a country as large as this with its wide local differences is absolutely necessary. However, this granting of discretion must be accompanied by the laying-down of general policy and with adequate field personnel to see that these policies are incorporated in the decisions made by the local boards. O.P.A. has never possessed such adequate field personnel.

Another serious disadvantage of a volunteer local-board system lies in the difficulty of securing adequate reports of what happens at the local-board level. Still using the gasoline program as an example, the rationing authority has never been able to get every board to make monthly reports of coupon issuance. Consequently, total rationed demand had to be estimated, and, of course, with a margin of error. In spite of this, however, it must be admitted that the estimates based on available reports did prove to be quite accurate.

In the early days of rationing no appreciable division of labor was possible within the local board. Each board consisted of from three to five members, aided by a number of volunteer helpers who performed clerical functions. As additional commodities were rationed, it was found necessary to increase the number of board members and to divide them into panels. Thus the original tire rationing board was divided into a tire panel and an automobile panel, soon to be followed by a food panel, a gasoline panel, a fuel-oil panel, and so on. Those boards which refused to be divided up into panels soon began to complain that they were being asked to carry out too many programs, that they had to read too much material, and that the regulations were too complicated. The board which operated in panels, however, so that each panel dealt with a specific commodity, found that it was able to handle its programs and still read the material sent to it. Moreover, because each panel had but one set of regulations to read, it did not find them too complicated.

Experience with a volunteer local-board organization emphasizes strongly the need to provide such a board with an adequate paid clerical staff. Experience indicates that it is simply too difficult to secure a sufficient number of well-trained volunteers to do the clerical work, and, unfortunately, rationing cannot be operated without an appreciable amount of clerical work in the local board. Too often the rationing authority in this country has been subjected to justified criticism because citizens have been asked to stand in line waiting their turn at the ration counter. Had adequate help been provided, work at the local-board level would have been expedited appreciably. Unfortunately, the rationing authority was never provided with adequate funds for the rationing job which it was called upon to perform. It would have taken many millions of dollars in excess of the amount spent by the rationing authority in 1942 and 1943 had it provided its local boards with adequate space, adequate equipment, and adequate personnel. This is an important point, since, as far as the majority of citizens are concerned, the local board is *the* rationing organization. Unless the local board is adequately staffed and equipped to do its job well, the rationing authority finds it impossible to build community-wide acceptance of its program.

Reliance upon a volunteer local-board organization proves especially unsatisfactory when the rationing authority experiences difficulty in getting complete explanations of its policy down to the local boards. In other words, while a paid organization may be expected to operate a program even if it does not know all the "whys," volunteers simply will not do this. They must be told the reasons and be convinced, or they will, in effect, make up their own programs. Without question, in this country the rationing authority has, to date, failed in keeping its local boards informed as to the reasons for its various decisions.

Likewise the rationing authority must see that material reaches its local boards in time. Regulations must be sent to the board appreciably before a program goes into effect. The same is true with the forms involved in the program. Otherwise the board is asked to institute and to answer questions on a program with which it is unfamiliar. The net result is that the boards become exasperated, as does the public in view of the poor service which it obtains from the boards. The long-run result, of course, is that the boards soon become the rationing authority's worst enemy, since there is nothing for them to do but to side with their local neighbors against the national office of the rationing authority. When such a point is reached, one of the main advantages of a nation-wide system of volunteer boards has disappeared.

The failure of the O.P.A. to provide its local boards with adequate information as to its policy and with regulations and forms sufficiently in advance of the introduction of a program is easily explained but cannot be justified. There were interminable delays in the clearing of forms not only within the organization but in securing the necessary clearances between O.P.A., other operating organizations, and the Bureau of the Budget. It was found impossible to avoid last-minute policy changes, especially in the early days of rationing when the staff of the rationing authority had had little experience with rationing. It was literally feeling its way. Such last-minute changes frequently resulted in form changes and in changes in the regulations themselves. The legal staff frequently found it necessary to debate at length as to the correct way of stating a certain policy in the regulations. As the printing requirements of rationing became greater, printing delays became unavoidable. Likewise, the rationing authority seemed to experience great difficulty in organizing a method of distributing material from Washington down through its regional and district offices to the local boards.

Many of the delays mentioned above may be explained in terms of the internal organization of the rationing authority in this country. The authority had a parallel legal and administrative organization, so that at each level of operation it was necessary for the executive and the legal counsel to agree on the matter, or each could effectively stymie the other. Likewise, the responsibility for printing was never centered in the rationing authority itself but existed in another department of the O.P.A. Finally, the rationing authority operated through two (at one time, three) layers of field offices —regional and district—and many delays were involved in the transmission of information through this overlapping field organization.

PUBLIC RELATIONS

Probably the most important step that the rationing authority may take as far as public relations are concerned is to establish a sound administration for its rationing programs. In practice this means several things, of which only a few will be mentioned. Local boards must be well housed, well equipped, well staffed, and fully informed as to the reasons back of each part of the program. It means that, so far as is possible, regulations must be in simple language rather than legalistic in character, although this need may be somewhat offset by developing board commodity panels.[6] In addition, a well-trained field staff, with adequate travel funds, must be developed to give the boards the supervision and aid which they need. Decisions must be made and adhered to, so that the public is not left in confusion as to what is expected of it.

Another important aspect of public relations is to keep both the trades involved and the general public fully informed. In the past the rationing au-

[6] In this country the first set of rationing regulations related to tires. While they were complicated and legalistic in nature, they did not weigh too heavily on the hands of the local boards since the boards had but one set of regulations with which to deal. However, by the time the board had also received a comparable set of regulations on gasoline, fuel oil, automobiles, and so on, it became evident that the regulations would have to be simplified if they were to be administered by part-time local-board volunteers. Consequently, a new trend was established toward simplifying orders, of which the Automobile Rationing Order No. 2B is typical.

thority has been forced to rely, in the main, upon press releases to give information to the public. Press releases are quite unsatisfactory for this purpose. Not only does the rationing authority have to write its releases in such a way that they will be acceptable to the press, thereby omitting much valuable information, but the authority is never sure of how his releases will actually appear in the press or even if they will appear at all. To some degree direct mailing to the trade has been possible, but printing delays have made this a slow and difficult process. Probably the most satisfactory medium would be paid newspaper advertising such as has been used in England and Canada. In this way the exact situation with the reasons involved might be put before the entire public in a minimum of time.

Although it is impossible to ration without causing the general public certain inconveniences, successful rationing depends upon minimizing these inconveniences. In the early days of rationing the rationing authority frequently found it necessary to cause people to stand in line in an effort to secure their rations. Gradually, however, steps have been taken to change this situation. For example, local war price and rationing boards are currently in a position to mail all gasoline rations to individuals rather than asking them to call at local boards for them. Likewise, in contrast to the way in which War Ration Books I and II were issued, it has been found possible to distribute War Ration Book III by mail. The steady minimizing of the inconveniences to which the public was originally subjected will be an important step in improving the public's acceptance of rationing.

Another major step in securing public acceptance of rationing involves assuring the entire public that a diligent effort is being made to treat people in the same circumstances in the same way. One important element in such a program must consist of continued work on the part of an adequate field staff with the local boards to see that regulations are being interpreted as uniformly as is possible. Another step in the same direction is for the rationing authority, whenever exceptions are made, to see that the exceptions, together with full explanations, are printed and made public.

No rationing program can secure public acceptance unless it is adequately enforced. However, there is a wide difference of opinions as to how the rationing authority should attempt to enforce its program. Experience indicates—and this experience is world-wide—that rationing cannot be in effect without having a black market. There are those, however, who insist that "by telling the people what is expected of them" this black market will be minimized, even in the absence of other enforcement measures. On the other hand, there are those who insist that the rationing authority must take on the function of a policeman.

Experience indicates that the vast majority of people will give fair compliance with a rationing program even in the absence of all-out enforcement activity. In other words, an honor system will have some influence, especially if a good job has been done in convincing the public and the trade of the necessity for the program. In practice, however, an honor system tends to break down if flagrant violators are not caught and penalized. When Mr. Jones finds that his neighbor is able to purchase any rationed commodity in the black market and that he is not penalized for this, Mr. Jones soon decides that his neighbor is merely smarter than he is. Consequently,

Mr. Jones enters the black market. Mr. Brown, Mr. Jones's neighbor, is the next victim of the same set of circumstances, with the result that the black market continues to grow.

The Canadian rationing authority, which for some time rationed some commodities under the honor system without coupons, has this to say about that system.

The advantage of this system is that it avoids the drain on manpower involved in coupon rationing, but it has the drawback of causing bitterness on the part of those who believe that their neighbours are less law-abiding than themselves. It can work satisfactorily only when a fairly intense publicity campaign can be focussed upon it. As soon as several commodities come under it, or when "honour" rationing is combined with coupon rationing, the public gets both confused and apathetic. Moreover, it can be effective only so long as the ration is not greatly below the normal consumption. Beyond these limits "honour" rationing tends to break down.[7]

We have also had some experience with the honor system in this country. Early in March, 1943, the rationing authority announced that the so-called ban on nonessential driving on the East Coast would remain in effect but that, instead of being actively enforced against all violators, only "criminals" would be punished for violations. Prior to this announcement, traffic figures compiled by automatic recording machines had fallen to a low level. Immediately following this announcement, however, traffic began to increase. This experience seems to indicate that honor systems are not too effective, or at least are not so effective as systems under which the violator runs the risk of penalties.

As a matter of fact, unless the rationing authority penalizes members of the trade who violate rationing orders, it cannot expect even honest dealers to have any respect for its programs, since what rationing without adequate enforcement does is to transfer business from the honest to the dishonest dealer. In other words, if for no other reason than to protect the honest dealer, the dishonest dealer must be penalized. No rationing authority can expect to have the respect of any trade, which is necessary if satisfactory general public acceptance is to be secured, unless it makes an effort to go after violators.

Of course, the difficulty of securing compliance varies markedly from commodity to commodity. In the case of those commodities in which the extent of the cut imposed by the rationing system is not great, compliance can be secured with a minimum of punishment for violations. On the other hand, where the cut imposed is deep, the number of people who seek to secure additional supplies in the black market tends to increase. By the same token, however, it is in just such programs that the rationing authority cannot afford to have any significant black-market operations.

CONCLUSIONS

In review, it seems that our experience with rationing over the last thirty months justifies the following conclusions:

1. While some wartime rationing is essential even in this country, the manpower costs of rationing are so great that any commodity should be well scrutinized before it is placed under rationing.

2. Durable goods, particularly those with active secondhand markets and with supplies available for rationing equal to a small part of the total supply in use, are not good candidates for rationing.

---

[7] *Report of the Wartime Prices and Trade Board, September 3, 1939, to March 31, 1943*, p. 59.

3. In wartime it is frequently difficult to forecast far in advance which commodities will need to be rationed.

4. The desirability of rationing a commodity as soon as possible after a decision to ration has been made frequently makes it impossible for the rationing authority to take the time required to produce a "perfect" rationing program.

5. Neither sprung nor preannounced rationing programs are always best; which to use depends upon the circumstances facing the rationing authority.

6. The immediate impact of a rationing program on demand is frequently to reduce demand below the level expected by the rationing authority and even below that called for by the regulations.

7. Much remains to be done both to and for the local rationing boards if they are to be integrated into a well-working local rationing organization. The rationing authority must secure a more adequate paid field personnel and organize it more effectively to supervise local boards; it is doubtful whether the local boards are effectively organized to handle rationing programs which require (*a*) supplemental rations and (*b*) a significant reduction in consumption.

8. Public relations of the rationing authority may be greatly improved by such measures as (*a*) improving local-board operation, (*b*) developing better communication channels to the trades and to the public, and (*c*) increasing the effectiveness of its enforcement activities.

CHARTING A COURSE BETWEEN INFLATION AND DEPRESSION

Secretary of the Treasury Fred Vinson and the Truman Administration's Tax Bill

By BARTON J. BERNSTEIN

DURING PRESIDENT HARRY S. TRUMAN'S early months in the White House, he received the benefits of the bi-partisan, wartime foreign policy and enjoyed the customary honeymoon with Congress on domestic policy. But as World War II drew to to an end, national agreement on domestic issues crumbled under the assault of rival and anxious interest groups seeking consolidation and advancement of their wartime gains.[1]

---

[1] On the early difficulties of the administration, see Barton J. Bernstein: "The Truman Administration and the Politics of Inflation" (Harvard, unpublished Ph.D. thesis, 1963); "The Postwar Famine and Price Control, 1946," *Agricultural History*, XXXVIII (1964), 235-40; "The Presidency Under Truman," *Yale Political*, IV (Fall, 1964), 8 ff.; "The Removal of War Production Board Controls on Business, 1944-1946," *Business History Review*, XXXIX (Summer, 1965), 243-60; "The Truman Administration and Its Reconversion Wage Policy," *Labor History*, VI (Fall, 1965), 214-31; "The Truman Administration and the Steel Strike of 1946," *Journal of American History*, LII (1966), 791-803; "Clash of Interests: The Postwar Battle Between the Office of Price Administration and the Department of Agriculture," *Agricultural History*, XL (Jan., 1967), 45-57; "Reluctance and Resistance: Wilson Wyatt and Veteran's Housing in the Truman Administration," *Register of the Kentucky Historical Society*, LXV (1967), 47-66; "The Ambiguous Legacy: The Truman Administration and Civil Rights" (paper given at the A.H.A., 1966). Also see: Allen J. Matusow, "Food and Farm Policies During the First Truman Administration, 1945-1948" (Harvard, unpublished Ph.D. thesis, 1963); Richard O. Davies, *Housing Reform During the Truman Administration* (Columbia, Mo., 1966), ix-59; Tris Coffin, *Missouri Compromise* (Boston, 1947); Mary Hinchey, "The Frustration of the New Deal Revival, 1944-1946" (Univ. of Missouri unpublished Ph.D., 1965); Samuel Lubell, *The Future of American Politics* (N.Y., 1951); Richard Kirkendall, ed., *The Truman Period as a Research Field* (Columbia, Mo., 1967); Eric Goldman, *The Crucial Decade: America, 1945-1955* (N.Y., 1956); William Berman, "The Politics of Civil Rights in the Truman Administration" (Ohio State unpublished Ph.D., 1963).

THE REGISTER OF THE KENTUCKY HISTORICAL SOCIETY, 1968, vol. 66, pp. 53-64.

Despite their differences, however, most groups did expect the federal government to act to avoid repetition of the disastrous inflation and the painful depression after World War I. In 1945, few suggested that the government should simply withdraw from the economy and allow demobilized soldiers to join the expected millions of unemployed. Though most agreed upon the necessity for some federal activity, they disagreed on its nature, and that issue sparked a debate by major interest groups and within the councils of government.[2]

Many Americans feared that, without federal activity, the economy would sink back into depression, from which the war, not the New Deal, had rescued the nation. During the war, economists and leading members of the administration looked ahead uneasily to the months following V-J Day. Based partly on the experience of the period after World War I, they predicted jobless millions and another depression, unless the government intervened to create substantial aggregate demand. So pessimistic were most experts that they differed only on the severity of the expected economic decline. There are few better measures of the prevailing fear than one business journal's self-declared "optimism": the editors rejected estimates of eight million unemployed by Christmas and offered instead, as evidence of their faith in the economy, predictions of only about three million by January and no more than five or six million by the spring.[3]

[2]For the best analysis of economic developments after World War I, see Paul Samuelson and Everett Hagen, *After the War—1918-1920* (Washington: National Resources Planning Board, 1945). Also see: John Hicks, *Rehearsal for Disaster* (Gainesville, Fla., 1961); and George Soule, *Prosperity Decade: From War to Depression, 1917-1929* (N.Y., 1947), 81-113; vol. VIII in *The Economic History of the United States*.

[3]Most of the estimates are mentioned, and many are discussed, by Everett Hagen, "The Reconversion Period: Reflections of a Forecaster," *Rev. Eco. Stat.* (May, 1947), 69-73. Also see: Hagen, assisted by Nora Kirkpatrick, "Forecasting Gross National Product and Employment during the Transition Period: An Example of the 'Nation's Budget' Method," *Studies In Income and Wealth*, XI (1949), 275-351; and W. S. Woytinsky, "What Was Wrong in Forecasts of Postwar Depression?" *J. Pol. Econ.* (April, 1947), 142-151. Woytinsky, who was not a Keynesian, had been more accurate in his forecasts. *The Magazine of Wall Street*, Sept. 1, 1945, 575, offered the more "sanguine" estimate. However, the liberal Committee for Economic Development, a business group, as well as some other business organizations, had forecast substantial business investment and strong demand, and they did not fear unemployment. *National City Bank Letter*, Sept., 1945; *Commercial and Financial Chronicle*, June 7, Aug. 16, 1945. For a wartime discussion of fiscal policy, see *American Economic Review*, (May, 1945), 329-54. Also see A. E. Holmans, *United States Fiscal Policy, 1945-1959* (Oxford, 1961), 14-44. On the generally neglected issue of expanded foreign trade as an anti-depression

To avoid widespread unemployment, most concerned economists urged government activity for creating jobs. By following a Keynesian fiscal policy and by developing related welfare programs, they hoped to avert or to minimize the imminent depression: they recommended reduced taxes, deficit spending, public works, and expanded unemployment benefits. While most businessmen and many Congressmen could endorse the idea of tax reductions, they opposed increased federal spending and enlarged public works. Indeed, few seemed to understand modern economics, and most were still reluctant to accept the *principle* of compensatory fiscal policy. The debate on the Employment Act revealed the level and quality of economic thought. Introduced as "The Full Employment Act" (which directed the federal government to provide the necessary investment to guarantee continuing full employment), the measure fell under attack by the self-proclaimed defenders of fiscal orthodoxy, who were usually enemies of an expanded federal government. Amended beyond recognition, the final law was merely a pious hope that the nation could achieve substantial employment. (Senator Alben Barkley of Kentucky, the Majority Leader, remarked during the legislative wrangling that the bill "promised anyone needing a job the right to go out and look for one.")[4]

policy, see: Committee on Banking and Currency, *Hearings, Export-Import Bank Act of 1945*, 79 Cong., 1 Sess.; Carl Parrini, "The Export-Import Bank and United States Government Foreign Investment" (University of Wisconsin unpub. M.A. thesis, 1957); Lloyd Gardner, *Economic Aspects of New Deal Diplomacy* (Madison, Wis., 1964), 261-329; William Appleman Williams, *The Tragedy of American Diplomacy* (rev. ed., N.Y., 1962), 204-309.

[4] Stephen Bailey, *Congress Makes a Law: The Story Behind the Employment Act of 1946* (N.Y., 1950), 9-30, 243-48. The Barkley quote is from Herbert Stein, "Twenty Years of the Employment Act" (unpub. ms., 1965), 2. Aside from authorizing three institutions—the Council of Economic Advisers, an annual economic report by the President, and the Joint Congressional Committee on the Economic Report—the act accomplished little. It did institutionalize a new source of economic advice to the President, but it did not mark a new point in the murky national dialogue on fiscal policy, nor did it even foreclose the continued use of outmoded and ill-conceived arguments. (When the Congress passed the act in early 1946, inflation and near full employment had refuted the earlier pessimistic forecasts and provided specific evidence to support doubts about the predictive skills of economists and the wisdom of their future recommendations.) Stein, "Twenty Years of the Employment Act," 6, accords limited significance to the measure, and then chiefly because it confirmed "the existing state of affairs," what he calls the postwar consensus, and then "permitted us to move on." In contrast, Edward S. Flash, *Economic Advice and Presidential Leadership: The Council of Economic Advisers* (N.Y., 1965), 9, finds that the act made "explicit the federal government's responsibility for using its resources to help avoid recession . . . and transformed the issue of conscious governmental intervention in the nation's economic welfare from 'whether' to 'how'."

The assault upon Keynesian theory and the expressed fears of big government would continue as part of the postwar dialogue, but major interest groups and government leaders *did agree* on some fiscal policies. It would be a mistake to assume, as have some historians, that the so-called conservatives of the postwar period accorded to fiscal orthodoxy a central place in their economic and political faith. Despite the Treasury's prediction of a deficit of about 30.5 billion (plus the cost of a tax reduction), even non-Keynesians did not recommend balancing the budget in fiscal 1946. Indeed, Keynesians and non-Keynesians, alike, agreed on the need for a tax reduction, but their reasoning differed greatly. Primarily concerned with aggregate demand, the Keynesians expected that the resulting deficit would contribute to higher spending, substantial investment and greater national income. In contrast, many non- or anti-Keynesians simply felt that the removal of some taxes would inspire business, boost its confidence and spark investment. To them, a tax cut was important as a stimulant to incentive, not as an addition to demand. Most believed that spending by government could not be as effective, nor as desirable morally, as spending by business. Indeed, many believed that large federal expenditures, and the resulting big government, threatened freedom, endangered the economy and impaired private initiative.[5]

II

Despite the fears of depression, neither Congress nor the administration, in preparing the new tax bill, emphasized consumer purchasing power. Truman's new Secretary of the Treasury, Fred M. Vinson, who was influential in formulating the administration's tax policy, was not concerned primarily with boosting consumer demand. A former seven-term Congressman from Kentucky, Vinson had won a reputation as a tax expert during his service on the Ways and Means Committee. After being elevated to a federal judgeship by President Franklin D. Roosevelt, he had left the Court of Appeals in 1943 at the President's request to take the helm of the Office of Economic Stabilization. An effective administrator who capably resolved conflicts between agencies, he was promoted by Roosevelt in 1945 to the directorship of the top domestic agency, the Office of War Mobilization and Reconversion. When Truman

---

[5]For the attitudes of businessmen and their organizations, see Senate Committee on Finance, *Hearings, Revenue Act of 1945*, 79 Cong., 1 Sess., *passim*. (Hereafter cited as *Revenue Hearings*.)

moved to rid the cabinet of strangers he had inherited, he fired Roosevelt's Secretary of the Treasury, Henry Morgenthau, Jr., in the summer of 1945 and advanced Vinson to the post. As an able leader and a man of great charm, Vinson was also the most skilled politician in the new cabinet (aside perhaps from James Byrnes, Secretary of State), and Truman wanted his counsel. Mixing conviviality with integrity, the new Secretary of the Treasury impressed political opponents and won allies. He commanded the respect of Congress and the affection and admiration of his friend, the President.[6]

Following the advice of Vinson and meeting popular expectations, the President in early September recommended what he called "limited" tax reduction. Probably because advisers were still laboring on specific measures, Truman was vague and did not even hint at his future proposals. However, he did express a general position: though a tax cut should remove barriers to speedy reconversion and to economic expansion, the new program must not neglect the estimated deficit of $30 billion, the federal debt of $263 billion and the government's responsibility to 85 million holders of federal bonds. Fearful of debt and worried about the immorality and danger of government deficits, Truman was eager to pay off the government's obligations.[7]

Even before Truman had endorsed the first tax cut in fifteen years, Congressmen had been urging a reduction. As early as June, Representative A. Willis Robertson, a Virginia Democrat, had proposed elimination of the wartime-imposed, corporate excess-profits levy following Japan's defeat. In late August, nearly two weeks after V-J Day, Representative Harold Knutson of Minnesota, the Senior Republican on the Ways and Means Committee, and his second-ranking Republican colleague, Representative Daniel Reed

---

[6]Vinson, "Taxation" (undated but in late August or early September), box 2, Samuel Rosenman Papers, Truman Library. On Vinson's earlier career, see James Bolner, "Fred M. Vinson: The Years of Relative Obscurity," *Register of the Kentucky Historical Society*, LXIII (1965), 3-16. Vinson's papers, still in his son's possession and unavailable to researchers, will soon be deposited in the library of the University of Kentucky. For Senator Robert Taft's acceptance of budget-balancing over a period longer than a year, see *Cong. Rec.* (Sept. 28, 1945), 9133.

[7]Vinson, "Taxation"; Truman's address of September 6, 1945, is reprinted in *Public Papers of the Presidents of the United States: Harry S. Truman, 1945* (Washington, 1961), 294-95. Truman also expressed the need for basic reform of the tax structure. Calling the message "dull and flat," Samuel Lubell advised Bernard Baruch that the tax recommendations were so vague that they were meaningless. Letter of Sept. 6, 1945, Baruch Papers, Princeton University.

of New York, backed by the House Minority Leader, Joseph Martin of Massachusetts, had endorsed a twenty per cent slash in individual and corporate tax liabilities. Senator Walter George of Georgia, the Democratic chairman of the Finance Committee, suggested that the administration would reduce taxes by about fifteen per cent, eliminating the excess profits tax and cutting levies on personal income.[8]

For nearly a month after Truman's speech, Congressmen continued to speculate publicly about the administration's tax program. Not until October did Vinson unveil the government's plan, and then to a closed session of the Ways and Means Committee. Pointing to lurking economic perils, he warned that taxation must not impede expansion of business, creation of jobs or maintenance of mass markets and large purchasing power. He focused on business investment, urged repeal (effective January 1, 1946) of the excess-profits tax, and recommended allowing losses incurred in 1946 to be carried back against excess profits assessed on corporations during the war years. To benefit private citizens, he proposed elimination of the three per cent normal personal income tax, thus exempting about twelve million tax payers; and reduction of the excise tax (by July 1) to the levels of June of 1942. The Treasury estimated the total cost of this tax program at $5.187 billion, including $2.55 billion in excess profits, $2.085 billion in personal income levies, and $547 million in excise taxes.[9]

In discussing the proposed tax cut, Vinson was more optimistic than many economists about the economy. He found a mixture of inflationary and deflationary currents: reduced federal expenditures, unemployment and downgrading were "deflationary factors" producing lower incomes; however, these conditions did not result from "any fundamental deflationary situation— . . . from a deficiency of total purchasing power—but simply from the transition from wartime to peacetime production." At first demand would fall, reasoned Vinson, but tax reductions would aid recon-

---

[8]*New York Times,* June 4, 1945, 1; Aug. 28, 1945, 1. On Senator George's views, see *New York Times,* Sept. 8, 1945, 1.

[9]*New York Times,* Sept. 7, 1945, 1; Sept. 9, 1945, 38; Sept. 17, 1945, 11; Sept. 20, 1945, 22; *Washington Post,* Oct. 1, 1945, 1; Oct. 2, 1945, 1. Vinson's statement of October 1 is printed in *Annual Report of the Secretary of the Treasury on the State of the Finance, 1946* (Washington, 1947), 326-32. (Hereafter cited as *Treasury Report.*) The net yield on the excess-profits tax was simply the difference between the 85½ per cent excess-profits rate and the lower corporate normal and surtax rates. The provision for a carry-back was in the revenue law of 1942 and Congress simply continued it. Vinson had suggested that the tax cut be restricted to $5 billion, but his program exceeded that.

version and keep the economy healthy. At the same time, the budgetary deficit and pent-up demand for capital and consumer goods would require a cautious tax policy and continued protection against inflation.[10]

III

Within the administration and Congress, the debate focused on the proposed removal of the excess-profits tax on corporations. Vinson charged that the tax was "too erratic a tax engine" in peacetime and called it an obstacle to prompt reconversion and business expansion, on which employment and reconversion would largely depend. Though Truman later confided that Vinson had not cleared this decision with him, the President backed his Secretary. It was Marriner Eccles, *bete noire* of Wall Street and chairman of the Federal Reserve Board, who dissented vigorously and counseled against withdrawal of the tax. Unlike Vinson and most government advisers, Eccles did not believe that elimination of the tax was necessary to induce production and investment. "The war," he later wrote, "demonstrated that if business had orders it would go ahead producing and furnishing employment notwithstanding high taxes." Foreseeing an extensive demand for goods, he thought that the serious problems of reconversion would be disruptions in the supply of labor and shortages of materials. Where Vinson had argued that the tax was dangerous because it would injure new businesses and small firms, Eccles contended that the recent exemption (of the first $25,000 of corporate income from the levy) had already given small business a "decided boon." Removal of the excess-profits tax, according to a Treasury analysis, would increase the profits of big business; it did not need the additional incentive, concluded Eccles.[11]

Taking a middle way in the dispute between Eccles and Vinson, the House Ways and Means Committee recommended a smaller excess-profits tax. As part of its plan for chopping taxes

[10]*Treasury Report, 1946,* 326-32.
[11]*Treasury Report, 1946,* 326-32; Office of War Mobilization and Reconversion, *Quarterly Report,* No. 4 (Oct., 1945), 24-26; Marriner Eccles, *Beckoning Frontiers: Public and Personal Recollections* (N.Y., 1951), 412-13. Excess-profits liability was concentrated in a few corporations: about $1.8 billion in 900 corporations; another $600 million in 6,000; and $100 million in another 12,000. *Cong. Rec.* (Oct. 11, 1945), 9762. Truman later implied to his Director of the Budget that the planned removal of the excess-profits tax had not been cleared very well with him. Harold Smith Conference (with Truman), Feb. 8, 1946, Smith Papers, Bureau of the Budget Library, Bureau of the Budget (Washington).

by about $5.35 billion (as opposed to Vinson's proposal of $5.187 billion), the committee slashed the excess-profits levy by sixty per cent for 1946, which would cost the government only $1.38 billion, but abolished the tax for 1947. The committee also cut the combined normal and surtax rates on corporate incomes by four percentage points, which would reduce government revenue by $405 million. In slicing ten per cent from the personal income tax, the committee exempted about 12 million families from income taxes and cut federal income by $2.6 billion. The bill also trimmed excise taxes by $535 million.[12]

Despite the disagreement with the Treasury on the excess-profits tax, the committee's unanimous report expressed the general priorities of most members of Congress and the views of the administration's advisers: the need to smooth the transition from war to peace by providing incentives for business expansion and, as a second thought, to support consumer purchasing power. Debate in the House revealed quickly that most objections to the bill were not severe, and that the legislators could compromise without much difficulty. Almost none seemed alarmed that the tax reduction would mean a larger federal debt. Despite the frequently stated fears of government debt and deficit spending which characterized many public discussions, balanced budgets and reduction of the debt did not occupy for most Congressmen a high place in their hierarchy of valued public policies. Though Congressmen might rail against government spending when opposing particular programs or resisting expansion of the federal bureaucracy, and their assaults relied upon the rhetoric of fiscal orthodoxy, budget balancing was not central to their political faith.[13]

A few legislators attacked the measure for providing too much aid for the wealthy and insufficient relief for small-income earners. But the House disregarded their protests and rushed the bill to the Senate, and in mid-October its Finance Committee held the first open hearings on the tax measure. Testifying before the committee, Vinson opposed the House draft and continued to labor for outright repeal of the excess-profits tax. He also objected to the flat cut of ten per cent in personal taxes as a bonus to high-income recipients. Because the House bill would slice almost $2 billion more from tax revenues in calendar 1947 than Vinson regarded

[12]The Revenue Bill of 1945, *H. Report* 1106, 79 Cong., 1 Sess. The committee did not hold hearings on the bill.
[13]*Cong. Rec.* (Oct. 11, 1945), 9618, 9631-33; *New York Times*, Oct. 8, 1945, 16.

as safe, he urged the committee to accept the administration's program and revise the House version.[14]

While Vinson feared that the House's tax reduction was excessive and favored the wealthy, the United States Chamber of Commerce, the State Councils of Chambers of Commerce and the National Association of Manufacturers, all sharing the sentiments of leading Congressional Republicans, demanded additional tax relief. Though rejecting Keynesian economics, business leaders, like most Congressmen, did not recommend a balanced budget for 1946. They recognized that the government had huge fixed expenditures resulting from the war, and they knew that the administration could balance the budget only by greatly raising taxes. But higher taxes threatened their welfare and collided with one of their basic economic beliefs—that taxes weaken incentive. Without difficulty, then, they chose the policy consonant with their interests and their larger view of the economy—lower taxes on business. They appealed for abolition of the excess-profits tax (by January 1), a twenty per cent reduction in personal-income levies, and other cuts in corporate taxes. Without these benefits, they reasoned, production would drop and reconversion would lag; the combination of taxes on the local, state and federal levels, warned one spokesman, would impose "tax burdens at the dangerous point of 30 per cent of national income."[15]

[14]*New York Times*, Oct. 12, 1945. The vote was 343-10. *Revenue Hearings*, 17 ff. Bernard Baruch, foreseeing years of post-war prosperity, warned that the reduction was inflationary, and opposed repeal of the excess-profits tax on the grounds that there was no need for additional stimulation of business. Baruch to Sen. George March 17, 1944, Baruch Papers. Sen. George wanted to repeal the excess-profits tax because "new business could not compete with old well established business with a substantial credit base against the excess profits tax." He did not want to cut individual rates so much but the committee went beyond his views. Sen. George to Baruch, Oct. 22, 1945, Baruch Papers.

[15]*Revenue Hearings*, 171 ff., 140 ff., 196 ff. Recent popular expositions of the reception of the "new economics" have failed to understand that opposition to the use of fiscal policy as a counter-cyclical weapon has seldom been a central tenet in the faith of most businessmen or Congressmen. At least since Hoover's Presidency, American history reveals their willingness to depart from practices dictated by the principles of fiscal orthodoxy whenever other, more important goals, of which there were many, could be served. However, most were not prepared to generalize their behavior to a new principle, but instead, for various reasons, clung to the old beliefs, or at least expressed them when convenient. Many undoubtedly did not understand their own priorities in the area of economic policy, and they did not realize that a balanced budget was low in their scale of values. It remained low until at least the early fifties. Certainly, despite the World War II experience and a half decade of large budget deficits, most continued to

In opposition, the Congress of Industrial Organizations (CIO), representing a number of consumer groups and labor unions, urged retention of the excess-profits tax and greater relief to low-income groups. "The prime objective of the ... bill should be to counteract the decline in purchasing power in the critical transition period," asserted a CIO lobbyist. His analysis emphasized that the savings of 38 million workers averaged only about $350 (for a total of about $13 billion), which was inadequate to maintain full employment. To create the mass purchasing power which would mean prosperity, the CIO recommended a three-point program: a repeal of the three per cent tax, freeing about 12 million families; a doubling of exemptions for single persons and couples, costing nearly $4 billion; and an allowance to individuals to carry back or forward unused exemptions for two years, estimated at $675 million.[16]

Disregarding labor's plan, the Finance Committee proposed a tax cut of nearly $5.8 billion—about $430 million more than the House measure and nearly $600 above the administration's recommendation. Under pressure from Vinson, the committe had acceded to the administration's plan and repealed the excess-profits tax. The committee also had removed the normal income tax, reduced

---

fear *years* of unbalanced budgets and suspected that the prolonged deficit spending might bleed the economy and contribute to its collapse. In 1945 some businessmen emphasized the need for a balanced budget—but by 1947 or 1948. By 1947 and 1948, some proponents of the balanced budget were so confused in their thinking that they could not often distinguish between their fear of a cumulative imbalance (over the course of years) and their fear of a significant annual imbalance. [See for example, "Statement of James V. Forrestal" before the President's Air Policy Commission, Dec. 3, 1947, Air Policy Commission Papers #1, Truman Library; Walter Millis, ed., *The Forrestal Diaries* (N.Y., 1951), *passim.*] In general, tax reduction should be more popular than spending, for advantages accrue directly to a larger group and there is no likelihood in such cases of evoking the fear of big government. Yet, as America demonstrated in the Kennedy years, tax reductions (more than deficits) also can unnerve a people who have learned to believe the rhetoric of sacrifice for the national interest; they fear that tax cuts may reveal or nurture a softness in the national character. For more traditional interpretations of these issues, see: Francis Sutton, *et al.*, *The American Business Creed* (Cambridge, 1956), particularly 202-205, 220-23, 367-79; and *Time* (Dec. 31, 1965), 64-69. Lamentably, there is still no adequate study on the reception of the "new economics" among economists, in the many layers of the federal government, or in the public dialogue. However, see Robert Lekachman, ed., *Keynes' General Theory: Reports of Three Decades* (N.Y., 1964); and his *The Age of Keynes* (N.Y., 1966).

[16]*Revenue Hearings*, 91 ff., 118 ff. Without expending much effort, almost as a token action, the AFL supported the CIO program. *American Federation of Labor Weekly News Service*, Oct. 23, 1945.

personal liability by ten per cent and relaxed rates on small businesses.[17]

In the Senate the doubts of a few could not puncture the consensus. Fearing profiteering and inflation, Senator Tom Connally, a Texas Democrat, vigorously fought repeal of the excess-profits tax; Senator J. William Fulbright, an Arkansas Democrat, was also skeptical about its abolition. But neither Congressmen nor labor leaders or administration advisors foresaw that continuation of the carry back-carry forward provision of the excess-profits tax might invite industry to resist peaceful settlement of wage disputes, encourage strikes and support union-busting by making recalcitrance financially possible. (Losses incurred from a shutdown, or any other reason, could be compensated by tax rebates from accumulated excess-profits taxes during the war years.) Nor did any realize that the repeal of the tax in 1946 might induce corporations to delay placing reconversion goods on the market until 1946, when their profits would be free of the tax.[18]

On October 24 the bill easily passed the Senate. The conference version generally followed the Senate measure. The Revenue Act repealed the excess-profits tax and left intact excise taxes. A compromise on the two bills extended relief to corporations previously free of the excess profits levy. Although the cost of the final bill was estimated at $700 million above the $5 billion Vinson had suggested as a limit for tax relief, and the annual deficit might reach $36 billion, neither Congressmen nor prominent businessmen seemed troubled. The Revenue Act was a popular measure, and the public considered it a victory for Vinson and Truman.[19]

---

[17] *S. Report* 655, 79 Cong., 1 Sess.

[18] *Wall Street Journal*, Oct. 25, 1945, 3; *CIO News*, Oct. 22, 1945. Already union leaders were accusing producers of holding back goods to break the pricing formula, and perhaps even controls, of the Office of Price Administration. After an investigation, the government concluded that at least one industry—the washing machine producers—was keeping products from the market until OPA granted a larger profit margin. Charles Hitch to Robert Nathan, Nov. 5, 1945, files of Deputy Director for Reconversion, OWMR Records, RG 240, National Archives.

[19] *Cong. Rec.* (Nov. 1, 1945), 9983, 10261. Aside from Senator Joseph O'Mahoney, a Wyoming Democrat, who believed that total demand would be so high that only repeal of the excess-profits tax was necessary, none raised serious objections about the *amount* of the tax cut. In the Senate, the measure passed on a voice vote. Where the Senate had provided benefits of $63 million for corporations and the House $405 million, the conference agreed on $347 million. The law reduced by four per cent the tax on corporations earning less than $25,000, and dropped by two per cent the rate on corporate profits above $50,000.

But what few knew was that President Truman had viewed the bill as a legislative defeat and had seriously considered a veto. Deterred by Vinson's pleading and reluctant to risk his prestige in another direct confrontation with Congress, when relations with the legislature were souring, Truman gracefully yielded. Still, he remained unhappy about the excessive tax cut, which he feared would leave the budget too unbalanced. Yet, despite the President's objections, the 1945 tax bill followed basically the analysis of his Secretary of the Treasury and also conformed closely to the general wishes of the business community. The act was designed to promote business expansion, which was expected to offset the deflationary effect of falling war expenditures. In formulating tax policy, Vinson and Truman had listened to predictions of disaster but had not followed those economists who had emphasized directly strengthening consumer demand. But as later events revealed, pessimistic forecasts of dwindling consumer demand proved grossly mistaken, and Vinson's policy was wiser than many economists would have judged in the early autumn of 1945.[20]

[Research for a study of the Truman administration, of which this article is one result, was assisted by grants from the Harry S. Truman Library Institute, the Samuel S. Fels Fund, and the American Council of Learned Societies.]

[20] Harold Smith Conference (with Truman), Jan. 15, 1946. The major errors in the estimates were in private capital formation and consumer purchasing. They understated the value of construction and, to some extent, producers durable equipment and net exports. In consumer expenditures, the most important underestimates were in non-durables and services. On economic forecasts and developments, see n. 3. In his diary of January 15, Smith also reported the discussion on proposing in the budget message a $1 billion reduction in excess taxes. Vinson opposed it (recorded Smith) "in the name of political strategy, his acquaintances on the Hill and so forth." He wanted to keep the options open until the pattern of reconversion was clearer. John Snyder, director of OWMR, opposed a reduction, but Samuel Rosenman, presidential adviser, thought that it was better political strategy to recommend the $1 billion cut and then hold the line there. In his message of Jan. 21, 1946, Truman omitted the reduction and called for the continuation of established taxes. *Public Papers of Truman, 1946,* 71-73.

EDWARD L. SCHAPSMEIER
FREDERICK H. SCHAPSMEIER

EISENHOWER AND EZRA TAFT BENSON: FARM POLICY IN THE 1950s

Speaking to a large rural audience at the National Plow-in Contest in Kasson, Minnesota, on 6 September 1952, General Dwight D. Eisenhower made this significant campaign promise:

And here and now, without any "ifs" or "buts," I say to you that I stand behind—and the Republican Party stands behind—the price support laws now on the books. This includes the amendment to the basic Farm Act, passed by votes of both parties in Congress, to continue through 1954 the price supports on basic commodities at 90 percent of parity.[1]

This unequivocal but carefully delimited pledge fostered the erroneous impression among many farmers that Ike favored permanent retention of the price support program. Rural Republicans, especially congressional candidates running in the farm belt, recalling the GOP debacle of 1948, also presumed their candidate understood the political ramifications of the farm issue. They did not want Adlai Stevenson doing what Harry Truman had accomplished —convincing voters in the countryside that a Republican victory meant the end of federal assistance to the rural sector. What was not really known was the extent of Eisenhower's disapproval of the scope of government intervention in agriculture.[2]

In view of his own convictions it was not surprising that Ike would select a man of similar persuasion to be his Secretary of Agriculture. Ezra Taft Benson was in all ways a true believer in "free agriculture."[3] This Mormon leader (he

---

[1] As quoted in Ezra Taft Benson, *Cross Fire: The Eight Years With Eisenhower* (Garden City: Doubleday, 1962), p. 39.

[2] Dwight D. Eisenhower, *The White House Years: Mandate for Change, 1953–56* (Garden City: Doubleday, 1963), p. 287; Sherman Adams, *First-Hand Report: The Story of the Eisenhower Administration* (New York: Harper, 1961), pp. 202–203; See also Clifford Hope's letters to Milton R. Young and Frank Carlson, both dated 8 September 1952, Papers of Clifford Hope (Kansas State Historical Society, Topeka).

[3] Ezra Taft Benson, "The Challenge For Cooperatives Today," Address to American Institute of Cooperatives, Logan, Utah, 26 August 1951, in Papers of Ezra Taft Benson (Archives of the Church of the Latter-Day Saints of Jesus Christ, Salt Lake City, Utah). See also an elaboration of this concept in Ezra Taft Benson (as told to Carlisle Bargeron), *Farmers At The Crossroads* (New York: Devin-Adair, 1956).

was one of the church's Council of Twelve Apostles) was then serving as Executive Secretary of the National Council of Farm Cooperatives. With a master's degree in agricultural economics from Iowa State College and field experience in the Idaho Extension Service, his qualifications seemed excellent.[4] He had endorsements from Thomas E. Dewey, whom he advised on agricultural matters in 1948, and from Senator Robert A. Taft, his first choice for the 1952 presidential nomination. Also recommending him was Allan Kline, the President of the powerful American Farm Bureau Federation.[5]

Only after his confirmation by the Senate, however, did it become widely known that Ezra Taft Benson adhered to the precepts of a doctrinaire religious-economic philosophy. His overall orientation stemmed, as one associate phrased it, "from two Smiths, Joseph and Adam, and from Thomas Jefferson."[6] This became evident at a press conference on 5 February 1953 when the new Secretary issued a controversial "General Statement on Agricultural Policy." It declared forthrightly: "Freedom is a God-given, eternal principle vouchsafed to us under the Constitution.... It is doubtful if any man can be politically free who depends upon the state for sustenance. A completely planned and subsidized economy weakens initiative, discourages industry, destroys character, and demoralizes the people."[7]

Benson's moralistic pronouncement about the virtues of laissez-faire economics sounded like an ideological edict from a spokesman of the radical right. Immediately dubbed "the epistle from the apostle," this dogmatic declaration of principles caused grave concern among bureau chiefs in the USDA.[8] Also apprehensive were leaders of the Farmers Union (the Farm Bureau's chief rival), certain Midwest Republicans, and some Democrats who now looked upon Benson as their mortal enemy. In one stroke the new Secretary had created a politically damaging image of himself as an uncompromising disciple of a Hoover-type individualism and unregulated free enterprise.[9]

Coming from South Dakota, where his constituents favored retention of high price supports, veteran GOP Senator Karl Mundt quickly let White House aides know that "Members of Congress on both sides of the Capitol are disgusted and mad at the apparent lack of political savvy on the part of the Secretary of

---

[4] *Iowa State Daily*, 4 June 1953; *Iowa State College Alumnus*, January 1953. His thesis was entitled "The Beef Cattle Situation in the Northern Range Area in its Relation to the Iowa Feeder" (1927).

[5] Robert Taft, Jr. to authors, 19 March 1970; Paul Friggens, "Meet the new Secretary and his Family," *Farm Journal* 77 (January 1953): 28–29, 97.

[6] Don Paarlberg (Director, Agricultural Economics, USDA), to authors, 3 September 1969. Benson once asserted, "Our marketing system is intricate.... When one first looks at such a complex system, he may easily get the impression of disorder in it. Yet there is a guiding principle. Adam Smith ... pointed out that individual producers and businessmen, acting in their own self-interest as they make their countless separate decisions to buy or sell or hold or ship, are led as if by an invisible hand to benefit the general public." "Foreword" to *Marketing: The Yearbook Of Agriculture 1954* (Washington, D.C.: GPO, 1954), p. vii. See also comments on Benson's religious philosophy in Oral History Memoirs of Don Paarlberg and Earl L. Butz, Dwight D. Eisenhower Library, Abilene, Kansas.

[7] "General Statement On Agricultural Policy by Ezra Taft Benson," 5 February 1953, Benson Papers, Eisenhower Library.

[8] Interview with Chester C. Davis, 9 August 1963.

[9] Interview with Sherman Adams, 13 August 1969.

Agriculture...."[10] When Sherman Adams and General Wilton Persons subsequently sent an emissary to assess rural sentiment, they were dismayed to hear the news that Benson was "not too popular."[11] The Administration was forced to defend an Agriculture Secretary even before a farm program had been sent to Congress.[12]

More criticism was hurled at Benson after he reorganized the Department of Agriculture. The Bureau of Agricultural Economics, the stronghold of liberal policy planners, was deliberately dismantled.[13] Ousted were such notable Democratic appointees as M. L. Wilson, Louis Bean, and Claude Wickard.[14] Displeased Democrats claimed the purge constituted an example of pure ideological zealotry. Benson fell into even more disfavor with longtime USDA employees when he lectured them on their responsibilities to the taxpaying public. Intending to encourage efficiency the Secretary asserted rather innocently, "The people of this country have a right to expect that everyone will give a full day's work for a day's pay."[15] The remark appeared to imply that all bureaucrats were inherently lazy, especially those hired by Democrats.

Since Benson was extremely eager to alter the farm statutes then on the books, even before their expiration, he took the initiative in an attempt to force quick repeal. At a Cabinet meeting on 19 June 1953 he suggested seeking a congressional resolution permitting immediate lowering of price supports. Though he approved of the principle of flexibility, Eisenhower nevertheless stated that under no circumstances would he refuse to fulfill his election commitment. On several occasions thereafter the President had to reaffirm this decision when Benson pressed for speedy action.[16] Clarifying his own position in terms of military strategy, the President explained that attainment of an objective sometimes involved delicate maneuvering—even temporary retreat. Politically speaking, Ike's announced path of "gradualism" was much more practicable. To reverse instantaneously an agricultural policy that reflected the trend of two decades would have evoked bitter opposition from many quarters and much internal dissention within the GOP.[17]

On 3 June, when the President and key Cabinet members appeared on nationwide television, Secretary Benson received an opportunity to elaborate publicly on his views. In response to Eisenhower's leading question, Benson

[10] Karl Mundt to Wilton B. Persons, 26 March 1953, in Papers of Sherman Adams, Baker Library, Dartmouth College, Hanover, New Hampshire.
[11] Homer Gruenther to Wilton B. Persons, 30 October 1953 and Gruenther to Sherman Adams, 11 November 1953, in Staff Files, Box 11, Eisenhower Library.
[12] The President was forced to ask friendly senators to defend his Agriculture Secretary. See Eisenhower to Hugh Butler, 9 November 1953; to Kenneth Keating, 30 November 1953; and to Arthur V. Watkins [circa November 1953], Central File, OF 1-Agriculture Department, Eisenhower Library.
[13] Howard R. Tolley, "Dismemberment of the BAE," *Journal of Farm Economics* 36 (February 1954): 14–16.
[14] Interview with M. L. Wilson, 17 July 1966; Dean Albertson, *Roosevelt's Farmer: Claude R. Wickard in the New Deal* (New York: Columbia University Press, 1961), p. 400.
[15] Quoted in Benson, *Cross Fire*, pp. 52–53.
[16] Benson to Eisenhower, 30 April 1953, Adams Papers, "Minutes of Cabinet Meetings," 19 June and 11 December 1953, Adams Papers.
[17] Memoirs of Paarlberg and Butz, Eisenhower Library.

announced the basic outline of the Administration's agricultural policy. Six objectives were enumerated: to build up markets; to allow for adjustments in production; to avoid pricing commodities "out of the world or domestic markets"; to take care not to hold an "umbrella over synthetic and competing products"; to emphasize consumer research; and to encourage a "self-help program for the farmers."[18]

With these objectives translated into specific recommendations, President Eisenhower presented a farm program to Congress on 11 January 1954. Combining gradualism with flexibility, he called for distinct but slow steps in reducing the level of price supports. This was to be accomplished by relating commodity loans to market needs and by using a "modernized parity . . . in steps of five percentage points of the old parity per year [starting 1 January 1955] until the change from old to modernized parity had been accomplished." While allowing for temporary set-asides of surpluses, Ike made no secret of the fact that "the key element of the new program is gradual adjustment to new circumstances and conditions."[19]

Moderate as it was, the Eisenhower farm measure caused consternation among members of both parties and in farm circles where the general idea of government management of agriculture was accepted without a questioning of overall efficacy. Yet times had changed drastically. Huge stocks of surpluses, particularly wheat, cotton, and dairy products, had accumulated in Commodity Credit Corporation warehouses. High storage costs (including spoilage) were incurred in addition to the depressing effect the existence of these excess stocks had on price levels. Too many producers had become accustomed to unloading their food and fiber on the government without any consideration for actual consumer preference. Some varieties of wheat and cotton had long been out of demand; synthetic fiber and substitute products, like nylon and oleomargarine, were far cheaper than those commodities still being produced in large quantities. Too many hogs were being raised with a fat content far exceeding the desires of diet-conscious consumers. Some industries, dairying for one, had become exceedingly lax and lazy; no advertising was done to woo the "Pepsi generation" into drinking milk. Certainly for the long-range benefit of agriculture some accommodation to the realities of the market place had to be achieved.[20]

The concept of flexibility in price supports was not new. Henry A. Wallace had introduced it into the Agricultural Adjustment Act of 1938 and it was also incorporated into the Acts of 1948 and 1949.[21] But World War II and military

[18] Transcript of television broadcast, 3 June 1953, Adams Papers.

[19] President's Message to the Congress of the United States, 11 January 1954, copy in Adams Papers. In a memorandum for Sherman Adams dated 18 December 1953 (probably sent by Gabriel Hauge), it was indicated that "modernized parity" meant reducing price supports 14 percent. Memo. in OF 110-N, Eisenhower Library.

[20] Oral History Memoir of Ezra Taft Benson, Eisenhower Library; Statement by Secretary of Agriculture, Ezra Taft Benson, Before the Senate Committee on Agriculture and Forestry, 18 January 1954, Benson Papers, LDS Archives; Statement by Secretary of Agriculture, Ezra Taft Benson, Before the House Committee on Agriculture, 10 March 1954, Benson Papers-LDS Archives; Ezra Taft Benson, "New Tasks and New Tools," address at Oregon State Fair, Salem, 10 September 1954, Benson Papers, LDS Archives.

[21] Henry A. Wallace, "Background of the Farm Program," NBC, 1 April 1954, transcript in Papers of Henry Agard Wallace, University of Iowa, Iowa City.

action in Korea had prevented meaningful reductions from taking effect so that, by 1954, no substantive lowering of price supports had taken place. Congress was loathe to reduce parity levels or stringently to enforce acreage allotments.[22] Living in retirement in the New York hamlet of South Salem, Henry Wallace voiced his personal approval of the Eisenhower–Benson plan to make flexibility an operative device. As a major architect of the original price support system, he claimed it was a "foolish waste" to produce for nonexistent markets. To the surprise of James Farley, who had asked him to denounce Ike's Agriculture Secretary, Wallace defended Benson even to the extent of supporting reorganization of the USDA. What the old New Dealer did criticize was Benson's peculiar penchant for getting himself pegged as an enemy of the dirt farmer. Wallace confided to Farley, "For Benson's sake I am sorry that he should have created this impression because I think the flexible price idea so far as corn is concerned is probably good."[23]

Despite GOP control of Congress, it took much persuasion by Benson and considerable pressure from Eisenhower to get a modicum of flexibility into the Agricultural Act of 1954. When finally passed after numerous compromises had been made, price supports for basic crops (except for tobacco) were set at a rate from $82\frac{1}{2}$ to 90 percent of parity for the first year, 1955; thereafter the rate could be lowered gradually to a minimum of 75 percent. Congress insisted on surplus set-asides as cushions and afforded wool special treatment for a four-year period. Only the threat of a presidential veto thwarted attempts to establish a two-price system for wheat.

Having displayed considerable political skill in securing adoption of his basic proposal, Eisenhower stated optimistically: "This new law—the central core of a vigorous, progressive agricultural program—will bring substantial lasting benefits to our farmers, our consumers, and our entire economy."[24] But Benson's attitude reflected his disappointment at not attaining total legislative success: "We had to take what we could get. We *had* to make a beginning somewhere. We had to start at least to reverse the 20-year trend toward socialism in agriculture."[25]

Because Benson was so intent upon discrediting the entire concept of price supports, his testimony before congressional committees along with his public statements tended to emphasize the negative side of the agricultural picture. Sherman Adams, Gabriel Hauge (Eisenhower's economic adviser), and Vice President Richard Nixon were among those who on more than one occasion urged the Secretary to publicize positive aspects of the Administration's farm program. Insiders on the White House Staff and certain Republican state chairmen disliked Benson's habit of exhorting farmers to stand on their own feet without seeming to sympathize with those who strove to earn a living under

---

[22] Earl L. Butz, "The Agricultural Economist In the Political Environment Of Policy Making," *Journal of Farm Economics* 37 (May 1955): 189–196.

[23] Henry A. Wallace to James A. Farley, 1 April 1954, Wallace Papers.

[24] Statement by the President Upon Signing the Agricultural Act of 1954, 28 August 1954, *Public Papers of the Presidents of the United States, Dwight D. Eisenhower, 1954* (Washington, D.C., 1960), p. 772.

[25] Benson, *Cross Fire*, pp. 254–255.

difficult circumstances.[26] By condemning federal assistance as a wasteful use of taxpayers' money, the farm bloc began to lose allies among urban representatives. Louisiana's Allen J. Ellender, Chairman of the Senate Committee on Agriculture and Forestry decried this tactic: "It is my belief that Mr. Benson tried to create friction between the consumers and producers. A lot of publicity was made available by the Secretary to show that these price support programs tended to raise the cost of living among the consumers."[27]

Prior to the mid-term elections of 1954 a considerable amount of legislation beneficial to agriculture was actually enacted. Social security was extended to farmers, tax depreciation benefits were granted, and lower freight rates resulted from approval of the St. Lawrence Seaway Project. Other helpful measures included establishment of an Agricultural Marketing Service; intensification of consumer-use programs within the Agricultural Research Service; passage of Public Law 480 (Agricultural Trade Development and Assistance Act) authorizing sale or barter of surpluses abroad; the Mutual Security Act making excess stocks of food and fiber available for purposes of national defense; reinauguration of a school milk program; and an experimental extension of crop insurance. Conservation was fostered by the Water Facilities Act, encouraging wise storage and use of water, and the Watershed and Flood Prevention Act, which provided matching funds for construction of local dams. Fifteen million dollars was also made available for loans to farmers living in areas stricken by natural disasters.

Doubtless, Benson's marketing expertise accounted for the considerable success in disposing of accumulated surpluses. With the help of the Francis Committee, an interdepartmental organization, various international roadblocks and state department restrictions were overcome in order to channel excess American holdings into overseas markets. Believing he must fight for American markets abroad, Benson traveled widely, even to the Soviet Union, to gain new outlets for American farm products. He denied the United States was either dumping or stealing markets from other nations. All types of sales (often for foreign currency to be spent in that country), barter, and outright gifts in conjunction with foreign aid programs helped get rid of four billion bushels of wheat, two and a half billion pounds of dairy products, and four and a half billion pounds of cottonseed products. In addition to government activities, Benson encouraged agricultural cooperatives to seek actively new markets abroad. Not only was this done as a result of the Secretary's urging, but many updated merchandizing techniques were adopted to enlarge domestic consumption.[28]

---

[26] Notes on Staff Meeting with President by Sherman Adams, 1 February 1953, Adams Papers; Minutes of the Cabinet, 23 October 1953 and 7 October 1955, Adams Papers; Gabriel Hauge to True D. Morse, 23 August 1955, Copy in Files of Howard Pyle, Box 34-Agriculture, Eisenhower Library. For an example of how Hauge edited a Benson speech to be delivered at the annual convention of the Farm Bureau, see Memorandum and text speech alterations, Hauge to General Persons, 11 December 1953, OF 1, Eisenhower Library.

[27] Allen J. Ellender to authors, 6 October 1969. See also Wesley McCune, *Ezra Taft Benson, Man With A Mission* (Washington, D.C.: Public Affairs Press, 1958), pp. 29–32.

[28] Clarence Francis to Sherman Adams, 16 June 1954, Files of Clarence Francis, Box 4, Eisenhower Library; Arthur E. Burns to authors, 24 September 1969; Benson, *Cross Fire*, pp. 254–256.

Having made a supreme effort to empty Commodity Credit Corporation storage bins, Secretary Benson took a stern stand against another influx of surpluses. Relentlessly he preached the necessity for adjusting production downward to conform more realistically to market demands. This cold, businesslike attitude antagonized farmers within inelastic markets, such as wheat growers, since they felt they were being sacrificed to economic efficiency. Ironically then, Benson's remarkable accomplishments increased his unpopularity with a considerable number of rural people.[29]

In spite of farmers being driven off the land, rapid technological improvements kept increasing crop yields. To cope with increasing surpluses, Secretary Benson adopted the Soil Bank to curtail overall production. This plan was originally presented to the House Committee on Agriculture in 1953 by Melvin P. Gehlbach, chairman of a group calling itself the Soil Bank Association (headquartered in Lincoln, Illinois). After the National Agricultural Advisory Commission studied it and gave their approval, USDA officials worked out the specific mechanism of operation.[30] By use of monetary incentives, both productive and marginal land was to be retired in either acreage or conservation reserves. In asking for enactment of this measure, President Eisenhower claimed:

It will help remove the crushing burdens of surpluses. . . .
It will reduce the massive and unproductive storage costs on government holdings. . . .
It will provide an element of insurance since farmers are assured income from the reserve acres. . . .
It will ease apprehension among our friends abroad over our surplus-disposal program. . . .
It will harmonize agricultural production with peacetime markets.[31]

With the 1956 election approaching both parties were receptive to a program aimed at helping farmers, especially in a period when rural income was on the decline. Congress passed the Soil Bank but added much more than the President had bargained for. Tacked on were provisions for support of feed grains; a two-price system for wheat and rice; a dual method of computing parity for

---

[29] Benson to K. B. Cornell (Manager, Cornell Accounting Firm, Clinton, Oklahoma), 21 October 1954, Correspondence of Secretary of Agriculture, RG (National Archives, Washington, D.C.); Press Release, 1 November 1954, Benson Papers, Eisenhower Library; Statement by Secretary Benson to the Senate Committee on Agriculture and Forestry, 19 January 1955 and to the House Committee on Agriculture, 27 March 1956, Benson Papers, Eisenhower Library; a confidential report, "Elements of Production Research," 12 April 1957, concluded: "Ill-advised public pressures operating through the legislative branch are chiefly responsible for our departure from a free agricultural economy into one of increasing complexity." OF 110-N, Eisenhower Library; Senator George D. Aiken to authors, 27 September 1969; Meade Alcorn to authors, 12 November 1969.

[30] Don Paarlberg informed Gabriel Hauge on 15 August 1955 "We have these plans [Soil Bank and Acres for Tomorrow] under analysis and will review them with the National Agricultural Advisory Commission in September." Correspondence File, Secretary of Agriculture, RG 16, National Archives. The National Agricultural Advisory Commission was established by Executive Order No. 10472 on 20 July 1953. It consisted of 18 members (9 of each party) with William I. Myers serving as Chairman and Don Paarlberg as Secretary. William I. Myers to authors, 15 December 1969.

[31] Special Message to the Congress on Agriculture, 9 January 1956, *Public Papers of the Presidents of the United States, Dwight D. Eisenhower, 1956*, pp. 386–388.

wheat, peanuts, cotton, and corn; 80 percent parity for dairy products; set-asides for wheat, cotton, and corn; and compulsory participation in acreage withdrawal programs for eligibility in the price support system. Eisenhower was advised by staff aides to sign it. Many Republicans from rural areas also pleaded with him to accept the legislative package as it stood. But the President, with Benson's counsel prevailing, sent a firm veto message to Congress. He contended:

(1) To return now to wartime 90 percent supports would be wrong....
(2) The provisions for dual parity would result in a permanent double standard ... for determining price supports....
(3) The provision for mandatory supports on the feed grains would create more problems for farmers...
(4) The multiple price plans for wheat and rice would have adverse effects upon producers of other crops, upon our relations with friendly foreign nations, and upon our consumers.[32]

The core of Eisenhower's logic, which represented Benson's thinking, was good economics only if one discounted the human element. Political considerations do take into account the needs of people and cost-price squeezes affected farmers more than many other segments of the population. Ike won out, however, and Congress gave him pretty much what he wanted.[33] Other agricultural legislation that session included a Great Plains program to combat effects of drought; increased appropriations for utilization research; refunds to farmers on taxes collected for gasoline used in the field; and passage of a Rural Development Program for assisting farm families earning less than $1,000 per year. This fledgling anti-poverty program did not get much beyond the pilot stage because Congress never gave it the financial support needed for necessary expansion. Given less need for marginal producers, this program might well have been the means to train farmers for nonagricultural vocations.[34]

Eisenhower won reelection easily but the Republican party, which had lost control of Congress in 1954, could not regain its majority. There were those from rural regions who blamed Benson's farm program for this. The Secretary of Agriculture denied it and always had sufficient statistics to prove his point. Nonetheless, Republicans in districts where farmers were in revolt thought it foolish to lose elections in order to remain ideologically pure.[35]

The last four years of Eisenhower's term constituted a period of mixed concepts and muddled improvisations. Expectations for the Soil Bank did not fully materialize and by 1960 the Commodity Credit Corporation again possessed large amounts of food and fiber. Costs exceeded those of any other program

---

[32] Veto of the Farm Bill, 16 April 1956, *Public Papers of Dwight D. Eisenhower, 1956*, pp. 386–388.

[33] Statement by the President on Signing Farm Bill, 28 May 1956, Copy in Adam Papers.

[34] A Report on the Working Conference, Rural Development Program, Washington, D.C., 11–12 July 1956, OF 106 F-N, Eisenhower Library; True D. Morse to authors, 14 January 1970.

[35] Adams, *First-Hand Report*, pp. 218–219; Marion B. Folsom to authors, 25 September 1969; Charles A. Halleck to authors, 28 October 1969.

(even those of the Truman years).³⁶ Executive power was used to thwart all attempts to restore full parity and in turn Democrats, aided by some Republicans, prevented Benson from achieving his goal of a totally free agriculture. Congressman W. R. Poage, Chairman of the House Committee on Agriculture and one of Benson's bitterest antagonists, admitted "there was never any kind of very sound over-all alternatives offered and ... Democratic leadership did resort to a great deal of day to day opposition."³⁷ Amid an atmosphere of rancorous congressional hearings and exchanges of public charges, agricultural policy soon degenerated into an incongruous combination of open production and continued price supports.³⁸

Although Farm Bureau leaders remained staunchly loyal to Benson, the President was constantly urged by others to drop the Secretary from his Cabinet.³⁹ After William Proxmire won a special Senate election in Wisconsin, held in 1957 to fill the vacant seat of Joseph McCarthy, requests for Benson's removal intensified.⁴⁰ Because Proxmire based his campaign on an anti-Benson theme, Wisconsin's Melvin Laird complained to Eisenhower that the Secretary of Agriculture "has become the issue, not the program which he stands for."⁴¹ Congressman A. L. Miller of Nebraska asserted angrily to Sherman Adams that Benson "talked too much" and was now without doubt a "real liability."⁴² Senator Karl Mundt informed Adams: "I am writing to tell you candidly ... [that] we cannot come close to electing a Republican House of Representatives or a Republican Senate in 1958 unless Secretary of Agriculture Benson is replaced by somebody who is personally acceptable to the farmers of this country."⁴³ Another communication to Adams from Representative H. Carl Anderson of Minnesota inquired bitterly, "Why must the Midwest Republicans be sacrificed so needlessly?"⁴⁴

³⁶ "Facts About Price Supports," 10 December 1957, OF 1, Eisenhower Library. This report concluded: "Heavy costs would be justified if they lead to a solution of the problem. Such is not the case." See also "Report to the Congress of the United States, Review of the 1959 Conservation Reserve Program, Commodity Stabilization Service, USDA," June, 1959, OF Q-T, Eisenhower Library, and Don Paarlberg's fine study, *American Farm Policy: A Case Study of Centralized Decision-Making* (New York: John Wiley & Sons, 1964), passim.
³⁷ W. R. Poage to authors, 24 September 1969.
³⁸ In a letter to the authors dated 13 October 1969, Orville L. Freeman described the situation at the onset of his tenure as follows: "When Mr. Benson's term came to a close, the Department of Agriculture not only was disorganized—it was demoralized."
³⁹ On the influence of the Farm Bureau see Charles M. Hardin, "The Republican Department of Agriculture—A Political Interpretation," *Journal of Farm Economics* 36 (May 1954): 210–227.
⁴⁰ In a letter to the authors dated 22 September 1969, William Proxmire said: "One of the principal reasons for my election was the obvious protest of Wisconsin farmers against the Benson policies, especially with respect to dairy prices which are so vital to the Wisconsin farm economy." After his victory, he wired President Eisenhower, "Respectfully but with great urgency, I appeal to you to take immediate action to replace Ezra Taft Benson as Secretary of Agriculture." Telegram, Proxmire to Eisenhower, 12 September 1957, Adams Papers.
⁴¹ Melvin Laird to Dwight D. Eisenhower, 22 October 1957, Copy in Adams Papers.
⁴² A. L. Miller to Sherman Adams, 26 October 1957, Adams Papers.
⁴³ Karl E. Mundt to Sherman Adams, 23 October 1957, Adams Papers.
⁴⁴ H. Carl Anderson to Sherman Adams, 28 August 1957, Adams Papers. Further evidence of Benson's unpopularity occurred when eggs were thrown at him while he was speaking at the National Corn Picking Contest near Sioux Falls, South Dakota. Aberdeen *American-News*, 22 October 1957. When Senator Edward J. Thye called for Benson's dismissal, he received this answer from Eisenhower: "Naturally I am not unaware of your strong feelings in this matter; yet

Protests or no, Eisenhower remained obdurate in his decision to keep Benson in the Cabinet even after the latter offered to resign. In his memoirs Ike did concede that once Benson arrived at a conclusion it was "earnestly held and argued, though not always with the maximum of tact." Aside from that mild criticism, Eisenhower defended his Agriculture Secretary in these words: "But he and I agreed that high, rigid governmental price supports could never solve the farm problem, and so I supported his every effort and they were honest efforts because he was and is a man of unimpeachable integrity—to make American agriculture more responsive to a free market."[45]

Since Eisenhower delegated considerable authority to subordinates, he considered it his duty to support them when they came under fire for discharging their responsibilities,[46] and he never gave serious thought to dropping Benson from his administrative team.[47] Ike's political instincts were sharper than his Agriculture Secretary's and he, as well as his top assistants and upper echelon USDA officials, sought to mitigate the more rigid and implacable stance assumed by Benson. In the final analysis some steps were taken by the Eisenhower administration to resolve perplexing problems stemming from technological change and the transformation of family-oriented farming into modern agribusiness. Although Benson was perceptive and courageous, he seemed overly motivated by doctrinaire principles at a time when hard-pressed farmers needed sympathetic help and encouragement. This sincere man, who truly loved the land and those who tilled it, never fully realized that his political rhetoric sounded too much like didactic sermons from Salt Lake City's Temple Square. Rural America liked Ike but at best only respected the hard-working Secretary of Agriculture who kept on his desk the stern prayer: "O God give us men with a mandate higher than the ballot box."[48]

---

it does seem to me that upon reflection you will concede to Mr. Benson not merely the right but more importantly the obligation vigorously to set forth the programs and concepts which, in his best judgment, are essential to the well being of our farm people. It is my opinion that if he failed to do so, he would be derelict in his responsibility, and though so doing may understandably create some difficulties, I hardly see how he could effectively carry out his responsibilities in any other manner." Eisenhower to Thye, 13 March 1958, OF 1, Eisenhower Library.

[45] Eisenhower, *Mandate for Change*, p. 354.

[46] Walter Bedell Smith to Maxwell D. Taylor, 25 January 1946. This letter contains a superb analysis of the Eisenhower staff method. For instance, Smith said of Ike: "He is an exponent of decentralization. He selects commanders and staff with care. He then accords them a full measure of confidence and a very large measure of independent authority. After this he backs them up completely, however, as a rule he allows only one serious mistake per officer." Papers of Walter Bedell Smith, Eisenhower Library. Evidently in the President's judgment Sherman Adams, and not Ezra Taft Benson, made the one mistake serious enough to warrant dismissal.

[47] Milton S. Eisenhower to authors, 23 September 1969.

[48] Ezra Taft Benson, *The Red Carpet, A Forthright Evaluation of Socialism—The Royal Road to Communism* (Derby, Conn.: Monarch Books, 1963), p. 171. The Elder Harold B. Lee offered the following remarks after Benson's installation as Secretary of Agriculture: "When a man is ordained to the apostleship in this church, that is not just a job, but it is a power from Almighty God, and that power will remain in him so long as he knows that down in his soul there is a fire of testimony and a determination to serve God at all hazard and keep His commandments, and then there will be given inspiration and revelation. There will be given the power to act in the world of men in a manner that only one so possessed might act and receive." Washington Stake Conference, 1 March 1953, Benson Papers, LDS Archives.